PHASE-LOCKED LOOPS

PHASE-LOCKED LOOPS

APPLICATION TO COHERENT RECEIVER DESIGN

ALAIN BLANCHARD

A Wiley-Interscience Publication
John Wiley & Sons   New York / London / Sydney / Toronto

Library of Congress Cataloging in Publication Data:

Blanchard, Alain, 1938-
  Phase-locked loops.

  "A Wiley-Interscience publication."
  Includes bibliographical references and index.
  1. Phase-locked loops.   I.   Title.

TK7872.P38B56   621.38′043   75-30941
ISBN 0-471-07941-3

Printed in the United States of America

10 9 8 7 6 5 4 3 2 1

PREFACE

The phase-locked loop has been in use for over fifteen years in the domain of space telecommunications and for several years in the far wider fields of telecommunications in general, radiolocalization, and measurement and frequency synthesis instrumentation.

One of the essential properties of the phase-locked loop is the restitution of the phase, and thereby the frequency, of a signal immersed in noise, which can then be coherently detected. The coherent detection method fully justifies the many surveys that have been devoted to this subject over the last 20 or 30 years. On the one hand, emitter power is restricted by various practical considerations, such as weight, power consumption, price, interference, or even international power radiation regulations. On the other hand, we require communication over ever-growing distances, implying the constant reduction of link power budgets. Consequently, in many applications, the signal received is very low and the standard reception and detection techniques fail, whereas coherent techniques are remarkably efficient.

Phase-locked loops are thus of considerable interest in that they are basic elements in current top-performance reception and detection equipment. Initially studied and developed in connection with space telecommunications, they are more and more widely used in all fields of telecommunications and electronics for, apart from performance considerations, they frequently represent the optimum solution from the points of view of price, simplicity of design, and facility of adjustment.

The present book was originally intended as a means for students who have already completed a course in general telecommunications and electronics to acquire the necessary basis for the conception of equipment using coherent techniques.

In the course of writing, it seemed clear that with a few additions to the main text, this book could form a most useful handbook for engineers and technicians in industry, responsible for the design, construction, or maintenance of telecommunications equipment.

In order to comply with this twofold objective the book includes demonstrations and detailed explanations of the fundamental operating principles in sufficient number to constitute a theoretical approach to the subject. An attempt has also been made to present the material in order of increasing difficulty, without transgressing, insofar as possible, the three independent parts.

A large number of experimental results and numerical examples are also given, either to familiarize the reader with the magnitudes normally encountered in practice or to corroborate theoretical results in cases where the theory is approximate or inadequate. Finally, many curves and formulas that are generally dispersed in other books have been regrouped to facilitate the work of the reader.

Of course, the results given in Part III are sometimes incomplete because research on different aspects of the subject is still in progress in many parts of the world, which proves the importance attached to the theory of phase-locked loops and the corresponding practical applications. The present material may be considered as an attempted synthesis, regrouping all the essential results so far obtained forming the basic data necessary for the design and constructions of this equipment together with all the notions and demonstrations that one must have in mind to be able to understand the working principle.

Part I begins with an introduction to phase-locked loops, followed by a detailed description of the different elements, with particular emphasis on phase detectors, which often represent the main difficulty in the construction of equipment based on coherent techniques (Chapters 1 and 2). Chapter 3 contains all the equations governing the servo device behavior in the frequency or time domain, within or beyond the linear operating range. Chapter 4, which deals with phase-locked loop stability, provides information permitting a judicious choice of loop for a given practical problem.

Part II deals solely with the linear operation of the device. Chapters 5 and 6 group all the results derived from the equations given in Chapter 3, comprising all the formulas and curves necessary for the description of both transient and sinusoidal steady-state operating conditions. The use of the device as a coherent phase demodulator and frequency discriminator is discussed in this section. Chapter 7 is devoted to the behavior of the device in the presence of noise. It is shown how the input additive noise modifies phase detector operation; the consequences with respect to the servo device are described, and the formulas giving the output signal-to-noise ratio in demodulator applications are set forth. Chapter 8 deals with loop performances when the signals are modulated by random processes. Finally, in Chapter 9, we describe the operating modifications occurring when the device is preceded by an automatic gain-control device or a limiter, two cases frequently encountered in telecommunications.

In Part III we have assembled the main theoretical results and certain important experimental results concerning nonlinear operation. The problems of phase and frequency input signal acquisition by a first-order loop and by second-order loops are discussed in detail in Chapter 10. The possibilities of natural acquisition being limited, in Chapter 11 we describe the acquisition by auxiliary devices and discuss their performances. Finally, in Chapter 12, we deal with the question of the phase-locked loop operating threshold in the presence of noise, when the input signal is modulated or not and when the loops are used as phase demodulators and frequency discriminators.

Appendix A comprises several phase-locked loop circuits, of types frequently encountered in current equipment, while Appendix B is a preliminary coherent receiver project.

Many of the results mentioned in the present paper were obtained in the course of research work performed at the Centre National d'Etudes Spatiales (France). I express my gratitude to the Direction of the Centre National d'Etudes Spatiales, in the person of Mr. Michel Bignier, General Director, who gave me the opportunity of working on this subject.

I should also like to thank all my colleagues both at the Centre Spatial de Brétigny and the Centre Spatial de Toulouse whose work has contributed to our further understanding and knowledge of the subject. First of all, I would mention Mr. Serge Vialle, a friend whose loss is deeply regretted; he was a brilliant experimenter in these difficult techniques. I should also like to thank Mr. Alain Pouzet for the many fruitful discussions we had together and Mr. Daniel Ludwig for his encouragement during the writing of this book.

Finally, may I reserve a special acknowledgement for Mrs. Juliette Milano, who typed the first version of the French text together with the subsequent alterations, for Mrs. Jean Dalens, who undertook the English translation of the text, and for Mrs. Suzanne Steinthal, who typed the final English version.

ALAIN BLANCHARD

Toulouse, France
June 1975

CONTENTS

CHAPTER FOUR • PHASE-LOCKED LOOP STABILITY   60

PART TWO   LINEAR BEHAVIOR

CHAPTER FIVE • TRANSIENT RESPONSE   81

PHASE-LOCKED LOOPS

PART ONE
GENERALITIES

CHAPTER 1
WORKING PRINCIPLES

A phase-locked loop is a device by means of which the phase of a frequency-modulated oscillator output signal is obliged to follow that of the input signal. A diagram of this device is shown in Fig. 1.1.

The two signals are applied to a phase detector, the output of which is a function of the phase difference between the two signals applied. This error voltage, after low-pass filtering in the loop filter, is applied to the modulation input of the frequency-modulated oscillator (or VCO for voltage-controlled oscillator) in such a way that the oscillator signal phase must follow the input signal phase.

The quality of this servocontrol and the conditions conducive to its effectiveness or those rendering it inoperative, when the signals are affected by disturbances or various modulations, will be dealt with in subsequent chapters. However, it is possible, without further introduction, using a simple example, to describe how the device works.

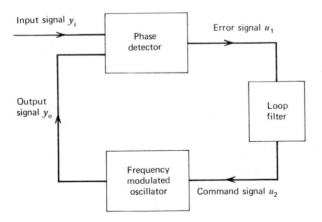

FIGURE 1.1.   General diagram of a phase-locked loop.

Let us suppose that the loop is unlocked, that is, that the command signal u_2 is not applied to the VCO modulation input. Let us then presume that y_i and y_0 are sinusoidal signals, which can be expressed as

$$y_i(t) = A \cos(\omega_i t + \theta_i)$$

$$y_0(t) = B \cos(\omega_0 t + \phi_0)$$

The angular frequency of the input signal is ω_i and ω_0 is the VCO central angular frequency. The values of the phase constants θ_i and ϕ_0 depend on the time origin selected. Generally speaking, ω_i differs from ω_0 and the signals are out of synchronism. If the phase detector has a sinusoidal characteristic (see Section 2.1), the output signal u_1 is expressed as

$$u_1(t) = K_1 \cos\left[(\omega_i - \omega_0)t + \theta_i - \phi_0\right] \tag{1.1}$$

Since signals y_i and y_0 are out of synchronism, the output signal of the phase detector u_1 is a sinusoidal signal of peak amplitude K_1, the angular frequency of which is equal to the difference in angular frequency between signals y_i and y_0.

At a given instant t, signal u_2 is applied to the VCO modulation input. After a period of time sufficiently long for transient phenomena to vanish and providing that the angular frequency difference $(\omega_i - \omega_0)$ does not exceed a certain value (see Chapters 10 and 11), it will be observed that the VCO output signal y_0 has become synchronous with the input signal y_i. Signal y_0 can then be expressed as

$$y_0(t) = B \cos(\omega_i t + \psi_0)$$

In other words, the quantity ϕ_0 has become a linear function of time expressed as

$$\phi_0 = (\omega_i - \omega_0)t + \psi_0 \tag{1.2}$$

and the phase detector output signal, or error signal, has become a dc signal, the value of which is

$$u_1 = K_1 \cos(\theta_i - \psi_0) \tag{1.3}$$

The loop filter is of the low-pass type, consequently letting through the dc signal u_1, so that the command signal u_2 is expressed as

$$u_2 = u_1 = K_1 \cos(\theta_i - \psi_0)$$

The VCO is a frequency-modulated oscillator. Its instantaneous angular frequency ω_{inst} is a linear function of the command signal, around the central angular frequency ω_0:

$$\omega_{\text{inst}} = \frac{d}{dt}(\omega_0 t + \phi_0) = \omega_0 + K_3 u_2$$

thus

$$\frac{d\phi_0}{dt} = K_3 u_2 \tag{1.4}$$

K_3, the proportionality constant, represents the VCO modulation sensitivity.*

If we insert Eqs. 1.2 and 1.3 in Eq. 1.4, we obtain

$$\omega_i - \omega_0 = K_1 K_3 \cos(\theta_i - \psi_0)$$

from which we derive

$$\psi_0 = \theta_i - \text{Arc} \cos \frac{\omega_i - \omega_0}{K_1 K_3} \tag{1.5}$$

The phase detector error signal u_1 may thus be written

$$u_1 = \frac{\omega_i - \omega_0}{K_3}$$

We can now recapitulate. Initially, signals y_i and y_0 were not synchronous: signal y_i was characterized by a frequency ω_i and a phase θ_i, and signal y_0 was characterized by a frequency ω_0 and a phase ϕ_0, ω_0 and ϕ_0 being completely independent of ω_i and θ_i. The loop was then said to be out of lock.

When the device is operating, signal y_0 becomes synchronous with signal y_i. The loop is then said to be in lock. The signals have the same angular frequency ω_i, but a difference in phase $\theta_i - \psi_0$ given by Eq. 1.5 exists between the two signals. This difference in phase produces in the phase detector output a dc component that goes through the filter to be supplied as a command signal to the VCO modulation input:

$$u_2 = u_1 = \frac{\omega_i - \omega_0}{K_3}$$

From the above equation it is clear that it is, in fact, by means of the dc signal u_2 that the VCO angular frequency changes from its central value ω_0 to the input signal angular frequency ω_i; that is to say,

$$\omega_{\text{inst}} = \omega_0 + K_3 u_2 = \omega_0 + \omega_i - \omega_0 = \omega_i$$

If the initial angular frequency difference $\omega_i - \omega_0$ is much lower than the product $K_1 K_3$, Eq. 1.5 becomes

$$\theta_i - \psi_0 \cong \text{Arc} \cos 0 = \frac{\pi}{2}$$

This means that if the frequency offset between the input signal and the VCO signal is slight when the loop is out of lock, the VCO signal is

*We reserve the notation K_2 for the gain of a possible amplifier situated between the phase detector and the VCO (see Section 2.3).

practically in phase quadrature with the input signal when the loop is in lock. Strictly speaking, the phase quadrature actually corresponds to $\omega_i = \omega_0$. It is for this reason that we tend to substitute the phase constant ψ_0 for the constant θ_0, so that $\theta_0 = \psi_0 - \pi/2$. Then

$$u_1 = K_1 \cos(\theta_i - \psi_0) = K_1 \sin(\theta_i - \theta_0)$$

The difference $\theta_i - \theta_0$ is often considered as the phase error between the two signals, which is null when the initial frequency offset is null. This is consequently only true to within $\pi/2$. Another method consists in representing the signals in conventional form:

$$y_i(t) = A \sin(\omega_i t + \theta_i)$$

$$y_0(t) = B \cos(\omega_0 t + \phi_0)$$

Using this notation, the phase detector output signal is written $u_1 = K_1 \sin[(\omega_i - \omega_0)t + \theta_i - \phi_0]$ when the loop is out of lock and $u_1 = K_1 \sin(\theta_i - \theta_0)$ when the device is working, with

$$\theta_i - \theta_0 = \text{Arc} \sin \frac{\omega_i - \omega_0}{K_1 K_3} \tag{1.6}$$

When the difference $\theta_i - \theta_0$ is sufficiently small, the following approximation is used:

$$u_1 \cong K_1(\theta_i - \theta_0)$$

The constant K_1 in this expression then represents the phase detector sensitivity and is measured in volts per radian (V/rad). The VCO modulation sensitivity K_3 is measured in hertz per volt (Hz/V) or in radians per second per volt (rad/s/V). The dimension of the product $K = K_1 K_3$, which is the servo device open-loop gain, usually designated as the "loop gain," is a time reciprocal, expressed in radians per second, when K_3 is expressed in radians per second per volt, or in hertz, when K_3 is expressed in hertz per volt (bear in mind that the radian, which is an angle unit, is dimensionless).

Equation 1.6 provides a basis for a preliminary conclusion regarding the conditions under which the servo device functions. This equation, as has been observed, is valid for the loop in lock. Let us suppose that, very gradually, so as to avoid generating transient disturbances, we now deviate the angular frequency ω_i from the VCO central angular frequency. When the difference $|\omega_i - \omega_0|$ exceeds the loop gain K, a θ_0 solution can no longer be found by means of Eq. 1.6. The synchronism equilibrium conditions no longer apply and the loop falls out of lock: the VCO angular frequency returns towards its central value and u_1 resumes its initial ac form. In physical terms, this means that the phase detector is no longer capable of producing the dc component necessary to the maintenance of synchronism. The values of ω_i corresponding to the appearance of this phenomena depend on the type of phase detector used.

For a phase detector of the sinusoidal type, the limits within which synchronization is possible, that is, within which it is possible to vary, very gradually, the input signal angular frequency of a previously locked loop, ranges from $\omega_0 - K$ to $\omega_0 + K$, ω_0 being the VCO central angular frequency and K the loop gain.

NUMERICAL EXAMPLE. Let us take $K_1 = 2$ V/rad as the sensitivity of a sinusoidal type phase detector and $K_3 = 10^4$ Hz/V (or $K_3 = 2\pi \times 10^4$ rad/s/V), the VCO modulation sensitivity, the central frequency of which is $\omega_0/2\pi = 10^3$ kHz. The value of the loop gain $K = K_1 K_3$ is

$$K = 2 \times 10^4 \text{ Hz} = 4\pi 10^4 \text{ rad/s}$$

The loop is observed to be in lock, even though the input signal frequency is $\omega_i/2\pi = 1010$ kHz. The dc component u_1, coming from the phase detector, which is also the dc component applied to the VCO modulation input, is thus equal to

$$u_1 = u_2 = \frac{\omega_i - \omega_0}{K_3} = \frac{2\pi \times 10^4}{2\pi \times 10^4} = 1 \text{ V}$$

This dc component originates in the phase error $\theta_i - \theta_0$ existing between the input signal and the VCO signal:

$$\theta_i - \theta_0 = \text{Arc} \sin \frac{\omega_i - \omega_0}{K} = \text{Arc} \sin \frac{2\pi \times 10^4}{4\pi \times 10^4} = \frac{\pi}{6}$$

If then, taking infinite care not to put the loop out of lock, we gradually increase the frequency difference between the input signal and the VCO central value as far as 20 kHz, the dc component $u_1 = u_2$ increases to 2 V. The phase error between the input signal and the VCO signal is then equal to $\pi/2$.

If we continue to increase the frequency difference beyond 20 kHz, the phase detector is incapable of providing the dc-error signal above 2 V, which would be necessary for the VCO frequency to remain keyed to that of the input signal, with the result that the loop falls out of lock.

The synchronization range of the loop under consideration lies between 980 and 1020 kHz.

CHAPTER 2
LOOP COMPONENTS

The three essential elements of a phase-locked loop are the phase detector, the loop filter, and the frequency-modulated oscillator or VCO.

There are several variants of each of these elements and the choice depends on the imaginative resources of the design engineer concerned, within the limits imposed by the possibilities of the technique in the desired frequency range, the waveform of the signals to be used, and the ultimate ends to be attained by the use of the device. The types here described are among those most frequently encountered.

Although certain integrated circuit phase-locked loops are now available (1, 2), in many cases each separate element has still to be defined.

2.1 PHASE DETECTORS

As regards the phase detectors, we distinguish two categories: sinusoidal signal phase detectors and square signal phase detectors. In the latter case, the square signal may be the original waveform of the signals used or may have been produced by hard limiting followed by amplification of sinusoidal signals.

When the signals applied to the phase detector and, in particular, the input signal are sinusoidal, the phase detectors generally use two- or four-diode bridges driven in various ways by means of impedance-matching amplifiers or transformers. The phase detector, in this case, is usually of the sinusoidal type. When the waveform is square, either diode (or transistor) bridges can be used, as for the sinusoidal signals, or logic circuits or devices derived from data-processing techniques. The phase detector characteristic is here of the linear type over an interval $(0, \pi)$ ("triangular") or over an interval $(0, 2\pi)$ ("sawtooth").

2.1.1 TWO-DIODE PHASE DETECTOR (FIRST LAYOUT)

Let us consider the circuit represented in Fig. 2.1 where D_1 and D_2 are perfect diodes, that is to say, their resistance is infinite or equal to a constant

8

small value ρ, according to the polarity of the voltage applied to their terminals.

The internal resistance of generator e_R is assumed to be zero, while the internal resistance of generators e_S and e'_S is included in the resistors R. It is further assumed that $e'_S = -e_S$.

Before attempting to calculate the voltage at point S, four different cases have to be distinguished, according to the state of conduction of the diodes.

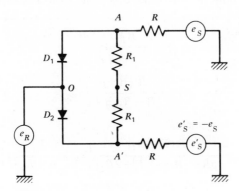

FIGURE 2.1. Two-diode phase detector circuit.

CASE 1 D_1 and D_2 do not conduct, which means that

$$v_A < v_O, \qquad v_{A'} > v_O$$

This twofold condition entails

$$\frac{R_1}{R+R_1} e_S < e_R < -\frac{R_1}{R+R_1} e_S$$

and furthermore $v_S = 0$.

CASE 2   Diode D_1 conducts and diode D_2 does not conduct, that is to say,

$$v_A \geqslant v_O, \qquad v_{A'} > v_O$$

which entails

$$e_R \leqslant \frac{R_1}{R+R_1} e_S$$

$$< -\frac{R_1(R+\rho)}{\rho R + R_1(R+\rho)} e_S$$

and furthermore,

$$v_S = \frac{R(R+R_1)}{\rho R + (R+\rho)(R+2R_1)}\left(e_R - \frac{R_1}{R+R_1}e_S\right)$$

CASE 3   Diode D_1 does not conduct and diode D_2 conducts, that is to say,

$$v_A < v_O, \qquad v_{A'} \leqslant v_O$$

This case is symmetrical with the preceding case. The following condition is fulfilled:

$$e_R \geqslant -\frac{R_1}{R+R_1}e_S$$

$$> \frac{R_1(R+\rho)}{\rho R + R_1(R+\rho)}e_S$$

while

$$v_S = \frac{R(R+R_1)}{\rho R + (R+\rho)(R+2R_1)}\left(e_R + \frac{R_1}{R+R_1}e_S\right)$$

CASE 4   Both diodes conduct, that is to say,

$$v_A \geqslant v_O, \qquad v_{A'} \leqslant v_O$$

which entails

$$-\frac{R_1(R+\rho)}{\rho R + R_1(R+\rho)}e_S \leqslant e_R \leqslant \frac{R_1(R+\rho)}{\rho R + R_1(R+\rho)}e_S$$

and

$$v_S = \frac{R}{R+\rho}e_R$$

If we call

$$\alpha = \frac{R_1}{R+R_1}$$

$$\beta = \frac{R_1(R+\rho)}{\rho R + R_1(R+\rho)}$$

$$\gamma = \frac{R(R+R_1)}{\rho R + (R+\rho)(R+2R_1)}$$

we observe that when the resistance ρ of the forward-biased diodes is very low as compared with R_1, $\beta \cong 1$. If, moreover, R is much smaller than R_1, then

$$\alpha \cong 1 \quad \text{and} \quad \gamma \cong \tfrac{1}{2}$$

These conditions may be summarized as in the table in Fig. 2.2.

The limit conditions for e_R and e_S obtained in this table are necessary conditions: if the state of conduction of the diodes is verified, the limit conditions are fulfilled. Bearing in mind the simplifying hypothesis referred to above, these limit conditions are mutually exclusive. They thus become sufficient conditions.

Let us consider, for instance, the following case:

$$e_R > -e_S$$

$$> e_S$$

Case	Diode D_1	Diode D_2	Limit Conditions	v_S
(1)	No conduct.	No conduct.	$e_R > e_S$ and $e_R < -e_S$	0
(2)	Conducting	No conduct.	$e_R \leqslant e_S$ and $e_R < -e_S$	$\frac{1}{2}(e_R - e_S)$
(3)	No conduct.	Conducting	$e_R > e_S$ and $e_R \geqslant -e_S$	$\frac{1}{2}(e_R + e_S)$
(4)	Conducting	Conducting	$e_R \leqslant e_S$ and $e_R \geqslant -e_S$	e_R

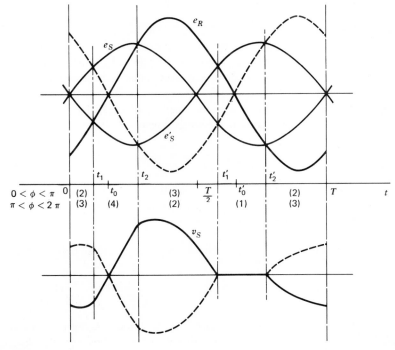

FIGURE 2.2. Limit conditions for the diode conduction and waveforms in a two-diode phase detector.

We can state positively that diode D_1 does not conduct and that diode D_2 conducts, and that consequently the corresponding expression of v_S is correct. For, if D_1 conducted, this would necessarily imply that $e_R < e_S$, whatever the state of conduction of D_2. Thus diode D_1 cannot conduct. In the same way, if D_2 did not conduct, this would necessarily imply that $e_R < -e_S$. Consequently, D_2 conducts. This means that:

$$v_S = \tfrac{1}{2}(e_R + e_S)$$

Subsequently, the left-hand columns of the table in Fig. 2.2 can be left aside and attention focussed on the two right-hand columns by means of which v_S is defined on the basis of conditions relating to e_R and e_S.

Let us now suppose that signals e_R, e_S and e'_S are sinusoidal signals of equal frequency, expressed as follows:

$$e_S = y_i(t) = A \sin \omega t$$

$$e'_S = -y_i(t) = -A \sin \omega t$$

$$e_R = y_R(t) = B \sin(\omega t - \phi)$$

These signals are represented in Fig. 2.2. Within a period T, the four cases described in the table can be distinguished and consequently the shape of the output signal $v_S(t)$ deduced. The e_R and v_S unbroken lines correspond to the case where $0 < \phi < \pi$, while the dashed lines correspond to the case where $\pi < \phi < 2\pi$.

The signal obtained at point S in the diagram of Fig. 2.1 is a signal with a period T and a relatively complicated waveform such that

$$v_S(t, \phi + \pi) = -v_S(t, \phi)$$

If this signal is sent through a low-pass filter, the dc component obtained is a dependent variable of ϕ, about which we have, generally speaking, little precise information. Let us examine two specific cases.

First of all, let us presume that the amplitudes of the signals applied are equal:

$$A = B$$

Under these conditions, the instants t_1, t_2, t'_1, and t'_2 corresponding to modifications in the conduction conditions of the diodes (Fig. 2.2) are respectively equal to

$$t_1 = \frac{t_0}{2}, \qquad t_1' = \frac{T}{2} + \frac{t_0}{2}$$

$$t_2 = \frac{T}{4} + \frac{t_0}{2}, \qquad t_2' = \frac{3T}{4} + \frac{t_0}{2}$$

the time t_0 being defined by $t_0 = \phi/\omega$.

Signal $v_S(t)$, in the case where $0 < \phi < \pi$, would then be expressed as

$$0 < t < \frac{t_0}{2}, \qquad\qquad v_S = \frac{A}{2}\left[\sin\omega(t - t_0) - \sin\omega t\right]$$

$$\frac{t_0}{2} < t < \frac{T}{4} + \frac{t_0}{2}, \qquad\qquad v_S = A\sin\omega(t - t_0)$$

$$\frac{T}{4} + \frac{t_0}{2} < t < \frac{T}{2} + \frac{t_0}{2}, \qquad v_S = \frac{A}{2}\left[\sin\omega(t - t_0) + \sin\omega t\right]$$

$$\frac{T}{2} + \frac{t_0}{2} < t < \frac{3T}{4} + \frac{t_0}{2}, \qquad v_S = 0$$

$$\frac{3T}{4} + \frac{t_0}{2} < t < T, \qquad\qquad v_S = \frac{A}{2}\left[\sin\omega(t - t_0) - \sin\omega t\right]$$

The dc component $\bar{v}_S$ of signal $v_S(t)$ is derived from the calculation of

$$\bar{v}_S = \frac{1}{T}\int_0^T v_S(t)\,dt$$

which leads to

$$\bar{v}_S = \frac{A}{\pi}\left(\cos\frac{\phi}{2} - \sin\frac{\phi}{2}\right)$$

that is,

$$\bar{v}_S = \frac{\sqrt{2}}{\pi} A \cos\left(\frac{\phi}{2} + \frac{\pi}{4}\right) \qquad \text{for } 0 < \phi < \pi$$

and

(2.1)

$$\bar{v}_S = -\frac{\sqrt{2}}{\pi} A \cos\left(\frac{\phi}{2} - \frac{\pi}{4}\right) \qquad \text{for } \pi < \phi < 2\pi$$

since $v_S(\phi + \pi) = -v_S(\phi)$.

The characteristic obtained is shown in Fig. 2.3, and could be called a "truncated" sinusoid. The phase detector sensitivity is obtained from

$$K_1 = \left| \frac{d\bar{v}_S}{d\phi} \right|_{\phi = \pi/2} = \frac{\sqrt{2}}{\pi} A \tfrac{1}{2} \sin\left(\frac{\pi}{4} + \frac{\pi}{4} \right)$$

that is,

$$K_1 = \frac{A}{\pi\sqrt{2}}$$

Let us now suppose that the amplitude of signal e_R is much greater than the amplitude of signals e_S and e'_S:

$$A \gg B$$

Instants t_1 and t_2 tend simultaneously towards $t_0 = \phi/\omega$, while t'_1 and t'_2 tend simultaneously towards $t'_0 = \phi/\omega + T/2$. The value of the dc component $\bar{v}_S$ of signal $v_S(t)$,

$$\bar{v}_S = \frac{1}{T} \int_0^T v_S(t)\, dt$$

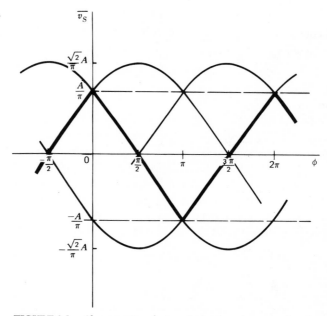

FIGURE 2.3.   Characteristic of a two-diode phase detector when $B = A$.

is then

$$\bar{v}_S = \frac{1}{2\omega T} 4A \cos \omega t_0$$

that is,

$$\bar{v}_S = \frac{A}{\pi} \cos \phi \tag{2.2}$$

Consequently, when $A \gg B$, the dc component obtained at point S of the circuit represented in Fig. 2.1 obeys a sinusoidal law of the difference in phase ϕ between signals e_R and e_S. This characteristic is represented in Fig. 2.4. The phase detector sensitivity,

$$K_1 = \left| \frac{d\bar{v}_S}{d\phi} \right|_{\phi = \pi/2} = \frac{A}{\pi}$$

is proportional to the amplitude of signal e_S. However, this sensitivity is limited, since A has to be smaller than B and for obvious technological reasons there are limits to the value that can be selected for B.*

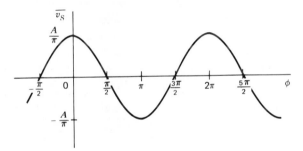

FIGURE 2.4. Characteristic of a two-diode phase detector when $B \gg A$.

On the other hand, we have indicated that the period of the signal $v_S(t)$ is equal to the period of the applied signals. Thus, the Fourier analysis of signal $v_S(t)$ produces, apart from the dc component $\bar{v}_S$, spectrum components at frequencies $f, 2f, 3f, \ldots$. Consequently, the attenuation of the low-pass filter designed to extract the dc component need only to be significant at frequency f. In particular, if the frequencies of signals e_S and e_R are not identical, the angular frequency difference Ω between the two can be interpreted as a linear variation of the difference in phase ϕ:

$$\phi = \Omega t + \theta$$

*However, it may be noted that if the waveform of signal e_R is square, the condition $B \gg A$ is replaced by $B > A$. It is a simple matter to check in this case that the same characteristic is obtained versus ϕ (ϕ being the difference in phase between the signal e_S and the fundamental of the square signal e_R).

If the angular frequency Ω is low and if the low-pass filter* does not attenuate this value, an output of the following type is obtained:

$$u_1 = \frac{A}{\pi} \cos(\Omega t + \theta) \tag{2.3}$$

This signal is identical, to within a quantity equal to the proportionality constant, with that which would be obtained by analog multiplication of signals e_S and e_R, followed by low-pass filtering to remove the high-frequency term. Thus, the analog multiplication would give

$$u = AB \sin(\omega t + \Omega t + \theta) \sin \omega t$$

that is,

$$u = \frac{AB}{2} \left[\cos(\Omega t + \theta) - \cos(2\omega t + \Omega t + \theta) \right]$$

After filtering out the angular frequency $(2\omega + \Omega)$ term, we get

$$u' = \frac{AB}{2} \cos(\Omega t + \theta)$$

which is identical with u_1, to within a quantity equal to the proportionality constant. It is for this reason that we tend to consider that the two-diode phase detector, when $B \gg A$, works like an analog multiplier, despite the fact that the waveform of the signal $v_S(t)$ in no way resembles what would be obtained by the analog multiplication of e_S by e_R, that is, the signal u above.

NUMERICAL EXAMPLE. The resistance ρ of the diodes when they are conducting is $\rho = 3\Omega$. The values of resistors R and R_1 of the Fig. 2.1 layout are, respectively, $R = 300$ Ω and $R_1 = 30$ kΩ. The values of the numerical coefficients α, β, and γ are consequently

$$\alpha = \frac{R_1}{R + R_1} = \frac{3 \, 10^4}{3.03 \, 10^4} \cong 1$$

$$\beta = \frac{R_1(R + \rho)}{\rho R + R_1(R + \rho)} = \frac{9.09 \, 10^6}{9.0909 \, 10^6} \cong 1$$

$$\gamma = \frac{R(R + R_1)}{\rho R + (R + \rho)(R + 2R_1)} = \frac{9.09 \, 10^6}{18.2718 \, 10^6} \cong \frac{1}{2}$$

The amplitude of signal e_S is equal to $A = 1$ V.

*The low-pass filtering required to produce the signal u_1 of Eq. 2.3 is not always easy to obtain. The equations, in fact, correspond to the presumption that there is no charge at point S. As the resistances R_1 have to be relatively large, compared with R, the signal at S has to be picked up at a very high impedance. If a high-input-impedance amplifier is used, care must be taken to ensure that there is no danger of this amplifier being saturated by the high amplitude signal (roughly $B/2$) present at S, for this would have disastrous consequences as regards the dc or low-frequency component.

(a) The amplitude of signal e_R is equal to $B = 1$ V. Then, the characteristic of the phase detector is a "truncated" sinusoid (Fig. 2.3) and the sensitivity is

$$K_1 = \frac{1}{\pi\sqrt{2}} = 0.225 \text{ V/rad}$$

(b) The amplitude of signal e_R is equal to $B = 10$ V. The characteristic is almost sinusoidal (Fig. 2.4) and the sensitivity is very close to $K_1 = 1/\pi = 0.318$ V/rad.

2.1.2 TWO-DIODE PHASE DETECTOR (SECOND LAYOUT)

Another possible two-diode phase detector layout is given in Fig. 2.5. Contrary to the preceding layout, we assume this time that the internal resistance of generators e_S and e_S' is zero and that the internal resistance of generator e_R is R.

Let us calculate the voltage at the point O. As in the previous example, four cases are to be distinguished, depending on the state of conduction of the diodes.

CASE 1 D_1 and D_2 do not conduct, corresponding to

$$e_S < v_O, \qquad e_S' > v_O$$

then $v_O = e_R$ and the twofold condition becomes

$$e_S < e_R, \qquad -e_S > e_R$$

CASE 2 D_1 conducts, D_2 does not conduct:

$$e_S \geqslant v_O, \qquad e_S' > v_O$$

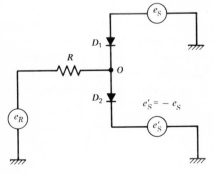

FIGURE 2.5. Second type two-diode phase detector circuit.

This case results in

$$v_O = \frac{Re_S + \rho e_R}{R + \rho}$$

the twofold condition becoming, in view of the fact that $e'_S = -e_S$,

$$e_S \geqslant e_R, \qquad -e_S > \frac{\rho}{2R + \rho} e_R$$

CASE 3 D_1 does not conduct and D_2 conducts ($e_S < v_O$ and $e'_S \leqslant v_O$), which gives

$$v_O = \frac{-Re_S + \rho e_R}{R + \rho}$$

the two fold condition becoming

$$e_S < \frac{\rho}{2R + \rho} e_R, \qquad -e_S \leqslant e_R$$

CASE 4 D_1 and D_2 conduct ($e_S \geqslant v_O$ and $e'_S \leqslant v_O$); then

$$v_O = \frac{\rho e_R}{2R + \rho}$$

the twofold condition becoming

$$e_S \geqslant \frac{\rho}{2R + \rho} e_R, \qquad -e_S \leqslant \frac{\rho}{2R + \rho} e_R$$

If the resistance ρ of the conducting diodes is presumed to be much lower than the resistance R, the four cases can be summarized as in the table of Fig. 2.6.

As in the case of the first phase detector considered, the limit conditions can be shown to be necessary and sufficient conditions. The case where signals e_S and e_R are sinusoidal signals is represented in Fig. 2.6. Unbroken lines represent $0 < \phi < \pi$ and dashed lines correspond to $\pi < \phi < 2\pi$.

The signal obtained at O is a signal of period T, such that $v_O(t, \phi + \pi) = -v_O(t, \phi)$, fairly complicated as regards waveform. If this signal is put through a low-pass filter, it is possible to pick up the dc component, which can easily be calculated.

When $A = B$, we can state that

$$\bar{v}_O = -\frac{\sqrt{2}}{\pi} A \cos\left(\frac{\phi}{2} + \frac{\pi}{4}\right), \qquad 0 < \phi < \pi$$

Case	Diode D_1	Diode D_2	Limit Conditions	v_O
(1)	No conduct.	No conduct.	$e_R > e_S$ and $e_R < -e_S$	e_R
(2)	Conducting	No conduct.	$e_R \leqslant e_S$ and $-e_S > 0$	e_S
(3)	No conduct.	Conducting	$e_S < 0$ and $e_R \geqslant -e_S$	$-e_S$
(4)	Conducting	Conducting	$e_S \geqslant 0$ (and $-e_S \leqslant 0$)	0

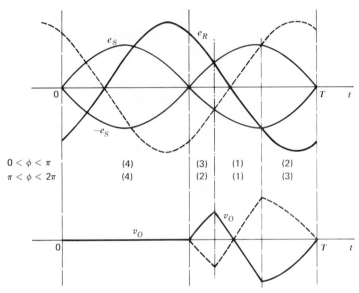

FIGURE 2.6. Limit conditions for the D_1 and D_2 diode conduction and waveforms in the second type phase detector.

and

$$\bar{v}_O = \frac{\sqrt{2}}{\pi} A \cos\left(\frac{\phi}{2} - \frac{\pi}{4}\right), \qquad \pi < \phi < 2\pi$$

The characteristic curve obtained is identical, within the sign, with that given by Eq. 2.1 and represented in Fig. 2.3.

When $B \gg A$, we get $\bar{v}_O = -(A/\pi)\cos\phi$. The characteristic curve obtained is identical, within the sign, with that given by Eq. 2.2 and represented in Fig. 2.4.

But the Fig. 2.5 diagram presents several advantages over the first layout considered. First of all, it is worth noting that resistance R is, this time, in series with the generator required to produce a high-amplitude signal (e_R). The internal impedance of generators e_S and e'_S has to be as low as possible, but the signals they are required to produce are of lower amplitude than the

e_R generator signal. The various generators will consequently be easier to construct and the power consumption will be lower. Furthermore, the amplitude of the signal at point O is $\leqslant A$, which is generally far below that of the signal at point S of the Fig. 2.1 layout. It will thus be easier to use a buffer amplifier to extract the signal at point O, with high-impedance conditions, before filtering.

2.1.3 FOUR-DIODE SINUSOIDAL PHASE DETECTOR

The waveform of the signal $v_O(t)$ of the previous layout leads us to think that if we could manage to substitute for the part where the signal is null $(0 < t < T/2)$ a signal identical with the part where it is not null $(T/2 < t < T)$, the waveform thus obtained would fairly closely approximate a sinusoidal signal, of frequency twice that of the signals applied. The behavior of the device would then be closer to that of an analog multiplier than can be obtained with a two-diode system. This result can be achieved by means of the circuit represented in Fig. 2.7 where four diodes are used in the standard configuration of the "ring modulator" or double-balanced modulator.

Since generators e_S and $e'_S = -e_S$ are presumed to have zero internal impedance, the potential at points A and A' is independent of the conduction conditions of the various diodes. In particular, the left-hand side of the circuit, by which the potential at point O is determined, is independent of the right-hand side and vice-versa.

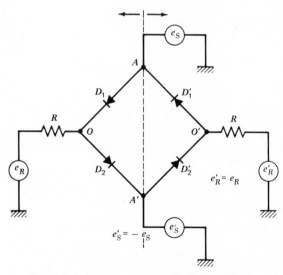

FIGURE 2.7. Four-diode phase detector circuit (double-balanced bridge or "ring modulator").

But the left-hand part of the circuit is in all respects identical with the two-diode layout of Fig. 2.5. As for the right-hand part of the circuit, it is also identical with that mentioned above, apart from the way in which diodes D_1' and D_2' are connected. The four cases corresponding to the conduction conditions of diodes D_1' and D_2' are summarized in the table in Fig. 2.8.

If $e_R' = e_R$ and if we examine the potential difference $v = v_{O'} - v_O$ (which can be extracted by means of a high-input-impedance differential amplifier, for instance) we get the signal represented in Fig. 2.8, within a period T (and in the case where $0 < \phi < \pi$), assuming signals e_S and e_R to be sinusoidal.

The signal $v(t)$ obtained is a signal of period $T/2$, consequently lacking a spectrum component at the frequency f of signals e_S and e_R. This signal is always such that $v(t, \phi + \pi) = -v(t, \phi)$ since this property holds for $v_O(t)$ and $v_{O'}(t)$. The dc component $\bar{v} = \bar{v}_{O'} - \bar{v}_O$, with $\bar{v}_{O'} = -\bar{v}_O$, is consequently

$$\bar{v} = -2\bar{v}_O$$

The characteristics of the four-diode detector are thus identical, apart

Diode D_1'	Diode D_2'	Limit Conditions	$v_{O'}$
No conduct.	No conduct.	$e_R < e_S$ and $e_R > -e_S$	e_R
Conducting	No conduct.	$e_R \geqslant e_S$ and $-e_S < 0$	e_S
No conduct.	Conducting	$e_S > 0$ and $e_R \leqslant -e_S$	$-e_S$
Conducting	Conducting	$e_S \leqslant 0$ (and $-e_S \geqslant 0$)	0

The limit conditions for the diodes D_1 and D_2 and for $v_{O'}$ are given in the Table of Figure 2.6.

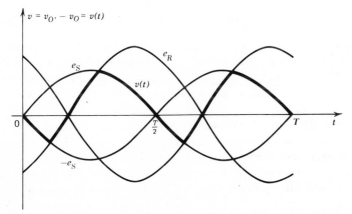

FIGURE 2.8. Limit conditions for the D_1' and D_2' diode conduction and output waveform in a four-diode phase detector (before filtering).

from sign and factor 2, with those of the two-diode detector.

In particular, if $B \gg A$, the dc output component is equal to:

$$\bar{v} = \frac{2A}{\pi} \cos\phi \tag{2.4}$$

By comparison with the dc component obtained at the output of a real analog multiplier, $u' = (AB/2)\cos\phi$, it will be noted that the law governing the variation versus ϕ and the proportionality to A are confirmed.* On the other hand, the proportionality to B and the value of the multiplying constant are not corroborated. Furthermore, a real analog multiplier only furnishes, apart from the dc component, a $2f$ frequency component, expressed as $-(AB/2)\cos(2\omega t - \phi)$. The four-diode layout produces the signal represented in Fig. 2.9 (still corresponding to the case where $B \gg A$) which comprises a $2f$ component, but also $4f, 6f, \ldots$ components.

It may also be noted that in this case the resulting signal is identical with that obtained by using a device ensuring the multiplication of signal e_S by the function $\text{Sign}[\sin(\omega t - \phi)] = \text{Sign}[e_R]$.[†] In this case, for the four-diode phase detector we can substitute the equivalent model shown in Fig. 2.10. The low-pass filtering necessary to extract the dc or low-frequency component is implied.

NUMERICAL EXAMPLE. Signal e_S is a sinusoidal signal of amplitude $A = 1$ V.

If signal e_R is a sinusoidal signal of amplitude $B = 10$ V, the characteristic is roughly sinusoidal and the sensitivity very close to $K_1 = 2/\pi = 0.636$ V/rad.

If signal e_R is a square signal of amplitude 1.1 V, the characteristic is strictly sinusoidal and the senstivity equal to $K_1 = 2/\pi = 0.636$ V/rad.

2.1.4 ANALOG MULTIPLIERS

The expressions and waveforms obtained in the preceding subsections for the phase detector output signal are only valid within the limits of application of the hypotheses: perfect diodes, perfectly sinusoidal generators having zero or very low internal resistance, and possibility of connecting generators in exact phase opposition.

In actual fact, the diode characteristic is more complex; it features a spurious capacitance at high frequencies and, finally, diodes are not all identical. Moreover, the generators do not supply perfectly sinusoidal signals, especially in cases where the internal impedance has to be very low. Finally, it is fairly difficult to obtain signals e_S and e_S' in strict phase opposition and

---

*Whence the name of pseudomultiplier sometimes used to designate this device.

[†]The function $\text{Sign}[\cdot]$ corresponds to the instantaneous sign of the quantity within the square brackets.

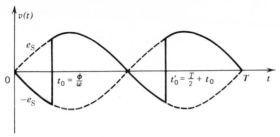

FIGURE 2.9. Output waveform in a four-diode phase detector (before filtering) when $B \gg A$.

with equal amplitude (as regards the latter point, there is no doubt that devices incorporating center-tap transformers are more efficient than those using amplifiers as phase shifters). The overall result is that the conditions of conduction of the diodes change more gradually, switching operations are less straightforward, sharp points are rounded off, and the final output signal is fairly close to that that would be obtained with a real multiplier (3).

On the other hand, devices of varying complexity, known as analog multipliers, are available. They comprise diodes or transistors, symmetrical amplifiers or center-tap transformers. Providing they are used with care, these devices actually operate like multipliers. In particular, when the operating frequency is not too high, certain devices on the market, presented in integrated circuit form, really work like a multiplier of signal $e_S(t)$ by signal $e_R(t)$ (4,5).

The main difficulty underlying the design and construction of a real multiplier is to obtain a device capable of multiplying effectively e_S by e_R in circumstances where the amplitudes of the signals involved vary considerably. If a device is available that is capable of performing the multiplication efficiently when the amplitudes of the applied signals are equal, this difficulty can be avoided. This is so, in particular, when "squarers" are used (device by which the input signal is squared), as shown in the diagram of Fig. 2.11. Since the signals applied to the two squarers are, respectively, $(e_S + e_R)$

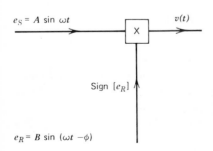

$e_S = A \sin \omega t$

$v(t)$

X

Sign $[e_R]$

$e_R = B \sin(\omega t - \phi)$

FIGURE 2.10. Model of a four-diode phase detector if $B \gg A$ (or if $B \geqslant A$, when e_R is a square-wave signal expressed

$$e_R = B \operatorname{Sign}[\cos(\omega t - \phi)]).$$

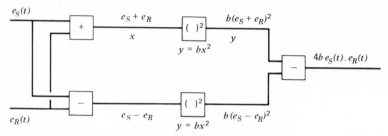

FIGURE 2.11.   Model of an analog multiplier using two "squarers."

and $(e_S - e_R)$, by means of a differential amplifier at the output, it is possible to pick up a signal directly proportional to $e_S \cdot e_R$.

If no perfect "squarers" are available, and if we make the layout slightly more complex, we can obtain a fairly close approximation to a multiplier. Let us consider, for example, the diagram of Fig. 2.12, where the input–output characteristic of the nonlinear elements A, B, C, and D is of the type

$$y = ax + bx^2 + cx^3$$

If the input of nonlinear element A is $x_A = e_S + e_R$, the output signal obtained is

$$y_A = a(e_S + e_R) + b(e_S + e_R)^2 + c(e_S + e_R)^3$$

Elements B, C, and D work with signals $x_B = e_S - e_R$, $x_C = -e_S + e_R$, and $x_D = -e_S - e_R$. By means of differential amplifiers the following operations can be performed: $y_A - y_B$ and $y_C - y_D$, then $(y_A - y_B) - (y_C - y_D)$. It is easy to check that:

$$(y_A - y_B) - (y_C - y_D) = 8 b e_S e_R$$

Therefore the output signal will be identical with an actual multiplier, providing the characteristic of the nonlinear elements used can be approximated by a third-degree polynomial.

2.1.5   LINEAR PHASE DETECTORS OVER AN INTERVAL $(0, \pi)$ ("TRIANGULAR" CHARACTERISTIC)

The diode phase detectors described in the preceding subsections produce a characteristic that is sinusoidal (if $B \gg A$) or formed of truncated sinusoids (if $B = A$). Consequently, the slope of the characteristic, that is, the sensitivity K_1 of the phase detector, is not constant. In particular, we note a change in sign corresponding to each interval of value π of the phase difference ϕ between the signals applied.

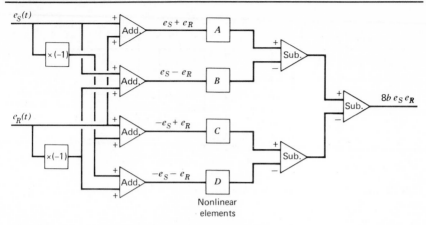

FIGURE 2.12.   Model of an analog multiplier using nonlinear elements.

It may be of interest, in certain cases, to linearize, insofar as possible, the characteristic within an interval $(0, \pi)$ or extend the interval where the slope sign is constant, or even attempt both these improvements simultaneously. Fairly complex analog devices have been proposed. We mention, in particular, the device known as "Tanlock" (6, 7), a detector having a wide dynamic range (8) or a phase-subtraction device (9). However, when the signals involved are square waves, or when it is possible to transform them into square waves, phase detectors featuring a linear characteristic over a certain interval are fairly easy to construct.

Let us consider the four-diode device shown in Fig. 2.7. We have seen that if the waveform of signal e_R is square and if its amplitude exceeds the peak value of signal e_S, the device operates like a multiplier of signal e_S by the function $\text{Sign}[e_R]$. Let us now suppose that signal e_S is also a square wave and has the same period as signal e_R. The signal $v(t)$ then obtained at the detector output is that shown in Fig. 2.13. Its value is $+A$ if e_S and e_R have the same sign and $-A$ when e_S and e_R have opposite signs. We can then write

$$v(t) = A \, \text{Sign}\left[\, e_S \,\right] \cdot \text{Sign}\left[\, e_R \,\right]$$

If we use the conventional term "phase difference" to designate the quantity $\phi = \omega t_0$, t_0 being the delay of signal e_R as compared with signal e_S and ω being the angular frequency corresponding to the frequency f of the signals (ϕ is the phase difference of the fundamental of signal e_R with regard to the fundamental of signal e_S), we can calculate the value of the dc

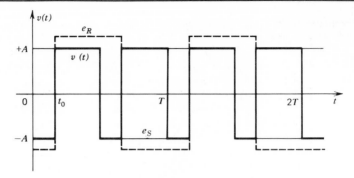

FIGURE 2.13. Output waveform in a four-diode phase detector when input signals are square waves.

component of the signal $v(t)$ as a function of ϕ:

$$\bar{v} = \frac{1}{T} \int_0^T v(t)\, dt$$

When $0 < \phi < \pi$, we get

$$\bar{v} = \frac{A}{T}\left[\int_0^{t_0} -dt + \int_{t_0}^{T/2} dt + \int_{T/2}^{T/2+t_0} -dt + \int_{T/2+t_0}^{T} dt \right] = \frac{A}{T}(T - 4t_0)$$

that is,

$$\bar{v} = \frac{2A}{\pi}\left(\frac{\pi}{2} - \phi\right) \qquad \text{for } 0 < \phi < \pi$$

and

$$\bar{v} = \frac{2A}{\pi}\left(\phi - \frac{3\pi}{2}\right) \qquad \text{for } \pi < \phi < 2\pi$$

(2.5)

The corresponding characteristic* is represented in Fig. 2.14, whilst the phase detector sensitivity value is

$$K_1 = \left| \frac{d\bar{v}}{d\phi} \right| = \frac{2A}{\pi}$$

NUMERICAL EXAMPLE. Signals e_S and e_R are square signals, respectively 1 V and 1.1 V in amplitude. The characteristic is triangular (Fig. 2.14) and the

---

*The characteristic can be made independent of signal e_S amplitude by including a limiter in the circuit preceding the phase detector, designed to deliver a signal of amplitude $\pm A'$ (according to the sign of e_S) and such that $A' < B$. The phase detector sensitivity value is then $2A'/\pi$.

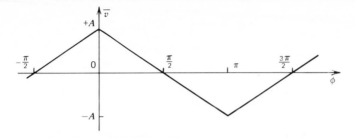

FIGURE 2.14. Characteristic of a four-diode phase detector when input signals are square waves (or characteristic of an EXCLUSIVE-OR logic circuit).

detector sensitivity value is

$$K_1 = \frac{2}{\pi} = 0.636 \text{ V/rad}$$

But signal $v(t)$ of Fig. 2.13, and consequently the characteristic $\bar{v}(\phi)$ of Fig. 2.14, can be obtained more simply than this four-diode device method, by using logical circuits. We can state in standard terms that

$$e_S \Leftrightarrow 1 \qquad \text{when } e_S = +A$$

$$e_S \Leftrightarrow 0 \qquad \text{when } e_S = -A$$

$$e_R \Leftrightarrow 1 \qquad \text{when } e_R = +B$$

$$e_R \Leftrightarrow 0 \qquad \text{when } e_R = -B$$

When the e_S and e_R signals are applied to an AND circuit, we get a 1 output when e_S and e_R are simultaneously positive and a 0 output otherwise. If the same operation is effected in another AND circuit, with the complementary signals $\overline{e_S}$ and $\overline{e_R}$, we get a 1 output when e_S and e_R are simultaneously negative and a 0 output otherwise. If the output signals of the two AND circuits are applied to an OR circuit (see diagram of Fig. 2.15) we get 1 when signals e_S and e_R have the same sign and 0 otherwise. We then constructed what the logical engineers call an EXCLUSIVE–OR circuit. The EXCLUSIVE–OR output signal is similar to that of the four-diode detector, except that the levels are $(0, 1)$ instead of $(-A, +A)$. If this signal is applied to a trigger producing $-A'$ or $+A'$ outputs, according to whether the signal is 0 or 1, the network implements the function

$$v(t) = A' \text{ Sign} \left[e_S \right] \cdot \text{Sign} \left[e_R \right]$$

If the $v(t)$ signal dc component $\bar{v}$ is extracted by low-pass filtering, we get

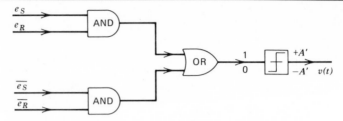

FIGURE 2.15.   EXCLUSIVE-OR logic circuit used as phase detector.

a characteristic similar to that represented in Fig. 2.14 when replacing A by A'. In particular, the sensitivity value of the phase detector is

$$K_1 = \frac{2A'}{\pi}$$

It is also easy to obtain a linear phase comparator characteristic over an interval $(0, \pi)$, when signals e_S and e_R have a square waveform, by using an AND circuit. Although in this case the operation $\text{Sign}[e_S] \cdot \text{Sign}[e_R]$ is omitted, it is a simple matter to check that if we filter the output signal of a straightforward AND circuit followed by a trigger supplying $+A'$ or $-A'$, according to whether the AND circuit output is equal to 1 or 0, we get the following dc-component values: 0 when the signals are in phase, $-A'/2$ when they are in quadrature, and $-A'$ if they are in opposite phase. As compared with the EXCLUSIVE–OR circuit, the AND circuit phase detector sensitivity is half that of the former system and, furthermore, the characteristic is centered round the value $-A'/2$.

2.1.6   LINEAR PHASE DETECTOR OVER AN INTERVAL $(0, 2\pi)$ ("SAWTOOTH" CHARACTERISTIC)

When the signals to be compared are naturally square waves or again when it is possible to transform sinusoidal signals into square signals by means of peak-clipping, it may be of interest to use as phase detector a network of the type shown in the circuit diagram of Fig. 2.16.

The sinusoidal signals e_S and e_R are respectively applied to a "limiter-differentiator-limiter" chain producing a sharp positive pulse at each zero-crossing with a positive slope of input signals e_S and e_R. If ϕ is the phase difference of signal e_R with respect to signal e_S, the pulse train obtained at the lower chain output is delayed by $t_0 = \phi/\omega$ with respect to the upper chain output pulse train.

The pulse trains are applied to a flip-flop circuit (bistable multivibrator)

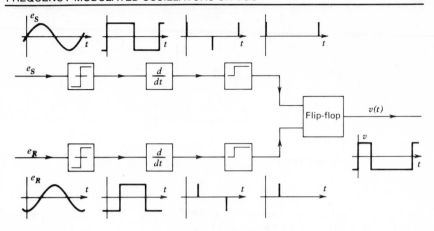

FIGURE 2.16. Circuits and waveforms in a linear over $(0, 2\pi)$ phase detector.

producing a $+A'$ output for the first pulse arrival and a $-A'$ output for the next one. Within a period T of the input signals, the output $v(t)$ obtained is thus equal to $+A'$ from $t=0$ to t_0 and $-A'$ from t_0 to T. The dc component of this signal is

$$\bar{v} = \frac{1}{T} \int_0^T v(t) \, dt$$

$$= \frac{1}{T} \left[\int_0^{t_0} A' \, dt + \int_{t_0}^T - A' \, dt \right] = \frac{A'}{T} (2t_0 - T)$$

that is,

$$\bar{v} = \frac{A'}{\pi} (\phi - \pi) \qquad \text{for } 0 < \phi < 2\pi \tag{2.6}$$

Function $\bar{v}(\phi)$, represented in Fig. 2.17, is monotonic within an interval $(0, 2\pi)$.* The slope representing the phase detector sensitivity is equal to $K_1 = A'/\pi$.

2.2 FREQUENCY-MODULATED OSCILLATORS OR VCO

The frequency-modulated oscillators used in the phase-locked loops are not fundamentally different from those employed for other uses (frequency

---

*When this type of detector is used in a phase-locked loop, the VCO signal is in opposite phase to the input signal (for an initial frequency difference of zero).

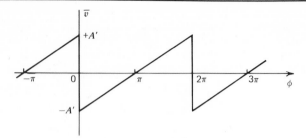

FIGURE 2.17.   Characteristic of a flip-flop phase detector.

modulation, automatic frequency control, etc.). The main properties generally required of a VCO are as follows:

• Adequate central frequency stability,
• A relatively high modulation sensitivity K_3 (in Hz/V or in rad/s/V). The latter point is important because K_3 enters directly in the expression of the open-loop gain K and the latter is generally fairly high. These first two properties are rather contradictory and, for the sake of stability, we sometimes have to make do with a fairly poor modulation sensitivity and use an operational gain amplifier K_2 preceding the VCO (see Section 2.3). The dc drift from this amplifier must of course remain compatible with the VCO inherent stability.
• The widest possible frequency deviation, theoretically $\pm K$, in order to derive maximum benefits from the whole synchronization range of the loop (Chapter 1). Generally speaking, a smaller frequency variation range is acceptable, but, in this case, care must be taken in the way in which the theoretical results are used, in particular those involving acquisition (Chapters 10 and 11), for they are all based on the assumption that the VCO linear frequency variation range is at least equal to the open-loop gain.
• A tolerance of 5% or 10% on the linearity of the modulation characteristic within the frequency variation range is more often than not acceptable, except for certain applications, such as the use of a phase-locked loop as a good linear frequency discriminator.
• Finally, the modulation bandwidth rarely needs to be very wide, except, once again, in the case of frequency demodulator uses (and for high-gain first-order loops).

The VCO technology depends on the values required for the above magnitudes and the frequency at which the loop is going to operate.

For relatively low-frequency applications and up to a few megahertz, providing the central frequency stability requirements are not too stringent, an astable multivibrator can be used. The frequency is varied by means of the capacitor charging current (using transistors Q_3 and Q_4) as in the diagram of Fig. 2.18. The essentially square-wave signal obtained (at transistor Q_1 and Q_2 collectors) is, as noted in Section 2.1, particularly well adapted to diode or logic circuit phase-detector operation.

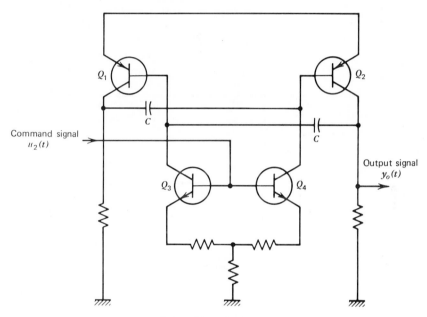

FIGURE 2.18.   Frequency-controlled astable multivibrator.

In fact, in this frequency range, the frequency of a large number of oscillator types can be varied by means of a command signal. In particular, all the oscillators usually frequency-adjusted by a judicious choice of resistor or capacitor values in the feedback network (double-T, bridged-T, Wien bridge, etc.) can be used, by replacing the resistors and capacitors by elements serving the same purpose but variable by means of a bias voltage. As an example, the diagram in Fig. 2.19 shows an RC circuit phase-shift oscillator, where the resistors are Field Effect Transistors.* By means of the

*Another solution would be to use resistances that vary according to the voltage applied to the resistor terminals (voltage-variable resistances), but this is rarely adopted in practice.

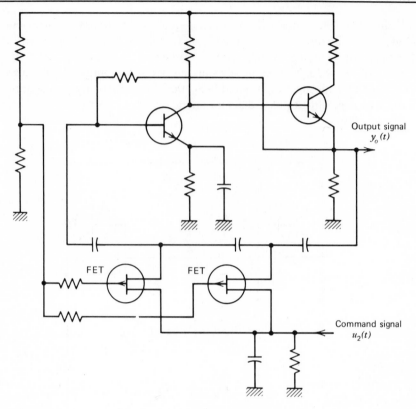

FIGURE 2.19.   Frequency-modulated phase-shift oscillator.

command signal, the resistance values can be varied and thus the oscillator frequency.

When the frequency is higher (up to a few hundred megahertz) we can consider the whole series of LC oscillators, where the value of one of the elements is controlled by means of the command signal. In most cases, the element selected will be a capacitor composed entirely or in part, of the capacitance of a reverse-biased semiconductor junction* (diode varactor or varicap). The simplest circuits for this purpose are the Hartley (Fig. 2.20a) and Clapp (Fig. 2.20b) type oscillators.

For the highest frequencies in the range, the oscillating circuits are replaced by resonant lines tuned by means of varactor diodes. In an extremely wide range of frequencies (from a few megahertz to a few

---

*Certain circuits use voltage-tunable ferroelectric capacitors.

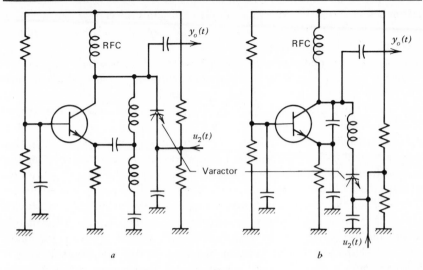

FIGURE 2.20. LC oscillators frequency controlled by using a varactor diode: (a) Hartley-type oscillator, (b) Clapp-type oscillator.

gigahertz), tunnel diode oscillators can also be used. In this case, the oscillator frequency is also varied by means of a varactor diode in the output oscillating circuit, as shown in the diagram of Fig. 2.21.

In certain very high-frequency applications (several gigahertz), the VCO could be a klystron reflex, where the reflector voltage is used to control the frequency; it could be a frequency-modulated magnetron, where the

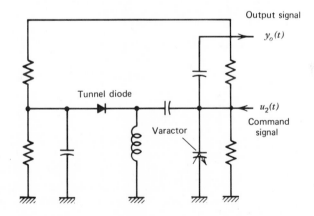

FIGURE 2.21. Tunnel diode frequency-controlled oscillator.

frequency is varied by adjusting the electron gun grid voltage; or, again, it could be a BWO (backward-wave oscillator) voltage-controlled by adjustment of the helix grid voltage.

However, as pointed out at the beginning of this section, one of the essential qualities of a VCO is its central frequency stability. When a very high degree of stability is required, the commonest solution is to use a quartz-stabilized oscillator (VCXO). Up to a certain point, the series-resonant frequency of a quartz crystal can be controlled by means of a variable capacitor (varactor diode) connected in series with the crystal. This gives a certain frequency variation range without too much deteriorating the central frequency stability. But stability, which can, in fact, be improved by placing the oscillator in a thermostatic container, is always obtained to the detriment of the frequency variation range and the modulation sensitivity (10). The systems most frequently used are either the Clapp-type oscillator, where the oscillating circuit is replaced by the crystal, the varactor diode and the characteristic linearization elements, or the oscillators having a feedback circuit comprising the quartz crystal, the varactor diode, and the linearization elements (Fig. 2.22). The quartz crystals used in the VCXO are generally AT-cut crystals (for stability), oscillating at fundamental frequency, which situates their oscillation frequency range between a few MHz and 20 or 30 MHz.

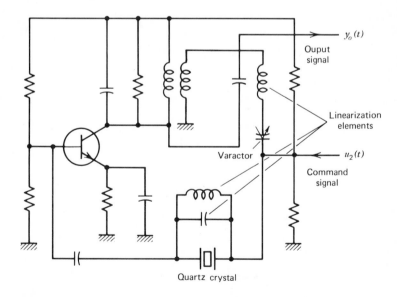

FIGURE 2.22. Quartz-crystal frequency-controlled oscillator.

When the VCO central frequency stability is required to be exceptionally high in very low or very high frequency loops, we use a VCXO followed either by a logic circuit frequency division chain or a frequency multiplication chain, composed of transistors or varactor diodes,* or step-recovery diodes.

NUMERICAL EXAMPLE.   We have to construct a 100-MHz phase-locked loop, having a synchronization range of ± 10 kHz. We have a phase detector of sinusoidal characteristic and slope $K_1 = 1$ V/rad. The VCO frequency stability must be such that the phase error involved will not exceed 0.1 rad. We are required to define the main characteristics of the VCO.

Since the synchronization range is equal to the loop gain (Chapter 1) we shall have

$$K = K_1 K_3 = 2\pi \times 10^4 \text{ rad/s}$$

The VCO modulation sensitivity will thus be

$$K_3 = 2\pi \times 10^4 \text{ rad/s/V} \quad \text{or} \quad K_3 = 10 \text{ kHz/V}$$

The VCO will also have to have a frequency variation range of ± 10 kHz around its central frequency (in order to be able to follow the input signal frequency within the synchronization range).

Let Δf_0 be the possible central frequency drift. The corresponding phase error is given by Eq. 1.6:

$$\Delta \phi = \text{Arc} \sin \frac{2\pi \Delta f_0}{K} = 0.1 \text{ rad}$$

which means that

$$\Delta f_0 \cong \frac{0.1 K}{2\pi} = 10^3 \text{ Hz}$$

The relative frequency stability required of the VCO is then

$$\frac{\Delta f_0}{f_0} = \frac{10^3}{100 \times 10^6} = 10^{-5}$$

Such a high degree of stability could only be achieved with a quartz-stabilized oscillator. Taking account of the loop frequency, we shall use a 20-MHz VCXO followed by a frequency multiplier by 5. The VCXO frequency variation range and modulation sensitivity required in this case will then be, respectively, ± 2 kHz and 2 kHz/V.

*In this case, the varactor diodes are not used as variable capacitors, as in the VCO, but as nonlinear elements favoring the production of high-order harmonics.

2.3 LOOP FILTERS

Loop filters are low-pass filters that are set between the phase detector output and the VCO modulation input. As will be seen in the subsequent chapters, the transfer function of the loop filter has a considerable influence on the properties of the loop and provides a means of modifying its performance by a judicious choice of the parameters introduced. The transfer functions most frequently used are very simple and obtained either by means of passive elements or by the use of passive elements used as feedback network for a high-gain amplifier. The latter type of set-up is sometimes referred to as an active filter.

In some cases, even when an entirely passive filter is required, a gain amplifier K_2 has to be fitted between the phase detector and the VCO. This is, in particular, the case when the phase detector sensitivity K_1 and the VCO modulation sensitivity K_3 are not high enough to produce a given loop gain K. The use of a gain amplifier K_2 enables us to overcome this difficulty, since the loop gain K then becomes

$$K = K_1 K_2 K_3$$

Generally speaking, the amplifier K_2 will be located between the loop filter and the VCO, but where this amplifier proves indispensable, it may be wiser to include it in an active filter, as discussed further on.

The most straightforward case we can conceive is the absence of a loop filter: the transfer function is $F(j\omega) = 1$ and the phase-locked loop is then of the first order (see Section 3.3).*

The simplest low-pass filter to construct is the RC filter of the transfer function:

$$F(j\omega) = \frac{1}{1 + j\omega RC}$$

that is,

$$F(j\omega) = \frac{1}{1 + j\omega\tau_1} \qquad \text{with } \tau_1 = RC \qquad (2.7)$$

The use of such a filter produces a second-order loop. However, the performances obtained are relatively restricted, mainly because only one

---

*In fact, we have seen that phase detectors require in all cases the use of a low-pass filter, if only to remove the components at frequency f and harmonic frequencies of f which accompany the dc component output. But it is always assumed that the bandwidth of this filter is sufficiently wide as not to influence the properties of the loop (see Chapter 4).

additional parameter is involved: the time constant τ_1. We shall see further on that this prevents an independent choice of the two essential parameters of a second-order loop, namely the natural angular frequency and the damping factor, when the loop gain K is otherwise given (see Subsection 3.6.2).

If we add a resistor in series with the capacitor C of the filter, we obtain the required additional parameter. The transfer function of the filter represented in Fig. 2.23 is given by

$$F(j\omega) = \frac{U_2(j\omega)}{U_1(j\omega)} = \frac{R_2 + 1/jC\omega}{R_1 + R_2 + 1/jC\omega}$$

$$= \frac{1 + j\omega R_2 C}{1 + j\omega(R_1 + R_2)C}$$

that is,

$$F(j\omega) = \frac{1 + j\omega\tau_2}{1 + j\omega\tau_1} \quad \text{with } \tau_2 = R_2 C \quad \text{and} \quad \tau_1 = (R_1 + R_2)C \quad (2.8)$$

By judicious choice of elements R_1, R_2 and C, we can obtain, independently, the time constants τ_1 and τ_2. The main technological limitation of such a filter becomes evident if we try to construct a very low natural angular frequency loop (known as a "narrow" loop). In this case, as will be seen further on, the natural angular frequency ω_n is related to the loop gain K and the time constant τ_1 by Eq. 3.41:

$$\omega_n^2 = \frac{K}{\tau_1}$$

If the loop gain K is high and if we require a relatively low natural angular

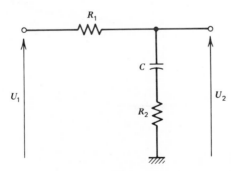

FIGURE 2.23. A very commonly used loop filter for a second-order phase-locked loop.

frequency ω_n, we are obliged to have a very high time constant τ_1: a few hundred, or even a few thousand seconds. It is not always easy to find, in compact form, capacitors C capable of producing such a time constant. Another solution is then to use an active filter of the type described below.

Let us consider the layout shown in Fig. 2.24, consisting of a resistor R_1 connected in series with an operational amplifier of gain $-G$ (the minus sign to indicate the fact that the output signal is in opposite phase to the input signal). A network comprising a resistor R_2 and a capacitor C supplies feedback to this amplifier. If we assume the input impedance of the amplifier to be far greater than the impedances R_1 and $R_2 + 1/jC\omega$, the filter transfer function is written

$$F(j\omega) = \frac{U_2(j\omega)}{U_1(j\omega)}$$

$$= -\frac{1}{R_1}\left[\frac{1}{R_2 + 1/jC\omega} + \frac{1}{G}\left(\frac{1}{R_1} + \frac{1}{R_2 + 1/jC\omega}\right)\right]^{-1}$$

which can also be expressed

$$F(j\omega) = -G\frac{1 + j\omega R_2 C}{1 + j\omega(R_1 + GR_1 + R_2)C}$$

that is,

$$F(j\omega) = -G\frac{1 + j\omega\tau_2}{1 + j\omega\tau_1} \quad \text{with } \tau_2 = R_2 C \quad \text{and} \quad \tau_1 = (R_1 + GR_1 + R_2)C \quad (2.9)$$

The transfer function of the resulting filter can be formulated in a manner similar to that of the passive filter, apart from the multiplying constant $-G$.

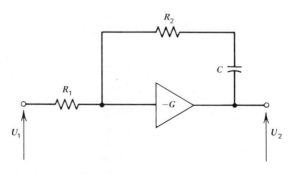

FIGURE 2.24.   Another type of loop filter for a second-order phase-locked loop.

The loop gain K is enhanced by this amplification and becomes

$$K = K_1 K_3 G$$

Two eventualities are possible.

In the first case, the product $K_1 K_3$ is inadequate and it is decided to use an active filter of gain G to increase the loop gain K. The natural angular frequency of the loop ω_n is always given by Eq. 3.41:

$$\omega_n^2 = \frac{K}{\tau_1} = \frac{K_1 K_3 G}{(R_1 + GR_1 + R_2)C}$$

If the gain G is sufficiently high, $\tau_1 \cong GR_1 C$, and consequently

$$\omega_n^2 \cong \frac{K_1 K_3 G}{GR_1 C} = \frac{K_1 K_3}{R_1 C}$$

To obtain a given natural angular frequency ω_n for a given $K_1 K_3$ product, the same time constant $R_1 C$ is required as for a passive filter: we have thus resolved the loop gain question, but realization of the time constant $R_1 C$ is still problematical.

NUMERICAL EXAMPLE. We have a phase detector of sensitivity $K_1 = 0.5$ V/rad and a VCO of sensitivity $K_3 = 2 \times 10^4$ rad/s/V. We have to construct a second-order loop of natural angular frequency $\omega_n = 10$ rad/s.

If a passive filter is used, the loop gain is equal to $K = 10^4$ rad/s and the time constant $R_1 C$ is equal to

$$R_1 C = \frac{K}{\omega_n^2} = \frac{10^4}{10^2} = 100 \text{ s}$$

If an active filter, with a gain $G = 100$ is used, the loop gain value is $K = 10^6$ rad/s while the product $R_1 C$ is still equal to 100 s.

In the second case, the product $K_1 K_3$ is adequate, but to obtain a given ω_n, the necessary time constant $R_1 C$ cannot easily be realized with a passive filter. An active filter of gain G can be used, but the product $K_1 K_3 G$ will then have to be divided by G, using, for instance, a potentiometric divider, to keep the loop gain at its value $K_1 K_3$. The time constant τ_1 of Eq. 2.9 to be achieved is still the same, but more easily obtained, the product $R_1 C$ being multiplied by G.

NUMERICAL EXAMPLE. We have a phase detector of sensitivity $K_1 = 4$ V/rad and a VCO of sensitivity $K_3 = 25 \times 10^4$ rad/s/V. We have to construct a second-order loop of loop gain $K = 10^6$ rad/s and of natural angular frequency $\omega_n = 10$ rad/s. It is observed that the value of the product $K_1 K_3$ is a means of obtaining the loop gain value without further amplification.

If a passive filter is used, the time constant τ_1 must be chosen so that:

$$\tau_1 = \frac{K}{\omega_n^2} = \frac{10^6}{10^2} = 10^4 \text{ s}$$

If an active filter of gain $G = 100$ is used, and a potentiometric divider of gain $1/100$ is placed between the detector and the loop filter, the $R_1 C$ elements should confirm

$$GR_1C = \frac{K}{\omega_n^2} = 10^4 \text{ s}$$

that is $R_1 C = 10^2$ s.

Remark 1   The transfer function, in the case of the active filter, is

$$F(j\omega) = -G\frac{1 + j\omega\tau_2}{1 + j\omega\tau_1}$$

We have seen that the constant G corresponds to the operational amplifier gain. The minus sign, necessary for the stability of the filter, will slightly alter the behavior of the loop as described briefly in Chapter 1, where we assumed the constants K_1 and K_3 to be positive. We have seen that most phase detectors exhibit an alternating positive-negative slope. We shall see in Chapter 10 that, of the two possible operating points, corresponding to the two sensitivities of equal absolute value but opposite sign, only one corresponds to stable conditions. According to the sign of K_1 and K_3, this may be, for example, for the VCO signal in lead-quadrature with respect to the input signal. If in the phase detector–VCO modulation input relationship, a change of sign is introduced, the stable operating position will correspond to the phase detector slope opposite in sign, as, for the same example, for the VCO signal in lag-quadrature with respect to the input signal. As the stable point is generally only of interest as a reference and not by reason of its exact value, the minus sign can be disregarded.

If the phase detector is of the linear type over an interval $(0, 2\pi)$, the sign of K_1 or K_3 has to be changed, in order to preserve an operating stability when a minus sign is introduced through the active filter.*

---

*It should be remembered that in this case (Subsection 2.1.6) the stable operating point corresponds to the VCO signal in opposite phase with respect to the input signal.

Remark 2. If the gain G of the operational amplifier is very high, the filter transfer function can be written

$$F(j\omega) = -G\frac{1+j\omega\tau_2}{1+j\omega\tau_1} \cong -G\frac{1+j\omega\tau_2}{1+j\omega GR_1C}$$

$$\cong -G\frac{1+j\omega\tau_2}{j\omega GR_1C}$$

If we ignore the minus sign (see Remark 1 above) we obtain the following transfer function:

$$F(j\omega) = \frac{1+j\omega\tau_2}{j\omega\tau_1'} \quad \text{with } \tau_2 = R_2C \quad \text{and} \quad \tau_1' = R_1C \qquad (2.10)$$

This approximate transfer function (strictly speaking, not applicable where very low frequencies are concerned) presents the advantage of simplifying certain calculations, while remaining a very close approximation to the active filter transfer function. It is for this reason that it is very often used in specialized technical literature. It should be noted that a loop filter with a transfer function corresponding to Eq. 2.10 would have an infinite gain with a dc current. The loop gain would then be infinite (we shall meet with this property again in the following chapters).

We should consequently bear in mind that we are dealing with only an approximation, the use of which is justified if the gain G is necessary to increase the loop gain K and if, to preserve the value of ω_n, we have increased simultaneously τ_1, keeping the value of the product R_1C.

But if the gain G is only used to facilitate realization of the time constant τ_1 (with, for example, the simultaneous use of a potentiometric divider of value $1/G$), the approximation is no longer acceptable and its use should only be decided after careful consideration.

NUMERICAL EXAMPLE. We have a phase detector and a VCO such that $K_1 = 0.5$ V/rad and $K_3 = 2 \cdot 10^4$ rad/s/V. We require a loop where $K = 10^6$ rad/s and $\omega_n = 10$ rad/s. If an active filter of gain $G = 100$ is used and if the elements R_1 and C of the filter are so selected that $R_1C = 100$ s, the approximation is justified since the product $GR_1C = 10^4$ s is very large.

Using the same elements, we require to construct a loop such that $K = 10^4$ rad/s and $\omega_n = 10$ rad/s. If we decide to use an active filter of gain $G = 200$, and elements R_1 and C chosen so that $R_1C = 0.5$ s (and a $1/200$ gain potentiometric divider), the approximation is not justified since the product $GR_1C = 100$ s is not large enough.

CHAPTER 3
GENERAL EQUATIONS

We have seen in Chapter 1 that the phase-locked loop (represented in Fig. 1.1) is a device by means of which a frequency-modulated oscillator (VCO) delivers an output signal y_0, in synchronism with the input signal y_i. In fact, when y_i and y_0 are sinusoidal signals, out of synchronism, the sinusoidal phase detector output signal is itself a sinusoidal signal, the frequency of which is equal to the difference in frequency between signals y_i and y_0 (Eq. 1.1). But when synchronism is obtained, after an acquisition period, which will be studied in Chapters 10 and 11, the two signals are only separated by a phase error given by Eq. 1.6.

We now have to consider the behavior of the device when, with the loop locked, a modulation or disturbance is applied to the input signal phase. To this purpose, we have to write the equations governing the behavior of each component part of the loop. In particular, the phase detector characteristic is not linear. It is generally a periodic function of period 2π of the instantaneous phase difference between the signals applied (Section 2.1). This feature considerably complicates the behavior analysis of the device since, as we shall see, it is governed by a nonlinear differential equation.

However, since we are attempting to build a servo device, it stands to reason that we want it to work properly. If it works well, the difference in phase between signals y_i and y_0 will remain slight. It will then be possible to substitute for the real characteristic of the phase detector an ideal linear characteristic and the loop will then be governed by a linear differential equation. In most of the fields where phase-locked loops are used, every effort is made to apply the linear behavior hypothesis. Where this hypothesis is no longer applicable, other methods have to be used (see Chapters 10, 11, and 12).

The characteristics of the phase detectors studied in Section 2.1 (Eqs. 2.1–2.6) are such that the output signal is not null when the phase difference between the signals applied is null. The output signal is null when the phase difference is equal to $\pi/2$ for sinusoidal and "triangular" phase detectors and when it is equal to π for "sawtooth" phase detectors.

42

Since the static operating point of the servo device corresponds to a null signal at the phase detector output, we shall presume that a constant phase difference exists between the signals applied, such that the comparator output signal is null at the static operating point. Using this hypothesis as a basis, we can now translate all the characteristics of an angle $\pi/2$ or π, so that the output signal is null for zero instantaneous phase difference. Consequently, the results that we shall obtain for the instantaneous phase error between y_i and y_0 will only be valid to within $\pi/2$ or π.

On the other hand, the slope K_1 of the characteristic around the static operating point is positive or negative, according to the case under consideration. We shall assume that it is always possible to choose a positive K_1, even if we have to change the sign of K_3 (or that of a possible additional amplifier K_2) to obtain this end.

In the light of these remarks, the various characteristics may be replaced by those represented in Fig. 3.1. In the linear domain, they will all be replaced by the ideal characteristic: the straight line $u_1 = K_1\phi$ (Fig. 3.1). Since this ideal characteristic is independent of the phase detector used, the

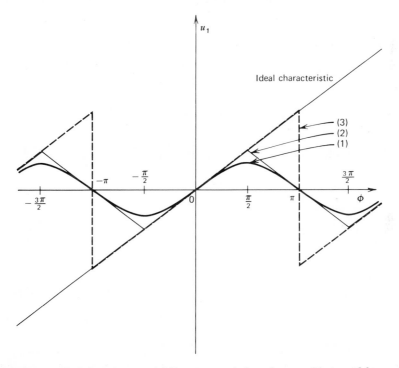

FIGURE 3.1.   Ideal characteristic of different types of phase detector: (1) sinusoidal type, (2) "triangular" type, (3) "sawtooth" type.

linear operating conditions are obviously independent of the type of phase detector. We shall establish the linear equations by "linearizing" the sinusoidal phase detector characteristic.

3.1 GENERAL TIME DOMAIN EQUATION FOR A LOOP USING A SINUSOIDAL CHARACTERISTIC PHASE DETECTOR

Let us consider the diagram shown in Fig. 3.2, for which the input signal y_i and the VCO output signal y_0 are expressed

$$y_i(t) = A \sin \left[\omega t + \phi_i(t) \right]$$

$$y_0(t) = B \cos \left[\omega t + \phi_0(t) \right]$$

These signals have not necessarily the same angular frequency, the difference being easily included in $\phi_i(t) - \phi_0(t)$. The phase detector used is of the sinusoidal type (see Section 2.1) and the error signal u_1 can be expressed

$$u_1(t) = K_1 \sin \left[\phi_i(t) - \phi_0(t) \right] \tag{3.1}$$

Let $F(j\omega)$ be the loop filter transfer function and $f(t)$ its impulse response. These two functions are both Fourier transforms, that is to say,

$$f(t) = \int_{-\infty}^{+\infty} F(j2\pi f) e^{j2\pi ft} \, df$$

$$F(j2\pi f) = \int_{-\infty}^{+\infty} f(t) e^{-j2\pi ft} \, dt$$

The loop filter output voltage is written, making allowances for a possible

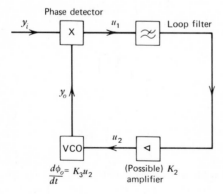

FIGURE 3.2. Block diagram of a phase-locked loop.

amplification of gain K_2 (see Section 2.3), as

$$u_2(t) = K_2 u_1(t) * f(t) \tag{3.2}$$

The symbol * represents the convolution product, that is to say,

$$u_2(t) = K_2 \int_{-\infty}^{+\infty} u_1(\tau) f(t - \tau) \, d\tau$$

Finally, since the VCO is a frequency-modulated oscillator, K_3 being its modulation sensitivity in rad/s/V,

$$\frac{d\phi_0}{dt} = K_3 u_2(t) \tag{3.3}$$

If we combine Eqs. 3.1, 3.2, and 3.3 we get the general time-domain equation that governs the behavior of a loop using a sinusoidal phase detector:

$$\frac{d\phi_0}{dt} = K_1 K_2 K_3 \{ \sin [\phi_i(t) - \phi_0(t)] * f(t) \}$$

The product $K_1 K_2 K_3$ is replaced by $K = K_1 K_2 K_3$. The constant K then represents the servo device open-loop gain (if there is no amplifier between the loop filter and the VCO modulation input, we can state simply $K_2 = 1$). The general equation is thus

$$\frac{d\phi_0}{dt} = K \{ \sin [\phi_i(t) - \phi_0(t)] * f(t) \} \tag{3.4}$$

3.2 GENERAL LINEARIZED EQUATIONS

Linearization of the problem consists in assuming the quantity $[\phi_i(t) - \phi_0(t)]$ to be small enough for the sine to be replaced by the corresponding angle. From this we derive the general linearized equation in the time domain:

$$\frac{d\phi_0}{dt} = K [\phi_i(t) - \phi_0(t)] * f(t) \tag{3.5}$$

The convolution product is not very easy to handle as an operator, mainly, perhaps, because we are unaccustomed to it. We generally prefer to reason in terms of frequency rather than time. But, when two magnitudes are related by a convolution product, their Fourier transforms are related by a simple product. If we take the Fourier transforms of each of the two parts of

Eq. 3.5 and call $\Phi_i(j\omega)$ and $\Phi_0(j\omega)$ the Fourier transforms of $\phi_i(t)$ and $\phi_0(t)$, we get

$$j\omega\Phi_0(j\omega) = K[\Phi_i(j\omega) - \Phi_0(j\omega)] \cdot F(j\omega)$$

from which we derive

$$[j\omega + KF(j\omega)]\Phi_0(j\omega) = KF(j\omega)\Phi_i(j\omega)$$

that is,

$$H(j\omega) = \frac{\Phi_0(j\omega)}{\Phi_i(j\omega)} = \frac{KF(j\omega)}{j\omega + KF(j\omega)} \tag{3.6}$$

The function $H(j\omega)$ is the general linearized transfer function of a phase-locked loop.

The instantaneous phase error is given by $\phi(t) = \phi_i(t) - \phi_0(t)$. The Fourier transform $\Phi(j\omega)$ is then given by

$$\Phi(j\omega) = \Phi_i(j\omega) - \Phi_0(j\omega)$$

from which we deduce that

$$\frac{\Phi(j\omega)}{\Phi_i(j\omega)} = \frac{j\omega}{j\omega + KF(j\omega)} \tag{3.7}$$

The quantity $\Phi(j\omega)/\Phi_i(j\omega) = 1 - H(j\omega)$ is the error function of the phase-locked loop.

If we wish to use operational notation, we can call s the operational variable, which gives for the transfer function:

$$H(s) = \frac{\Phi_0(s)}{\Phi_i(s)} = \frac{KF(s)}{s + KF(s)} \tag{3.8}$$

while the error function is given by

$$1 - H(s) = \frac{\Phi(s)}{\Phi_i(s)} = \frac{s}{s + KF(s)} \tag{3.9}$$

3.3 LINEARIZED EQUATIONS OF A FIRST-ORDER LOOP

A phase-locked loop is said to be of the first order when it comprises no loop filter (Fig. 3.3). The transfer function is then written

$$H(j\omega) = \frac{K}{j\omega + K} \tag{3.10}$$

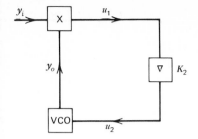

FIGURE 3.3.   First-order phase-locked loop.

while the error function is written

$$1 - H(j\omega) = \frac{j\omega}{j\omega + K} \tag{3.11}$$

On the basis of the transfer and error functions, it is very simple to obtain the time-domain equations describing the way in which the loop behaves linearly. For example, using Eq. 3.10 as a basis, we can write, in operational notation,

$$(s + K)\Phi_0(s) = K\Phi_i(s)$$

which enables us to find

$$\frac{d\phi_0}{dt} + K\phi_0(t) = K\phi_i(t)$$

In the same way, we can use Eq. 3.11 to obtain

$$\frac{d\phi}{dt} + K\phi(t) = \frac{d\phi_i}{dt}$$

The time-domain equations governing the loop operation are first-order linear differential equations. It is for this reason that this type of loop is said to be of the first order.

Remark.   In this particularly simple case, the general time-domain equations can be obtained directly, since we have simply to note that if there is no loop filter, Eq. 3.2 is written

$$u_2(t) = K_2 u_1(t)$$

which means that Eq. 3.4 becomes

$$\frac{d\phi_0}{dt} = K \sin\left[\phi_i(t) - \phi_0(t)\right] \tag{3.12}$$

Equation 3.12 is the time-domain equation corresponding to a first-order phase-locked loop having a sinusoidal phase detector. It is a nonlinear first-order differential equation, which we shall be using in Chapter 10. If we linearize the phase detector characteristic, we get back to

$$\frac{d\phi_0}{dt} + K\phi_0(t) = K\phi_i(t)$$

3.4 LINEARIZED EQUATIONS OF SECOND-ORDER LOOPS

When we study the phase-locked loop response to various disturbances, we shall see that the performances of the first-order loop are not particularly outstanding. It is consequently necessary, in many cases, to increase the order by inserting a loop filter in the loop. The most frequently used devices are those described in Section 2.3:

- The integrator, with phase-lead correction network, the transfer function of which is given by Eq. 2.10:

$$F(j\omega) = \frac{1 + j\omega\tau_2}{j\omega\tau_1}$$

- The one-pole low-pass filter, the transfer function of which is given by Eq. 2.7:

$$F(j\omega) = \frac{1}{1 + j\omega\tau_1}$$

- The one-pole low-pass filter, with phase-lead correction network, the transfer function of which corresponds to Eq. 2.8:

$$F(j\omega) = \frac{1 + j\omega\tau_2}{1 + j\omega\tau_1}$$

3.4.1 CASE OF THE LOOP FILTER HAVING A TRANSFER FUNCTION $(1 + j\omega\tau_2)/j\omega\tau_1$

The circuit diagram of the loop is shown in Fig. 3.4. It is here assumed that gain $-G$ of the loop filter operational amplifier is very large so that the filter transfer function (Eq. 2.9) is very closely approximated by Eq. 2.10 (the minus sign being omitted in accordance with Remark 1 of Section 2.3):

$$F(j\omega) = \frac{1 + j\omega\tau_2}{j\omega\tau_1}$$

$$\text{with } \tau_1 = R_1 C \quad \text{and} \quad \tau_2 = R_2 C$$

If we insert this expression of $F(j\omega)$ in Eqs. 3.6 and 3.7, we get the transfer and error functions:

$$H(j\omega) = \frac{K + j\omega K\tau_2}{K - \omega^2\tau_1 + j\omega K\tau_2} \tag{3.13}$$

$$1 - H(j\omega) = \frac{-\omega^2\tau_1}{K - \omega^2\tau_1 + j\omega K\tau_2} \tag{3.14}$$

If the operational notation is preferred, we get directly

$$H(s) = \frac{K\tau_2 s + K}{\tau_1 s^2 + K\tau_2 s + K} \tag{3.15}$$

$$1 - H(s) = \frac{\tau_1 s^2}{\tau_1 s^2 + K\tau_2 s + K} \tag{3.16}$$

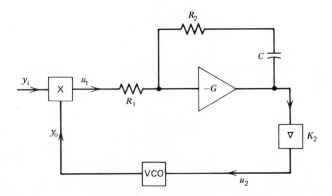

FIGURE 3.4. Second-order phase-locked loop with a loop filter of the type $F(s) = (1 + \tau_2 s)/\tau_1 s$ (active loop filter).

Using the transfer and error functions as a basis, it is very easy to obtain the linear time-domain equations governing the loop operation. In fact, Eq. 3.15 can be written

$$\tau_1 s^2 \Phi_0(s) + K\tau_2 s \Phi_0(s) + K\Phi_0(s) = K\tau_2 s \Phi_i(s) + K\Phi_i(s)$$

That is, reverting to the time variable,

$$\tau_1 \frac{d^2\phi_0}{dt^2} + K\tau_2 \frac{d\phi_0}{dt} + K\phi_0(t) = K\tau_2 \frac{d\phi_i}{dt} + K\phi_i(t) \tag{3.17}$$

In the same way, using Eq. 3.16 as a basis, we get

$$\tau_1 \frac{d^2\phi}{dt^2} + K\tau_2 \frac{d\phi}{dt} + K\phi(t) = \tau_1 \frac{d^2\phi_i}{dt^2} \tag{3.18}$$

These time-domain equations are second-order linear differential equations. It is for this reason that a loop using this type of filter is called a second-order loop.

3.4.2 CASE OF A LOOP FILTER HAVING A TRANSFER FUNCTION $1/(1 + j\omega\tau_1)$

The circuit diagram is that represented in Fig. 3.5. The loop filter transfer function is

$$F(j\omega) = \frac{1}{1 + j\omega\tau_1} \quad \text{with } \tau_1 = R_1 C$$

If we insert this expression in Eqs. 3.6 and 3.7 we obtain for the transfer and error functions, respectively,

$$H(j\omega) = \frac{K}{K - \omega^2\tau_1 + j\omega} \tag{3.19}$$

$$1 - H(j\omega) = \frac{-\omega^2\tau_1 + j\omega}{K - \omega^2\tau_1 + j\omega} \tag{3.20}$$

In operational notation, the transfer and error functions are easily

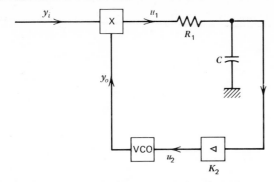

FIGURE 3.5. Second-order phase-locked loop with a loop filter of the type $F(s) = 1/(1 + \tau_1 s)$ (one-pole low-pass filter).

obtained from

$$H(s) = \frac{K}{\tau_1 s^2 + s + K} \tag{3.21}$$

$$1 - H(s) = \frac{\tau_1 s^2 + s}{\tau_1 s^2 + s + K} \tag{3.22}$$

The corresponding equations in terms of time are second-order linear differential equations:

$$\tau_1 \frac{d^2 \phi_0}{dt^2} + \frac{d\phi_0}{dt} + K\phi_0(t) = K\phi_i(t) \tag{3.23}$$

$$\tau_1 \frac{d^2 \phi}{dt^2} + \frac{d\phi}{dt} + K\phi(t) = \tau_1 \frac{d^2 \phi_i}{dt^2} + \frac{d\phi_i}{dt} \tag{3.24}$$

3.4.3   CASE OF A LOOP FILTER HAVING A TRANSFER FUNCTION $(1 + j\omega\tau_2)/(1 + j\omega\tau_1)$

The circuit diagram is that shown in Fig. 3.6. The corresponding loop filter transfer function is given by Eq. 2.8:

$$F(j\omega) = \frac{1 + j\omega\tau_2}{1 + j\omega\tau_1}$$

with $\tau_1 = (R_1 + R_2)C$ and $\tau_2 = R_2 C$

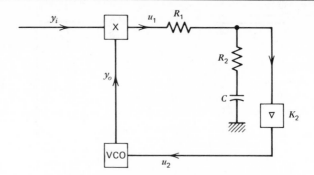

FIGURE 3.6. Second-order phase-locked loop with a loop filter of the type $F(s) = (1 + \tau_2 s)/(1 + \tau_1 s)$ (one-pole low-pass filter and phase-lead correction network).

If we insert Eq. 2.8 in Eqs. 3.6 and 3.7 we get the transfer and error function equations:

$$H(j\omega) = \frac{K + j\omega K\tau_2}{K - \omega^2\tau_1 + j\omega(1 + K\tau_2)} \qquad (3.25)$$

$$1 - H(j\omega) = \frac{-\omega^2\tau_1 + j\omega}{K - \omega^2\tau_1 + j\omega(1 + K\tau_2)} \qquad (3.26)$$

Using operational notation, we get, respectively,

$$H(s) = \frac{K\tau_2 s + K}{\tau_1 s^2 + (1 + K\tau_2)s + K} \qquad (3.27)$$

$$1 - H(s) = \frac{\tau_1 s^2 + s}{\tau_1 s^2 + (1 + K\tau_2)s + K} \qquad (3.28)$$

Using Eqs. 3.27 and 3.28 as a basis, we can easily derive the equations corresponding to the time domain:

$$\tau_1 \frac{d^2\phi_0}{dt^2} + (1 + K\tau_2)\frac{d\phi_0}{dt} + K\phi_0(t) = K\tau_2 \frac{d\phi_i}{dt} + K\phi_i(t) \qquad (3.29)$$

$$\tau_1 \frac{d^2\phi}{dt^2} + (1 + K\tau_2)\frac{d\phi}{dt} + K\phi(t) = \tau_1 \frac{d^2\phi_i}{dt^2} + \frac{d\phi_i}{dt} \qquad (3.30)$$

These are of course second-order linear differential equations, as in the two cases considered above.

3.5 GENERAL TIME DOMAIN EQUATIONS OF A SECOND-ORDER LOOP (FOR A PHASE DETECTOR WITH A SINUSOIDAL CHARACTERISTIC)

The equations given in Section 3.4 are only valid in the linear domain, in other words, so long as the phase error $\phi(t) = \phi_i(t) - \phi_0(t)$ remains small enough for the real phase detector characteristic to be replaced by the ideal characteristic. In the case of the sinusoidal characteristic detector, here used as an example, we can give general time equations that are simpler in form than Eq. 3.4. For this, we simply use directly Eqs. 3.1 and 3.3 and render explicit the time-domain equation representing the loop filter operation. We shall now consider this aspect for the types of loop filter discussed in the preceding section.

3.5.1 CASE OF A LOOP FILTER HAVING A TRANSFER FUNCTION $(1 + j\omega\tau_2)/j\omega\tau_1$

From the expression of the transfer function in operational notation $F(s)$ $= (1 + \tau_2 s)/\tau_1 s$, we derive the loop filter time equation,

$$\tau_1 \frac{du_2}{dt} = u_1 + \tau_2 \frac{du_1}{dt}$$

If we make allowances for the gain K_2 of an amplifier located, when necessary, between the filter and the VCO, we should get

$$\tau_1 \frac{du_2}{dt} = K_2 \left[u_1 + \tau_2 \frac{du_1}{dt} \right] \qquad (3.31)$$

Equations 3.1 and 3.3, together with the derived time equations, are written:

$$u_1(t) = K_1 \sin \left[\phi_i(t) - \phi_0(t) \right]$$

$$\frac{du_1}{dt} = K_1 \cos \left[\phi_i(t) - \phi_0(t) \right] \left(\frac{d\phi_i}{dt} - \frac{d\phi_0}{dt} \right)$$

$$\frac{d\phi_0}{dt} = K_3 u_2(t) \qquad (3.32)$$

$$\frac{d^2\phi_0}{dt^2} = K_3 \frac{du_2}{dt}$$

If we insert Eqs. 3.32 in Eq. 3.31, we get

$$\frac{\tau_1}{K_3}\frac{d^2\phi_0}{dt^2} = K_1 K_2 \sin\left[\phi_i(t)-\phi_0(t)\right]$$

$$+ K_1 K_2 \tau_2 \cos\left[\phi_i(t)-\phi_0(t)\right]\left(\frac{d\phi_i}{dt}-\frac{d\phi_0}{dt}\right)$$

that is, bearing in mind that $K_1 K_2 K_3 = K$, the loop gain

$$\tau_1\frac{d^2\phi_0}{dt^2} = K\sin\left[\phi_i(t)-\phi_0(t)\right]$$

$$+ K\tau_2 \cos\left[\phi_i(t)-\phi_0(t)\right]\left(\frac{d\phi_i}{dt}-\frac{d\phi_0}{dt}\right) \qquad (3.33)$$

As regards the instantaneous phase error $\phi(t)=\phi_i(t)-\phi_0(t)$,

$$\tau_1\frac{d^2\phi}{dt^2} + K\tau_2\cos\phi(t)\cdot\frac{d\phi}{dt} + K\sin\phi(t) = \tau_1\frac{d^2\phi_i}{dt^2} \qquad (3.34)$$

Equations 3.33 and 3.34 are second-order nonlinear differential equations, which cannot be solved by analytical methods. If we make the approximation

$$\sin\phi(t)\cong\phi(t)$$

$$\cos\phi(t)\cong 1$$

which is only valid if the instantaneous phase error $\phi(t)$ remains very small, whatever the value of t, we reencounter Eqs. 3.17 and 3.18 of the linear domain.

3.5.2   CASE OF A LOOP FILTER HAVING A TRANSFER FUNCTION $1/(1+j\omega\tau_1)$

The loop-filter time-domain equation, allowing for the possible gain K_2 amplifier, is

$$\tau_1\frac{du_2}{dt} + u_2 = K_2 u_1 \qquad (3.35)$$

If we insert in this equation Eqs. 3.32, we obtain the differential equation 3.36 relating $\phi_0(t)$ to $\phi_i(t)$:

$$\tau_1\frac{d^2\phi_0}{dt^2} + \frac{d\phi_0}{dt} = K\sin\left[\phi_i(t)-\phi_0(t)\right] \qquad (3.36)$$

The relationship between the instantaneous phase error $\phi(t) = \phi_i(t) - \phi_0(t)$ and $\phi_i(t)$ is given by Eq. 3.37:

$$\tau_1 \frac{d^2\phi}{dt^2} + \frac{d\phi}{dt} + K \sin\phi(t) = \tau_1 \frac{d^2\phi_i}{dt^2} + \frac{d\phi_i}{dt} \tag{3.37}$$

3.5.3 CASE OF A LOOP FILTER HAVING A TRANSFER FUNCTION $(1 + j\omega\tau_2)/(1 + j\omega\tau_1)$

In this case, the loop-filter time-domain equation, making allowances for a possible K_2 amplification, is given by

$$\tau_1 \frac{du_2}{dt} + u_2 = K_2 \left[u_1 + \tau_2 \frac{du_1}{dt} \right] \tag{3.38}$$

If we insert in Eq. (3.38) Eqs. 3.32, we obtain immediately

$$\tau_1 \frac{d^2\phi_0}{dt^2} + \frac{d\phi_0}{dt} = K \sin\left[\phi_i(t) - \phi_0(t) \right]$$

$$+ K\tau_2 \cos\left[\phi_i(t) - \phi_0(t) \right] \left(\frac{d\phi_i}{dt} - \frac{d\phi_0}{dt} \right) \tag{3.39}$$

As regards the instantaneous phase error $\phi(t) = \phi_i(t) - \phi_0(t)$,

$$\tau_1 \frac{d^2\phi}{dt^2} + \left[1 + K\tau_2 \cos\phi(t) \right] \frac{d\phi}{dt} + K \sin\phi(t) = \tau_1 \frac{d^2\phi_i}{dt^2} + \frac{d\phi_i}{dt} \tag{3.40}$$

By means of the approximation $\sin\phi(t) \cong \phi(t)$ and $\cos\phi(t) \cong 1$, we can get back to the linearized equations 3.29 and 3.30.

3.6 PARAMETERS OF A SECOND-ORDER LOOP

The denominator of second-order servo-control transfer and error functions, when operational notation is used, is generally formulated $s^2 + 2\zeta\omega_n s + \omega_n^2$. In order to preserve the analogy with second-order servo controls, we usually substitute for the time constants τ_1 and τ_2, which are physical parameters, the more mathematical parameters of natural angular frequency ω_n and damping factor ζ. We shall now see what becomes of the various equations involved when we apply this substitution to each of the three loop filter cases under consideration.

3.6.1 CASE OF A LOOP FILTER HAVING A TRANSFER FUNCTION $(1 + j\omega\tau_2)/j\omega\tau_1$

In this case, we can define

$$\omega_n^2 = \frac{K}{\tau_1} \tag{3.41}$$

$$2\zeta\omega_n = \frac{K\tau_2}{\tau_1} \tag{3.42}$$

If we divide by τ_1 the numerator and denominator of the transfer functions (Eqs. 3.13 and 3.15) and the error functions (Eqs. 3.14 and 3.16), and if we use Eqs. 3.41 and 3.42 notation, we obtain, respectively,

$$H(j\omega) = \frac{\omega_n^2 + j2\zeta\omega_n\omega}{\omega_n^2 - \omega^2 + j2\zeta\omega_n\omega} \tag{3.43}$$

$$1 - H(j\omega) = \frac{-\omega^2}{\omega_n^2 - \omega^2 + j2\zeta\omega_n\omega} \tag{3.44}$$

In operational notation, this gives

$$H(s) = \frac{2\zeta\omega_n s + \omega_n^2}{s^2 + 2\zeta\omega_n s + \omega_n^2} \tag{3.45}$$

$$1 - H(s) = \frac{s^2}{s^2 + 2\zeta\omega_n s + \omega_n^2} \tag{3.46}$$

3.6.2 CASE OF A LOOP FILTER HAVING A TRANSFER FUNCTION $1/(1 + j\omega\tau_1)$

To reduce the transfer and error function denominators to the canonical form, we state

$$\omega_n^2 = \frac{K}{\tau_1}$$

$$2\zeta\omega_n = \frac{1}{\tau_1} \tag{3.47}$$

With these notations, the transfer (Eqs. 3.19 and 3.21) and error (Eqs. 3.20

and 3.22) functions become

$$H(j\omega) = \frac{\omega_n^2}{\omega_n^2 - \omega^2 + j2\zeta\omega_n\omega} \qquad (3.48)$$

$$1 - H(j\omega) = \frac{-\omega^2 + j2\zeta\omega_n\omega}{\omega_n^2 - \omega^2 + j2\zeta\omega_n\omega} \qquad (3.49)$$

$$H(s) = \frac{\omega_n^2}{s^2 + 2\zeta\omega_n s + \omega_n^2} \qquad (3.50)$$

$$1 - H(s) = \frac{s^2 + 2\zeta\omega_n s}{s^2 + 2\zeta\omega_n s + \omega_n^2} \qquad (3.51)$$

3.6.3 CASE OF A LOOP FILTER HAVING A TRANSFER FUNCTION $(1 + j\omega\tau_2)/(1 + j\omega\tau_1)$

We state, by definition,

$$\omega_n^2 = \frac{K}{\tau_1}$$

$$2\zeta\omega_n = \frac{1 + K\tau_2}{\tau_1} \qquad (3.52)$$

With these notations, the quantity $K(\tau_2/\tau_1)$ can be expressed

$$K\frac{\tau_2}{\tau_1} = 2\zeta\omega_n - \frac{1}{\tau_1} = 2\zeta\omega_n - \frac{\omega_n^2}{K}$$

The transfer (Eqs. 3.25 and 3.27) and error (Eqs. 3.26 and 3.28) functions become

$$H(j\omega) = \frac{\omega_n^2 + j(2\zeta\omega_n - \omega_n^2/K)\omega}{\omega_n^2 - \omega^2 + j2\zeta\omega_n\omega} \qquad (3.53)$$

$$1 - H(j\omega) = \frac{-\omega^2 + j(\omega_n^2/K)\omega}{\omega_n^2 - \omega^2 + j2\zeta\omega_n\omega} \qquad (3.54)$$

$$H(s) = \frac{(2\zeta\omega_n - \omega_n^2/K)s + \omega_n^2}{s^2 + 2\zeta\omega_n s + \omega_n^2} \qquad (3.55)$$

$$1 - H(s) = \frac{s^2 + (\omega_n^2/K)s}{s^2 + 2\zeta\omega_n s + \omega_n^2} \qquad (3.56)$$

Remark 1. We pointed out in Section 2.3 that the transfer function $(1 + j\omega\tau_2)/j\omega\tau_1$ for a loop filter only corresponds to an approximation of the transfer function $(1 + j\omega\tau_2)/(1 + j\omega\tau_1)$ obtained by use of an active filter. The approximation is justified in a case where the gain of the operational amplifier used is very high, which implies that the time constant τ_1 is also very high (see Eqs. 2.9 and 2.10).

If we compare Eqs. 3.42 and 3.52, it is evident that the former is a good approximation of the latter when the quantity $K\tau_2$ is much larger than 1. However, it would be erroneous to conclude that, providing this condition is fulfilled, a loop using a filter $(1 + j\omega\tau_1)/(1 + j\omega\tau_2)$ is equivalent to a loop using a filter $(1 + j\omega\tau_2)/j\omega\tau_1$. If Eqs. 3.46 and 3.56 are compared, it will be seen that the equivalence is only obtained if ω_n^2/K is very small, in other words, if the time constant τ_1 is very large. The fact that the two conditions $(K\tau_2 \gg 1$ and $\omega_n^2/K \ll 1)$ are frequently fulfilled simultaneously, because the loop gain K is very high, may be a source of confusion.

NUMERICAL EXAMPLE. The time constant τ_2 of the filter of a loop such that $K = 5 \times 10^4$ rad/s and $\zeta = 2$ is equal to $\tau_2 = 20$ ms. Since the quantity $K\tau_2 = 10^3$ is much larger than 1, we can use Eq. 3.42, $2\zeta\omega_n = K(\tau_2/\tau_1)$, which gives $\omega_n\tau_1 = 250$. If we combine this relationship with Eq. 3.41, we get $\omega_n = K/250 = 200$ rad/s. Consequently, $\tau_1 = K/\omega_n^2 = 5 \times 10^4/4 \times 10^4 = 1.25$ s. This being so, whether the loop filter is of the active or passive type, the transfer and error functions are those given by Eqs. 3.53–3.56.

Remark 2. Of the three physical parameters K, τ_1, and τ_2, the loop gain K is liable to vary noticeably, at least in certain circumstances. This is because the phase detector sensitivity K_1 is an element in the expression of K. But, as we saw in Section 2.1, the phase detector sensitivity is generally proportional to the amplitude A of one of the two signals applied (it is proportional to the product AB of the amplitudes of the applied signals, in the case of a real multiplier). We generally use the VCO output as a reference signal e_R and the input signal as e_S. This is because the input signal is frequently accompanied by a relatively high noise power (see Chapter 7) and it is easier to obtain a high amplitude signal—or a square-wave signal—using the VCO output, which is a very clean signal (noise free). Then, when A varies, K_1 and K vary in the same proportions.

In all the formulas given up to and including Section 3.5, K is the only variable parameter. In the Section 3.6 formulas, on the other hand, K, ζ and ω_n vary simultaneously. If we use the subscript $_0$ to designate the values of these parameters when the input signal amplitude is equal to A_0, for any other value A of the amplitude, the parameters become

$$K = K_0 \frac{A}{A_0}$$

$$\omega_n = \omega_{n0} \sqrt{\frac{A}{A_0}}$$

For the damping factor ζ, we have to distinguish according to the loop filter type.

For a loop filter having a transfer function $(1 + j\omega\tau_2)/j\omega\tau_1$, ζ is given by Eq. 3.42, which becomes

$$2\zeta\omega_{n0}\sqrt{\frac{A}{A_0}} = K_0 \frac{A}{A_0}\frac{\tau_2}{\tau_1}$$

that is,

$$\zeta = \zeta_0 \sqrt{\frac{A}{A_0}}$$

For a loop filter having a transfer function $1/(1 + j\omega\tau_1)$, ζ is given by Eq. 3.47, which leads to

$$\zeta = \frac{\zeta_0}{\sqrt{A/A_0}}$$

Finally, for a loop filter having a transfer function $(1 + j\omega\tau_2)/(1 + j\omega\tau_1)$, ζ is given by Eq. 3.52, which gives

$$2\zeta\omega_{n0}\sqrt{\frac{A}{A_0}} = \frac{1 + K_0(A/A_0)\tau_2}{\tau_1}$$

that is,

$$\zeta = \frac{1 + K_0(A/A_0)\tau_2}{2\sqrt{K_0(A/A_0)\tau_1}}$$

When the quantity $K\tau_2 = K_0(A/A_0)\tau_2$ is much larger than 1, we can use the following formula:

$$\zeta \approx \zeta_0 \sqrt{\frac{A}{A_0}}$$

CHAPTER 4
PHASE-LOCKED
LOOP STABILITY

From the equations established in the previous chapter, it is clear that the behavior of a phase-locked loop is generally nonlinear. If the phase error is sufficiently low, however, the servo-loop behavior is linear. Using the corresponding equations, we can now examine the problem of stability, which is one of the essential considerations in any servo device. The conditions of stability that we shall obtain will thus be necessary conditions. There is no guarantee that they will be sufficient, since, strictly speaking, allowances should be made for the nonlinearity of the phase detector. One can always refer to specialized books, but experience shows that if the linear equation stability criteria are respected, the loop is generally stable.

We shall study the stability of the various loops by means of the Bode diagram, which is a representation of the amplitude and phase of the open-loop transfer function.

If K_1 is the phase detector sensitivity, K_2, the gain of a possible amplifier, $F(j\omega)$, the loop filter transfer function, and K_3, the VCO modulation sensitivity, the open-loop transfer function is expressed as

$$\mathcal{G}(j\omega) = K_1 K_2 F(j\omega) \frac{K_3}{j\omega} = K \frac{F(j\omega)}{j\omega}$$

4.1 FIRST-ORDER LOOP

The loop filter transfer function is $F(j\omega) = 1$; the open-loop transfer function is reduced to

$$\mathcal{G}(j\omega) = \frac{K}{j\omega}$$

The curve representing the gain in decibels, that is, the quantity $G = 20\log_{10}\text{Mod}[\mathcal{G}(j\omega)]$, versus ω on a logarithmic scale is a straight line of slope -1 (Fig. 4.1a). The phase shift $\theta = \text{Arg}[\mathcal{G}(j\omega)]$ is constant and equal to $-\pi/2$, whatever the value of ω (Fig. 4.1b). This type of loop is unconditionally stable.

Strictly speaking, a loop such as this does not really exist. It is perhaps easy to realize $F(j\omega) = 1$ (simply omit the filter), but the introduction of unwanted phase shifts is unavoidable: the phase detector output is always filtered to eliminate the 2ω angular frequency terms (unavoidable in a multiplier) and the residual ω terms (since diodes are never perfect). In the best case, the phase detector output filter is a one-pole low-pass filter, of time constant τ. The characteristic of the phase detector K_1 must consequently be replaced by the following transfer function:

$$\frac{K_1}{1+j\omega\tau}$$

On the other hand, the VCO is a frequency-modulated oscillator. That is to say, for the modulation path, as in any modulator, we always come to a point where the impedance has to be high for the modulation and low for the frequency to be modulated. In the best case, this takes place as if the modulation were sent through a one-pole low-pass filter, of time constant τ'. The modulation characteristic K_3 must thus be replaced by the following transfer function:

$$\frac{K_3}{1+j\omega\tau'}$$

Under these conditions, the open-loop transfer function becomes

$$\mathcal{G}(j\omega) = \frac{K}{j\omega(1+j\omega\tau)(1+j\omega\tau')}$$

The Bode diagram is represented in Fig. 4.2a and 4.2b. Since the asymptotical phase shift is equal to $-3\pi/2$, the loop is no longer unconditionally stable. On the diagram given, the phase shift is $-\pi$ for an angular frequency value such that the gain exceeds unity: the corresponding loop is not stable and begins to oscillate. To make it stable, we have to reduce the loop gain K. In a first approximation, the loop only restabilizes if K is below the smaller of the two angular frequencies $1/\tau$ and $1/\tau'$. This result is only approximate, since stability depends on the respective values of τ and τ', as well as on the safety margin we decide to allow. It should be remembered that this safety margin is reckoned in degrees of phase shift for the angular

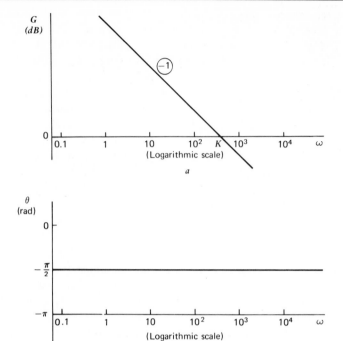

FIGURE 4.1. Bode diagram of a first-order loop: (a) Amplitude diagram; (b) phase diagram

frequency giving a unity gain or in decibels of gain for the angular frequency giving the phase shift of $-\pi$.

NUMERICAL EXAMPLE. We use a phase detector, of sensitivity $K_1 = 10$ V/rad and a VCO, the modulation sensitivity of which is $K_3 = 10$ kHz/V. We require to construct a first-order loop, such that $K = 6.28 \times 10^6$ rad/s.

Since the product $K_1 K_3$ is equal to 100 kHz, we have to add an operational amplifier of gain $K_2 = 10$.

The output filter of the phase detector is a one-pole low-pass filter, with a cutoff frequency of $f_c = 10$ MHz. The filtering for the modulation applied to the VCO is equivalent to that of a one-pole low-pass filter, with a cutoff frequency of $f_c' = 5$ MHz.

If there is no unwanted phase-shift from the operational amplifier, the frequency corresponding to an open-loop gain of unity is approximately $K/2\pi$. The phase-shift introduced by the phase detector and the VCO at this frequency is

$$\text{Arc} \tan \frac{K}{2\pi f_c} + \text{Arc} \tan \frac{K}{2\pi f_c'} = \text{Arc} \tan \frac{1}{10} + \text{Arc} \tan \frac{1}{5} \cong 17°$$

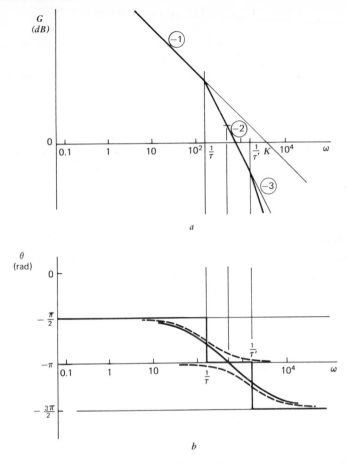

FIGURE 4.2. Bode diagram of a first-order loop with parasitic time-constant effects: (a) Amplitude diagram; (b) phase diagram.

If we wish to preserve a phase safety margin of 45°, the phase-shift introduced by the amplifier at frequency $K/2\pi$ must not exceed $45° - 17°$ $= 28°$. If the transfer function of this amplifier corresponds to a one-pole low-pass filter, with a cutoff frequency of F_c, we should obtain

$$\operatorname{Arc\,tan} \frac{K}{2\pi F_c} \leqslant 28°$$

Therefore $F_c \geqslant 1.88$ MHz.

In actual fact, if we take into account the different filtering operations, the phase-locked loop open-loop gain will be below 1 for the frequency $K/2\pi$, which intensifies the safety margin.

4.2 SECOND-ORDER LOOP WITH PERFECT INTEGRATOR

When we want to increase the order of the servo device, that is to say, when we want to increase the degree of the transfer function denominator by one unit, the first method that comes to mind consists in using as loop filter a filter having the following transfer function:

$$F(j\omega) = \frac{1}{j\omega}$$

This transfer function is the equivalent, in terms of frequency, of the time domain "integration" operation. The open-loop transfer function becomes

$$\mathcal{G}(j\omega) = \frac{K}{(j\omega)^2}$$

The gain curve is a straight line of slope -2, intersecting the axis 0 dB for an angular frequency value equal to $\sqrt{K}$. Since the phase-shift equals $-\pi$, whatever the angular frequency value, the loop is unstable and oscillates at angular frequency $\omega = \sqrt{K}$. A cross check consists in taking the transfer function expression given by Eq. 3.6, in which

$$F(j\omega) = \frac{1}{j\omega}$$

that is,

$$H(j\omega) = \frac{K}{(j\omega)^2 + K} = \frac{K}{K - \omega^2}$$

For $\omega = \sqrt{K}$, the transfer function denominator cancels out. An oscillation at this angular frequency is developed in the loop.

In fact, the presence of the spurious time constants τ and τ', related to the phase detector and VCO, modifies the phenomenon without improving the loop reliability. The phase-shift curve is then identical with that shown in Fig. 4.2b to within a $-\pi/2$ translation. The phase-shift is consequently invariably below $-\pi$. The Bode diagram does not, in this case, enable us to arrive at a conclusion, but the instability can be established by another method (see Section 4.6).

4.3 SECOND-ORDER LOOP WITH PERFECT INTEGRATOR AND PHASE-LEAD CORRECTION NETWORK

The loop described in the previous section can be stabilized by the use of a phase-lead correction network, having a transfer function $(1 + j\omega\tau_2)$. The loop filter transfer function becomes, after addition of a proportionality constant $1/\tau_1$ inserted for the sake of uniformity in notation,

$$F(j\omega) = \frac{1}{\tau_1} \frac{1 + j\omega\tau_2}{j\omega}$$

The open-loop transfer function is then written

$$\mathcal{G}(j\omega) = \frac{K(1 + j\omega\tau_2)}{\tau_1(j\omega)^2}$$

The corresponding Bode diagram is represented in Fig. 4.3. From this diagram, we can see that the entire phase curve is located above $-\pi$. In particular, if $1/\tau_2 < \omega_n$, for the value of ω corresponding to a gain of unity, the phase-shift falls between $-\pi/2$ and $-3\pi/4$, which corresponds to a

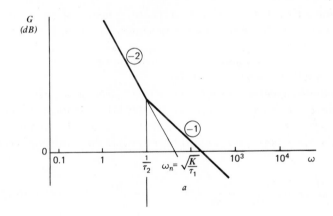

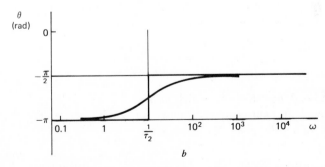

FIGURE 4.3. Bode diagram of a second-order loop with perfect integrator and phase-lead correction: (a) Amplitude diagram; (b) phase diagram.

very good stability. The condition $1/\tau_2 < \omega_n$ can be written $\omega_n \tau_2 > 1$. But for this type of loop according to Eqs. 3.41 and 3.42, $\omega_n \tau_2 = 2\zeta$, ζ being the damping factor. Consequently, as would be expected, the higher the damping factor, the greater the phase safety margin.

If we take into account the spurious time constants τ and τ' of Section 4.1, it will be noted that for the loop to remain stable, these time constants simply have to be such as not to introduce an additional phase shift between $\pi/4$ and $\pi/2$ for the angular frequency value corresponding to a gain of unity. When we use a second-order loop, we often aim at a relatively low value for the parameter ω_n. It follows that the angular frequency value corresponding to an open-loop gain of unity is also relatively low. The influence of the spurious time constants τ and τ' can consequently be rendered negligible, even if their values are not very low. Thus for a low ω_n second-order loop, the phase detector and VCO modulation bandwidths need not be of a particularly high value. This is not at all so in the case of a first-order loop with a high loop gain K.

NUMERICAL EXAMPLE. We have a phase detector of sensitivity $K_1 = 10$ V/rad and a VCO, the modulation sensitivity of which is $K_3 = 10$ kHz/V. The phase detector spurious time constant is $\tau = 15.9$ μs (-3 dB cutoff frequency $f_c = 10$ kHz) and the VCO spurious time constant is $\tau' = 31.8$ μs (-3 dB cutoff frequency $f_c' = 5$ kHz).

With these elements we cannot hope to construct a first-order loop, for its loop gain $K = K_1 K_3$ would be equal to 100 kHz and the device would be unstable. We can, for instance, check that for an angular frequency $\omega = 10^5$ rad/s, the open-loop gain is slightly over 1, while the open-loop phase-shift is equal to

$$90° + \text{Arc} \tan \frac{10^5}{2\pi \times 10^4} + \text{Arc} \tan \frac{10^5}{2\pi \times 5 \times 10^3} = 180° + 40.4°$$

To construct a second-order loop, with a loop filter having a transfer function $F(s) = (1 + \tau_2 s)/\tau_1 s$, such that $\omega_n = 100$ rad/s and $\zeta = 1$, the time constants τ_1 and τ_2 have to be selected so that

$$\tau_1 = \frac{K}{\omega_n^2} = \frac{2\pi \times 10^5}{10^4} = 62.8 \text{ s}$$

Furthermore, $2\zeta\omega_n = K\tau_2/\tau_1$, which implies that $\tau_2 = 2\zeta/\omega_n = 0.02$ s.

The open-loop gain is equal to unity for $\omega = 200$ rad/s. For this angular frequency value, the phase-shift is equal to

$$180° - \text{Arc} \tan (200 \times 0.02) + \text{Arc} \tan \frac{200}{2\pi \times 10^4} + \text{Arc} \tan \frac{200}{2\pi \times 5 \times 10^3}$$

that is,

$$180° - 76° + 0.18° + 0.36° \cong 180° - 75.5°$$

The phase safety margin, afforded by the correction network, is considerable. The phase shifts induced by the spurious time constants τ and τ' have no influence on the way in which the servo device operates.

4.4 SECOND-ORDER LOOP WITH LOW-PASS FILTER

Another way of increasing by one unit the degree of the transfer function denominator, is to use as loop filter, a low-pass filter having a transfer function

$$F(j\omega) = \frac{1}{1 + j\omega\tau_1}$$

In this case, the open-loop transfer function becomes

$$\mathcal{G}(j\omega) = \frac{K}{j\omega(1 + j\omega\tau_1)}$$

The corresponding Bode diagram is given in Fig. 4.4. If all undesirable factors are removed, the phase shift still falling between $-\pi/2$ and $-\pi$, a loop of this type will be stable. However, it should be noted that if the time constant τ_1 is high, the phase safety margin at angular frequency $\omega = \sqrt{K/\tau_1}$ is very small. If, for this angular frequency value, the phase shift induced by parasitic time constants equals or exceeds the safety margin, the loop becomes unstable.

For a loop of this type, Eqs. 3.41 and 3.47 show that $\omega_n = \sqrt{K/\tau_1}$, while the damping factor ζ is given by $2\zeta\omega_n = 1/\tau_1$, that is,

$$\zeta = \frac{1}{2\sqrt{K\tau_1}}$$

Consequently, when we require a large τ_1, the conditions being otherwise identical, we necessarily get a low damping factor. On the other hand, if τ_1 is small (τ_1 representing, for instance, a spurious time constant in a first order loop) the safety margin is large and the damping factor value is acceptable.

NUMERICAL EXAMPLE. The loop is such that $K = 10^5$ rad/s and $\omega_n = 100$ rad/s. Consequently, the time constant τ_1 is equal to $\tau_1 = 10^5/10^4 = 10$ s, while the damping factor is $\zeta = 1/2\omega_n\tau_1 = 5 \times 10^{-4}$.

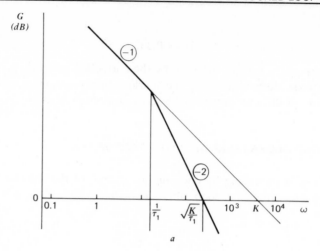

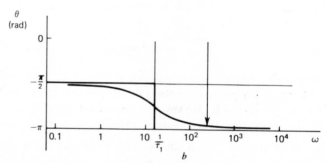

FIGURE 4.4. Bode diagram of a second-order loop with RC low-pass filter: (a) Amplitude diagram; (b) phase diagram.

For $\omega = \omega_n = 100$ rad/s, the phase shift is equal to $90° + \text{Arc} \tan 1000 \cong 179.94°$. The phase safety margin is insignificant and the slightest spurious phase shift will be sufficient to make the loop unstable.

4.5 SECOND-ORDER LOOP WITH LOW-PASS FILTER AND PHASE-LEAD CORRECTION

The use of a phase-lead correction network having a transfer function $(1 + j\omega\tau_2)$ enables us to increase the phase safety margin in the vicinity of the angular frequency corresponding to an open-loop gain of unity. This has the twofold advantage of corresponding to an acceptable damping factor value,

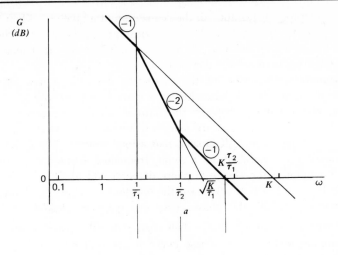

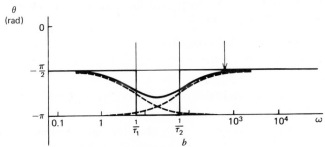

FIGURE 4.5. Bode diagram of a second-order loop with RC low-pass filter and phase-lead correction: (a) Amplitude diagram; (b) phase diagram.

even for low natural angular frequency ω_n loops, and permitting the use of phase detectors and VCOs with not very high modulation bandwidths, without compromising the loop stability.

With the correction network, the loop filter transfer function becomes

$$F(j\omega) = \frac{1+j\omega\tau_2}{1+j\omega\tau_1}$$

and the open-loop transfer function $\mathcal{G}(j\omega)$ is written

$$\mathcal{G}(j\omega) = \frac{K(1+j\omega\tau_2)}{j\omega(1+j\omega\tau_1)}$$

The corresponding Bode diagram is represented in Fig. 4.5. It will be noted that the phase safety margin is around $\pi/2$. For a given time constant

τ_1, corresponding for instance to the realization of a given ω_n when the loop gain K is fixed, the greater the time constant τ_2, the greater the safety margin. For predetermined ω_n, K, and τ_1, Eq. 3.52 shows that increasing τ_2 actually means increasing the damping factor ζ.

Consequently, by means of the correction network, which implies the insertion of an additional parameter τ_2, we are free to choose a satisfactory damping factor value, whatever the natural angular frequency value selected. Moreover, a phase safety margin of around $\pi/2$ means that we can use a phase detector and VCO such that a total spurious phase shift of $\pi/4$ (approximately) is induced at angular frequency $\omega = K\tau_2/\tau_1$. This considerably facilitates the choice of phase detector and VCO.

It will be noted in Fig. 4.5 that if the constant τ_1 is very large (or again if τ_2 is relatively small), the phase curve can get very close to $-\pi$ for an angular frequency $\omega = 1/\sqrt{\tau_1\tau_2}$. This case, naturally, corresponds to a low loop damping factor. If the phase curve is not sent over the $-\pi$ axis by spurious time constants, the loop remains stable, as is clear from the following method (Section 4.6).

NUMERICAL EXAMPLE. The parameters of a second-order loop, using a loop filter of the type $F(s) = (1 + \tau_2 s)/(1 + \tau_1 s)$, are as follows:

$$K = 10^5 \text{ rad/s}, \qquad \omega_n = 100 \text{ rad/s}, \qquad \zeta = 1$$

The time constant τ_1 is equal to 10 s, while the time constant τ_2 is given by Eq. 3.52:

$$2\zeta\omega_n = \frac{1 + K\tau_2}{\tau_1}$$

which leads to $1 + K\tau_2 = 2 \times 10^3$, that is, $K\tau_2 \cong 2 \times 10^3$ and

$$\tau_2 = \frac{2 \times 10^3}{10^5} = 2 \times 10^{-2} \text{ s}$$

For the angular frequency, $\omega = K\tau_2/\tau_1 = 200$ rad/s, corresponding to an open-loop gain equal to unity, the phase-shift is equal to

$$90° + \text{Arc} \tan(200 \times 10) - \text{Arc} \tan(200 \times 0.02) \cong 180° - 76°$$

The phase safety margin is very high. The use of a phase detector and VCO having not very high bandwidths (see Section 4.3), will consequently have no ill effects on the loop stability.

4.6 SUPERIOR-ORDER LOOPS

In the previous sections, we have discovered four operational types of loop:

- The first-order loop, which remains stable as long as the loop gain K is below a certain level; if we attempt to increase the loop gain beyond the limit, the stability may be affected by spurious time constants.
- The second-order loop with integrator and phase-lead correction.
- The second-order loop with low-pass filter, the disadvantage of which is the restrictive choice of K, ζ, and ω_n; this type of loop can also become unstable through parasitic time constant interference.
- The second-order loop with low-pass filter and phase-lead correction network.

Superior-order phase-locked loops have sometimes been proposed to improve the response to certain signals or to enhance the loop memory in a case of disappearance of the input signal (11, 12). On the other hand, if we wish to make allowances for a high spurious time constant, we should consider a second-order loop as a third-order loop. For this reason, it would be of interest to examine the stability problem in the case of loops of a higher order than 2. To take an example, we now analyze two particular cases corresponding to loop filters having transfer functions:

$$F(j\omega) = \frac{1}{j\omega} \cdot \frac{1}{1 + j\omega\tau}$$

and

$$F(j\omega) = \frac{(1 + j\omega\tau_2)^2}{(j\omega\tau_1)^2}$$

The first transfer function corresponds to the case of a perfect integrator, followed by a low-pass filter of time constant τ (which may be a parasitic time constant). The second corresponds to the use of a double integrator followed by a double phase-lead correction.

The phase-locked loop open-loop transfer function in the first case is

$$\mathcal{G}(j\omega) = \frac{K}{(j\omega)^2(1 + j\omega\tau)}$$

The corresponding Bode diagram is given in Fig. 4.6.

In the second case, shown in Fig. 4.7, the open-loop transfer function is

$$\mathcal{G}(j\omega) = \frac{K(1 + j\omega\tau_2)^2}{\tau_1^2(j\omega)^3}$$

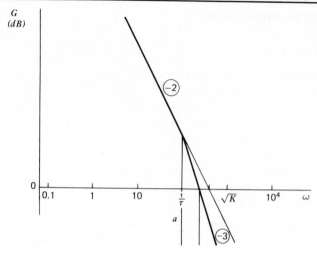

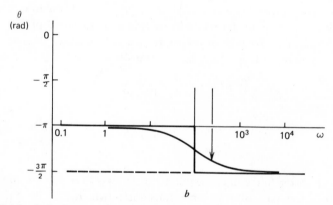

FIGURE 4.6. Bode diagram of a third-order loop with perfect integrator and RC low-pass filter: (a) Amplitude diagram; (b) phase diagram.

In Fig. 4.6, we indicate the phase security margin: when the gain is equal to unity, the phase shift is closer to $-3\pi/2$ than $-\pi$ and, strictly speaking, this phase shift is never equal to $-\pi$. Nevertheless, a loop of this type is unstable.

In Fig. 4.7, we have represented the fact that the gain is far higher than 1 for an angular frequency value corresponding to a phase shift of $-\pi$. But a loop of this type can be stable under certain conditions.

In these two cases, we consequently have to use another method, such as that of determining whether the closed-loop transfer function $H(s)$ poles

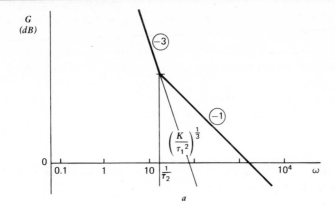

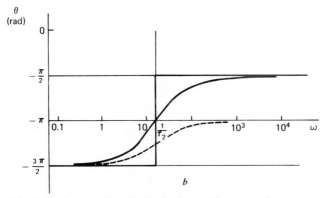

FIGURE 4.7. Bode diagram of a third-order loop with two perfect integrators and two phase-lead corrections: (a) Amplitude diagram; (b) phase diagram.

comprise a positive real component. Equation 3.8 shows that if the transfer function poles contain a positive real part, the general solution of the time differential equation [which will enable us to calculate the response $\phi_0(t)$ to an input signal $\phi_i(t)$], also has positive real exponents. The solution is thus an increasing function of time, which is a phenomenon characterizing instability.

Determination of poles with positive real parts can be undertaken in various fashions. The best-known method is a graphic method called the root locus method (13, 14), which consists in representing in the complex plane the $H(s)$ denominator root locus when the open-loop gain K varies. If, for certain values of K, the root locus cuts the imaginary axis and passes into the right-hand half-plane, the servo mechanism is unstable for these loop gain values.

For the two cases with which we are concerned, the graphical method can be replaced by an analytical method, since the $H(s)$ denominator is only of the third degree.

In the case of the integrator followed by a low-pass filter, the $H(s)$ denominator, designated as $D(s)$, is written

$$D(s) = \tau s^3 + s^2 + K$$

The problem is to determine the sign of the real part of the three complex solutions to the equation

$$D(s) = D(\sigma + j\omega) = 0$$

that is

$$\tau(\sigma + j\omega)^3 + (\sigma + j\omega)^2 + K = 0$$

This equation can be broken down into a system of two real equations, σ and ω being, in this case, real variables:

$$\tau\sigma^3 - 3\tau\sigma\omega^2 + \sigma^2 - \omega^2 + K = 0$$

$$3\tau\sigma^2\omega - \tau\omega^3 + 2\sigma\omega = 0$$

The solutions of the second equation of this system are as follows:

$$\omega = 0$$

$$\omega^2 = 3\sigma^2 + \frac{2}{\tau}\sigma$$

The first solution, $\omega = 0$, corresponds to the determination of the pure real roots of equation $D(s) = 0$. The equation

$$D(\sigma) = \tau\sigma^3 + \sigma^2 + K = 0$$

has no positive σ solution, since the τ and K constants are positive.

If we insert the value $\omega^2 = 3\sigma^2 + (2/\tau)\sigma$ of the second solution in the first equation of the system, we can then proceed to determine whether the σ equation thus obtained, $P(\sigma) = 0$, has positive roots:

$$P(\sigma) = 8\tau\sigma^3 + 8\sigma^2 + \frac{2}{\tau}\sigma - K = 0$$

The expression $P(\sigma)$ is an increasing function of σ when σ is positive. For $\sigma = 0$, $P(\sigma) = -K$. When σ approaches infinity, $P(\sigma)$ approaches infinity. We conclude that there is a positive root, and consequently the third-order loop using an integrator and a low-pass filter constitutes an unstable phase-locked loop.

In the case of the double integrator followed by a double phase-lead correction, the $H(s)$ denominator is written

$$D(s) = \tau_1^2 s^3 + K(1 + \tau_2 s)^2$$

that is,

$$D(s) = \tau_1^2 s^3 + K\tau_2^2 s^2 + 2K\tau_2 s + K$$

The problem is to determine whether the equation $D(s) = D(\sigma + j\omega) = 0$ possesses a complex root $\sigma + j\omega$ having a positive real part σ:

$$\tau_1^2(\sigma + j\omega)^3 + K\tau_2^2(\sigma + j\omega)^2 + 2K\tau_2(\sigma + j\omega) + K = 0$$

This complex equation can be broken down into a system of two real equations:

$$\tau_1^2 \sigma^3 - 3\tau_1^2 \sigma\omega^2 + K\tau_2^2 \sigma^2 - K\tau_2^2 \omega^2 + 2K\tau_2 \sigma + K = 0$$

$$3\tau_1^2 \sigma^2 \omega - \tau_1^2 \omega^3 + 2K\tau_2^2 \sigma\omega + 2K\tau_2 \omega = 0$$

The second equation of the system breaks down as follows:

$$\omega = 0$$

$$\omega^2 = 3\sigma^2 + 2K\frac{\tau_2^2}{\tau_1^2}\sigma + 2K\frac{\tau_2}{\tau_1^2}$$

The case $\omega = 0$ corresponds to determination of the real roots of $D(s) = 0$. The quantity

$$D(\sigma) = \tau_1^2 \sigma^3 + K\tau_2^2 \sigma^2 + 2K\tau_2 \sigma + K$$

is always positive when $\sigma > 0$, since the constants K and τ_2 are positive quantities.

If we insert the expression

$$\omega^2 = 3\sigma^2 + 2K\frac{\tau_2^2}{\tau_1^2}\sigma + 2K\frac{\tau_2}{\tau_1^2}$$

in the first equation of the system, we can then proceed to determine whether the equation $P(\sigma) = 0$ thus obtained has positive roots:

$$P(\sigma) = 8\tau_1^2\sigma^3 + 8K\tau_2^2\sigma^2 + \left(4K\tau_2 + 2K^2\frac{\tau_2^4}{\tau_1^2}\right)\sigma + 2K^2\frac{\tau_2^3}{\tau_1^2} - K = 0$$

The function $P(\sigma)$ is an increasing function for $\sigma > 0$. For $\sigma = 0$,

$$P(0) = 2K^2\frac{\tau_2^3}{\tau_1^2} - K; \qquad \text{when } \sigma \to +\infty, \quad P(\sigma) \to +\infty$$

Consequently, the equation $P(\sigma) = 0$ possesses a positive σ root if and only if $P(0)$ is negative. The loop is thus unstable if

$$K \leqslant \frac{1}{2}\frac{\tau_1^2}{\tau_2^3}$$

When this type of loop is used, it is important to make sure that the open-loop gain K remains in all circumstances higher than the minimum value defined above.

PART TWO
LINEAR BEHAVIOR

INTRODUCTION

We shall now analyze the phase-locked loop response to various modulations or disturbances under linear conditions (see the linear equations of Chapter 3).

The loops are assumed to be in lock to begin with, having a null phase error. This actually means that the VCO signal is in lead- or lag-quadrature with the input signal (for a sinusoidal or triangular detector) or in opposite phase (for a sawtooth detector).

Throughout this second part we assume the phase error to be low enough to verify the hypothesis of linear differential equations as the basic working principle. This has the tremendous advantage of facilitating calculation and especially prediction of loop performances. To put this into practice, we have adopted the following procedure:

- We assume a priori that the input signal — VCO signal phase discrepancy is low enough for the loop to function under linear conditions.
- This enables us to calculate the loop response, and in particular the phase variation between the input signal and VCO signal (known as the phase error).
- We subsequently check that the phase error obtained is small enough to guarantee the validity of the calculations.

If various disturbances or modulations appear simultaneously, it is the total phase error, calculated on the basis of the superposition principle, that must remain small.

Finally, if the phase error is low enough, this means that the VCO remains in synchronism with the input signal. We can thus conclude that the loop stays in lock.

The linearity hypothesis is not only applicable to the phase detector. When we use an amplifier with gain K_2 in the loop, it is important to avoid saturation; in other words, the amplifier output voltage must be constantly proportional to the input voltage. Finally, the VCO must be such that the instantaneous frequency swing is consistently proportional to the command signal u_2 and the VCO output amplitude is a constant for the whole frequency deviation range, at least if the phase detector used is sensitive to amplitude variations of the reference signal (as a multiplier would be).

CHAPTER 5
TRANSIENT RESPONSE

In this chapter we shall examine the response of the various loops to different disturbances occurring at instant $t=0$. The disturbances involved are

- Input signal phase step θ
- Input signal angular frequency step $\Delta\omega$
- Linear variation of slope R of the input signal frequency.

The loop is assumed to be in lock at instant $t=0$, with a null phase error. That is to say that for $t<0$, the signals are expressed

$$y_i(t) = A \sin \omega t$$

$$y_0(t) = B \cos \omega t$$

When $t=0$ and for $t>0$, the signals become

$$y_i(t) = A \sin \left[\omega t + \phi_i(t) \right]$$

$$y_0(t) = B \cos \left[\omega t + \phi_0(t) \right]$$

$\phi_0(t)$ being the loop response to excitation $\phi_i(t)$ and $\phi(t) = \phi_i(t) - \phi_0(t)$ is the phase error. We shall only calculate $\phi(t)$ as this is the basic term enabling us to check the validity of the linear operation hypothesis; $\phi_0(t)$ is easily deduced from the $\phi_i(t)$ and $\phi(t)$ expressions.

5.1 PHASE STEP RESPONSE

At instant $t=0$, a θ amplitude phase step is applied to the input signal:

$$\phi_i(t) = \theta \Upsilon(t)$$

where $\Upsilon(t)$ is the unit step function, that is, the time integral of the delta Dirac function.

In operational notation, we can write

$$\Phi_i(s) = \frac{\theta}{s}$$

The loop phase error is obtained from the error function given in Eq. 3.9:

$$\frac{\Phi(s)}{\Phi_i(s)} = 1 - H(s)$$

Consequently,

$$\Phi(s) = \left[1 - H(s) \right] \frac{\theta}{s} \tag{5.1}$$

5.1.1 FIRST-ORDER LOOP

The error function is given by Eq. 3.11 and the error is expressed as

$$\Phi(s) = \frac{\theta}{s + K}$$

We can use the final value theorem to determine immediately the final phase error, since

$$\lim_{t \to \infty} \phi(t) = \lim_{s \to 0} s \cdot \Phi(s)$$

When $s \to 0$, $s\Phi(s) \to 0$, there is no steady-state phase error. This can be checked by means of the expression of $\phi(t)$ obtained by computing the Laplace transform reciprocal of $\Phi(s)$:

$$\phi(t) = \theta e^{-Kt} \tag{5.2}$$

5.1.2 SECOND-ORDER LOOP WITH $F(s) = (1 + \tau_2 s)/\tau_1 s$

The error function, given by Eq. 3.16, leads to

$$\Phi(s) = \frac{\tau_1 s\, \theta}{\tau_1 s^2 + K\tau_2 s + K}$$

The final value theorem provides an instant cross check that $\phi(t)$ approaches zero when t tends towards infinity. We derive the $\phi(t)$ expression from the $\Phi(s)$ transform reciprocal. It is then simpler to use the notation of

Eq. 3.46:

$$\Phi(s) = \frac{s\,\theta}{s^2 + 2\zeta\omega_n s + \omega_n^2}$$

We can decompose the denominator of this expression into a product $(s + \alpha)(s + \beta)$, that is,

$$\Phi(s) = \frac{s\,\theta}{\left[s + \omega_n\left(\zeta + \sqrt{\zeta^2 - 1}\,\right)\right]\left[s + \omega_n\left(\zeta - \sqrt{\zeta^2 - 1}\,\right)\right]}$$

leading to

$$\phi(t) = \theta\,\frac{\omega_n\left(\zeta - \sqrt{\zeta^2 - 1}\,\right)e^{-\omega_n\left(\zeta - \sqrt{\zeta^2 - 1}\,\right)t} - \omega_n\left(\zeta + \sqrt{\zeta^2 - 1}\,\right)e^{-\omega_n\left(\zeta + \sqrt{\zeta^2 - 1}\,\right)t}}{\omega_n\left(\zeta - \sqrt{\zeta^2 - 1}\,\right) - \omega_n\left(\zeta + \sqrt{\zeta^2 - 1}\,\right)}$$

At this point, we have to distinguish three cases, according to the damping factor ζ:

$$\zeta > 1, \quad \phi(t) = \theta e^{-\zeta\omega_n t}\left[\cosh\omega_n\sqrt{\zeta^2 - 1}\;t - \frac{\zeta}{\sqrt{\zeta^2 - 1}}\sinh\omega_n\sqrt{\zeta^2 - 1}\;t\right]$$

$$\zeta = 1, \quad \phi(t) = \theta e^{-\omega_n t}\left(1 - \omega_n t\right) \tag{5.3}$$

$$\zeta < 1, \quad \phi(t) = \theta e^{-\zeta\omega_n t}\left[\cos\omega_n\sqrt{1 - \zeta^2}\;t - \frac{\zeta}{\sqrt{1 - \zeta^2}}\sin\omega_n\sqrt{1 - \zeta^2}\;t\right]$$

The curves representing $\phi(t)/\theta$ for different values of ζ are given in Fig. 5.1. The maximum phase error value, obtained for $t = 0$, is θ. The expressions of Eqs. 5.3 above and the curves shown are then only valid if $\theta < \frac{1}{2}$ rad (approximately) for a sinusoidal detector, if $\theta < \pi/2$ for a detector linear over $(-\pi/2, +\pi/2)$, or if $\theta < \pi$, when the detector is linear over $(-\pi, +\pi)$.

The curves in Fig. 5.2 provide a means of checking that the choice $\zeta = 1$ corresponds to the fastest elimination of the loop error.

NUMERICAL EXAMPLE. The signal applied to a phase-locked loop, of natural angular frequency $\omega_n = 200$ rad/s, undergoes a phase step $\theta = 1$ rad. What damping factor ζ should be chosen to give a phase error below 10^{-3} rad after a period of 100 ms?

The curves of Fig. 5.2 show that if ζ is chosen so that $1/2\sqrt{2} \leqslant \zeta \leqslant 2\sqrt{2}$,

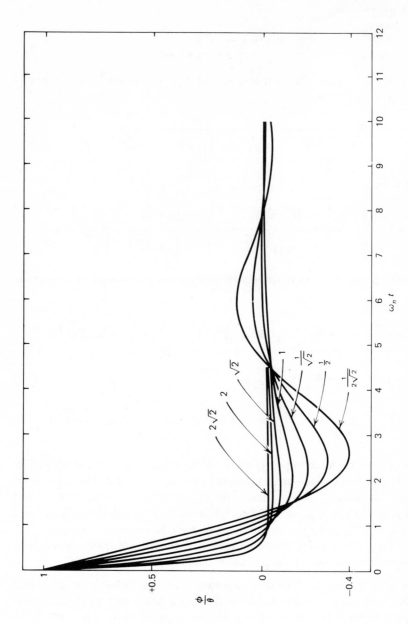

FIGURE 5.1. Phase error of a second-order loop $[F(s) = (1 + \tau_2 s)/(\tau_1 s)]$ for a phase step of the input signal.

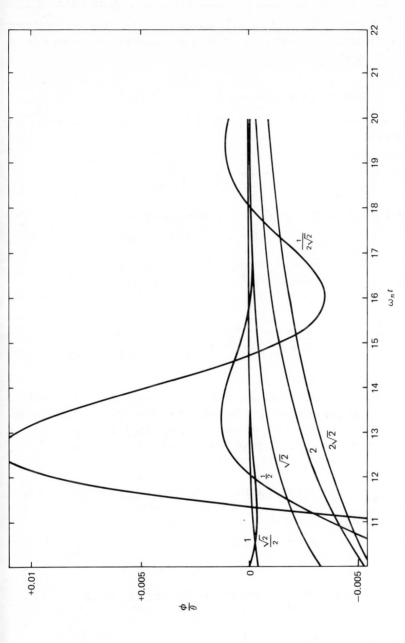

FIGURE 5.2. Tail of the phase error [second-order loop with $F(s) = (1 + \tau_2 s)/(\tau_1 s)$] for a phase step of the input signal.

85

the phase error is below 10^{-3} rad for $\omega_n t = 20$.

The choice of ζ in the range $1/\sqrt{2}$ to 1 guarantees a phase error below 4×10^{-4} rad after 50 ms.

5.1.3   SECOND-ORDER LOOP WITH $F(s) = 1/(1 + \tau_1 s)$

The error function is given by Eq. 3.51. Substituting this equation in Eq. 5.1, we get the following expression for the loop phase error:

$$\Phi(s) = \frac{(s + 2\zeta\omega_n)\theta}{s^2 + 2\zeta\omega_n s + \omega_n^2}$$

If we break down the denominator of $\Phi(s)$ into a product $(s + \alpha)(s + \beta)$, we get:

$$\Phi(s) = \frac{s\,\theta}{\left[s + \omega_n\left(\zeta + \sqrt{\zeta^2 - 1}\,\right)\right]\left[s + \omega_n\left(\zeta - \sqrt{\zeta^2 - 1}\,\right)\right]}$$

$$+ \frac{2\zeta\omega_n\theta}{\left[s + \omega_n\left(\zeta + \sqrt{\zeta^2 - 1}\,\right)\right]\left[s + \omega_n\left(\zeta - \sqrt{\zeta^2 - 1}\,\right)\right]}$$

The reciprocal transform of the first term of the right-hand member is identical with that calculated in Section 5.1.2. If we take into account the second term of the right-hand member, this leads simply to changing the sign of the second term of the $\phi(t)$ expressions given in Section 5.1.2:

$$\zeta > 1, \quad \phi(t) = \theta e^{-\zeta\omega_n t}\left[\cosh \omega_n \sqrt{\zeta^2 - 1}\; t + \frac{\zeta}{\sqrt{\zeta^2 - 1}}\sinh \omega_n \sqrt{\zeta^2 - 1}\; t\right]$$

$$\zeta = 1, \quad \phi(t) = \theta e^{-\omega_n t}\left(1 + \omega_n t\right) \qquad\qquad\qquad (5.4)$$

$$\zeta < 1, \quad \phi(t) = \theta e^{-\zeta\omega_n t}\left[\cos \omega_n \sqrt{1 - \zeta^2}\; t + \frac{\zeta}{\sqrt{1 - \zeta^2}}\sin \omega_n \sqrt{1 - \zeta^2}\; t\right]$$

The corresponding representative curves are given in Fig. 5.3.

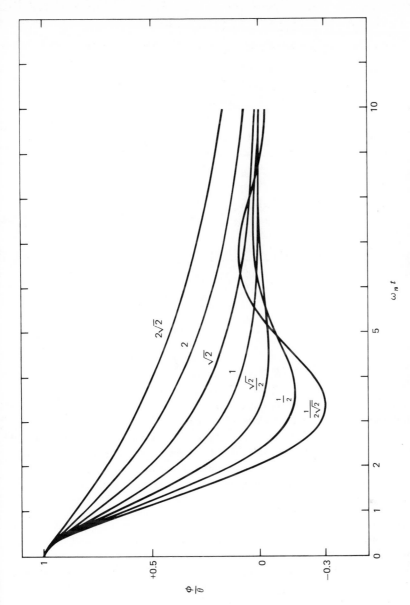

FIGURE 5.3. Phase error of a second-order loop $[F(s) = 1/(1 + \tau_1 s)]$ for a phase step of the input signal.

87

5.1.4   SECOND-ORDER LOOP WITH $F(s) = (1 + \tau_2 s)/(1 + \tau_1 s)$

The error function is given by Eq. 3.56 and leads to

$$\Phi(s) = \frac{\left(s + \omega_n^2/K\right)\theta}{s^2 + 2\zeta\omega_n s + \omega_n^2}$$

If we break down the denominator into a product $(s + \alpha)(s + \beta)$ and take the reciprocal transform, we get

$$\zeta > 1, \quad \phi(t) = \theta e^{-\zeta\omega_n t}\left[\cosh\omega_n\sqrt{\zeta^2 - 1}\ t + \frac{\omega_n/K - \zeta}{\sqrt{\zeta^2 - 1}}\sinh\omega_n\sqrt{\zeta^2 - 1}\ t\right]$$

$$\zeta = 1, \quad \phi(t) = \theta e^{-\omega_n t}\left[1 + \left(\frac{\omega_n}{K} - 1\right)\omega_n t\right] \tag{5.5}$$

$$\zeta < 1, \quad \phi(t) = \theta e^{-\zeta\omega_n t}\left[\cos\omega_n\sqrt{1 - \zeta^2}\ t + \frac{\omega_n/K - \zeta}{\sqrt{1 - \zeta^2}}\sin\omega_n\sqrt{1 - \zeta^2}\ t\right]$$

Generally speaking, ω_n/K is much smaller than ζ, so the curves representing Eqs. 5.5 differ only slightly from the curves of Figs. 5.1 and 5.2.

In light of the above remark a comparison of the curves of Fig. 5.3 (use of a simple low-pass RC filter as loop filter) with the curves of Fig. 5.1 (use of a low-pass filter with phase-lead correction network) shows the advantage of the correction network from the rapidity of response point of view. We can thus affirm that the phase correction network not only improves stability but also improves appreciably the loop response to an abrupt disturbance of the input signal phase.

NUMERICAL EXAMPLE. The damping factor is chosen equal to 2 ($\zeta = 2$). Without a correction network, for an input signal phase step θ, the phase error will be $\theta/10$ after a time of around $9/\omega_n$.

With a correction network, the phase error is $\theta/10$ after a time of only $0.5/\omega_n$.

5.2   FREQUENCY STEP RESPONSE

A frequency step $\Delta\omega$ is applied to the input signal at instant $t = 0$:

$$\phi_i(t) = \Delta\omega \cdot t \cdot \Upsilon(t)$$

In operational notation,

$$\Phi_i(s) = \frac{\Delta\omega}{s^2}$$

The phase error is consequently given by

$$\Phi(s) = [1 - H(s)]\frac{\Delta\omega}{s^2} \tag{5.6}$$

5.2.1   FIRST-ORDER LOOP

In operational notation, the phase error is written

$$\Phi(s) = \frac{\Delta\omega}{s(s+K)}$$

The Laplace transform reciprocal gives directly

$$\phi(t) = \frac{\Delta\omega}{K}(1 - e^{-Kt}) \tag{5.7}$$

The steady-state error is equal to $\Delta\omega/K$. If we make allowances for the linearization of the phase detector characteristic, we get back to the result obtained in Chapter 1 (within $\pi/2$, since this term disappears in the present notation).

5.2.2   SECOND-ORDER LOOP WITH $F(s) = (1 + \tau_2 s)/\tau_1 s$

The phase error is obtained by substituting Eq. 3.46 in Eq. 5.6:

$$\Phi(s) = \frac{\Delta\omega}{s^2 + 2\zeta\omega_n s + \omega_n^2}$$

As in the case of the phase step response, the denominator is broken down into a product $(s+\alpha)(s+\beta)$ before taking the reciprocal transform. We finally get the following results:

$$\zeta > 1, \quad \phi(t) = \frac{\Delta\omega}{\omega_n}e^{-\zeta\omega_n t}\frac{\sinh\omega_n\sqrt{\zeta^2-1}\ t}{\sqrt{\zeta^2-1}}$$

$$\zeta = 1, \quad \phi(t) = \frac{\Delta\omega}{\omega_n}e^{-\omega_n t}\cdot\omega_n t \tag{5.8}$$

$$\zeta < 1, \quad \phi(t) = \frac{\Delta\omega}{\omega_n}e^{-\zeta\omega_n t}\frac{\sin\omega_n\sqrt{1-\zeta^2}\ t}{\sqrt{1-\zeta^2}}$$

The curves representing $(\omega_n/\Delta\omega)\phi(t)$ for various values of the damping factor ζ are given in Figs. 5.4 and 5.5. The maximum phase error increases as ζ decreases.

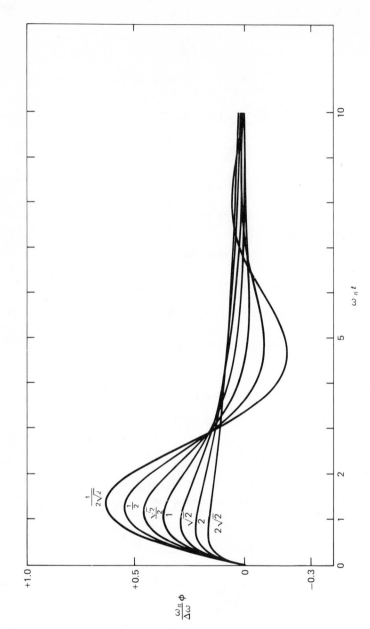

FIGURE 5.4. Phase error of a second-order loop $[F'(s) = (1 + \tau_2 s)/(\tau_1 s)]$ for a frequency step of the input signal.

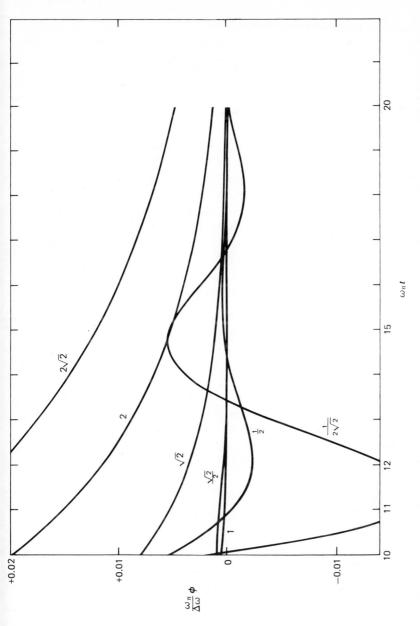

FIGURE 5.5. Tail of the phase error [second-order loop with $F(s) = (1 + \tau_2 s)/(\tau_1 s)$] for a frequency step of the input signal.

91

It will also be noted that the steady-state error is null, as would be expected when a perfect integrator is included in the loop filter. A cross check is provided by the final value theorem.

NUMERICAL EXAMPLE. Let us assume a 100-Hz frequency step on the loop input signal; the damping factor is $\zeta = 2$. An oscilloscope connected to the phase detector output gives a maximum phase error value of 0.44 rad. What will be the phase error after 40 ms?

We note on Fig. 5.4 that for $\zeta = 2$, the maximum phase error corresponds to $(\omega_n/\Delta\omega)\phi = 0.22$. Consequently, the loop natural angular frequency $\omega_n = \Delta\omega/2 = 314$ rad/s. After 40 ms, $\omega_n t = 12.6$ and $(\omega_n/\Delta\omega)\phi = 0.01$ (Fig. 5.5). The phase error after 40 ms is thus 2×10^{-2} rad.

If the loop damping factor had been fixed at $\sqrt{2}/2$, the maximum phase error, for the same ω_n, would have been 0.92 rad, but the phase error would have dropped much faster, reaching a value of only 2×10^{-3} rad after 31.4 ms.

5.2.3 SECOND-ORDER LOOP WITH $F(s) = 1/(1 + \tau_1 s)$

In this case the phase-locked loop error function is given by Eq. 3.51. The phase error due to the angular frequency step $\Delta\omega$ is found by substituting Eq. 3.51 in Eq. 5.6:

$$\Phi(s) = \frac{\Delta\omega}{s^2 + 2\zeta\omega_n s + \omega_n^2} + \frac{2\zeta\omega_n\Delta\omega}{s(s^2 + 2\zeta\omega_n s + \omega_n^2)}$$

The first term of the right-hand member corresponds to the response computed in Section 5.2.2. Making allowances for the additional term corresponding to the second term of the right-hand member, we get an overall response as follows:

$$\zeta > 1, \quad \phi(t) = 2\zeta\frac{\Delta\omega}{\omega_n} + \frac{\Delta\omega}{\omega_n}e^{-\zeta\omega_n t}$$

$$\times \left[\frac{1 - 2\zeta^2}{\sqrt{\zeta^2 - 1}} \sinh\omega_n\sqrt{\zeta^2 - 1}\,t - 2\zeta\cosh\omega_n\sqrt{\zeta^2 - 1}\,t \right]$$

$$\zeta = 1, \quad \phi(t) = 2\frac{\Delta\omega}{\omega_n} - \frac{\Delta\omega}{\omega_n}e^{-\omega_n t}(2 + \omega_n t) \qquad (5.9)$$

$$\zeta < 1, \quad \phi(t) = 2\zeta\frac{\Delta\omega}{\omega_n} + \frac{\Delta\omega}{\omega_n}e^{-\zeta\omega_n t}$$

$$\times \left[\frac{1 - 2\zeta^2}{\sqrt{1 - \zeta^2}} \sin\omega_n\sqrt{1 - \zeta^2}\,t - 2\zeta\cos\omega_n\sqrt{1 - \zeta^2}\,t \right]$$

The representative curves are given in Fig. 5.6. The steady-state error, corresponding to the first term of the right-hand member of the above expressions, is $2\zeta\Delta\omega/\omega_n$. With this type of loop, $2\zeta\omega_n=1/\tau_1$ (Eq. 3.47) and $\omega_n^2=K/\tau_1$ (Eq. 3.41) and consequently $\omega_n/2\zeta=K$. The steady-state error is thus in fact equal to $\Delta\omega/K$, as for the first-order loop. But having a high value for K infers a low value for the loop damping factor. The curves of Fig. 5.6 show clearly that to reduce the steady-state error, ζ has to be small; but we observe at the same time the appearance of a marked overshoot implying the loop's relative instability. The use of a phase-lead correction will remedy this defect, since ζ can then be chosen independently of K.

5.2.4 SECOND-ORDER LOOP WITH $F(s)=(1+\tau_2 s)/(1+\tau_1 s)$

In this case the phase error is derived in operational notation from the substitution of Eq. 3.56 in Eq. 5.6:

$$\Phi(s)=\frac{\Delta\omega}{s^2+2\zeta\omega_n s+\omega_n^2}+\frac{\omega_n^2}{K}\frac{\Delta\omega}{s\left(s^2+2\zeta\omega_n s+\omega_n^2\right)}$$

Using the same computation as in Section 5.2.3, we get

$$\zeta>1,\qquad \phi(t)=\frac{\Delta\omega}{K}+\frac{\Delta\omega}{\omega_n}e^{-\zeta\omega_n t}$$

$$\times\left[\frac{1-\zeta\omega_n/K}{\sqrt{\zeta^2-1}}\sinh\omega_n\sqrt{\zeta^2-1}\,t-\frac{\omega_n}{K}\cosh\omega_n\sqrt{\zeta^2-1}\,t\right]$$

$$\zeta=1,\qquad \phi(t)=\frac{\Delta\omega}{K}+\frac{\Delta\omega}{\omega_n}e^{-\omega_n t}\left(\omega_n t-\frac{\omega_n^2}{K}t-\frac{\omega_n}{K}\right) \qquad (5.10)$$

$$\zeta<1,\qquad \phi(t)=\frac{\Delta\omega}{K}+\frac{\Delta\omega}{\omega_n}e^{-\zeta\omega_n t}$$

$$\times\left[\frac{1-\zeta\omega_n/K}{\sqrt{1-\zeta^2}}\sin\omega_n\sqrt{1-\zeta^2}\,t-\frac{\omega_n}{K}\cos\omega_n\sqrt{1-\zeta^2}\,t\right]$$

The steady-state error is $\Delta\omega/K$ but, this time, a high value can be chosen for the loop gain, without reference to the damping factor value. When the loop gain K is very high, the above expressions are very close to equations represented in Figs. 5.4 and 5.5. This example further corroborates the fact

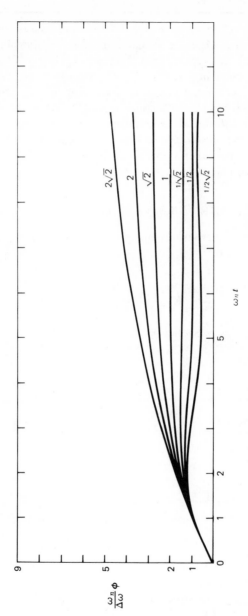

FIGURE 5.6. Phase error of a second-order loop $[F(s) = 1/(1 + \tau_1 s)]$ for a frequency step of the input signal.

that the loop with a filter $(1 + \tau_2 s)/\tau_1 s$, physically unrealizable, closely approximates the loop with filter $(1 + \tau_2 s)/(1 + \tau_1 s)$ when the open-loop gain K is high, and produces mathematical expressions that are easier to handle.

5.3 RESPONSE TO A LINEAR FREQUENCY VARIATION

When $t \geqslant 0$, the instantaneous frequency $f_{i_{\text{inst}}}$ of the input signal varies linearly versus time, with a slope R expressed in hertz per second:

$$f_{i_{\text{inst}}} = f + Rt\Upsilon(t)$$

The input signal phase variation (with respect to the normal phase $2\pi ft$) is written

$$\phi_i(t) = 2\pi R \frac{t^2}{2}\Upsilon(t) = \Re \frac{t^2}{2}\Upsilon(t)$$

$\Re = 2\pi R$ is the slope of the input angular frequency and is expressed in rad/s^2. The input phase Laplace transform is $\Phi_i(s) = \Re / s^3$, which gives a phase error

$$\Phi(s) = \left[1 - H(s) \right] \frac{\Re}{s^3} \qquad (5.11)$$

5.3.1 FIRST-ORDER LOOP

In operational notation, we get $\Phi(s) = \Re / s^2(s + K)$ that is,

$$\phi(t) = \frac{\Re t}{K} - \frac{\Re}{K^2}(1 - e^{-Kt}) \qquad (5.12)$$

The steady-state phase error (that is, where t is such that $e^{-Kt} \cong 0$) consists of a constant tracking error $-\Re / K^2$ and a term increasing linearly versus time, corresponding to the steady-state error for an angular frequency deviation $\Delta\omega = \Re t$. As t increases, this error can cause the loop to work outside the linear domain. When the error reaches $\pi/2$,* the loop falls out of lock.

5.3.2 SECOND-ORDER LOOP WITH $F(s) = (1 + \tau_2 s)/\tau_1 s$

The phase error in operational notation is

$$\Phi(s) = \frac{\Re}{s\left(s^2 + 2\zeta\omega_n s + \omega_n^2\right)}$$

*or π, if the detector characteristic range extends to π.

The reciprocal transform has the same form as that corresponding to the second term of the right-hand member of the expression in Section 5.2.3. This gives

$$\zeta > 1, \qquad \phi(t) = \frac{\mathcal{R}}{\omega_n^2} - \frac{\mathcal{R}}{\omega_n^2} e^{-\zeta \omega_n t}$$

$$\times \left[\cosh \omega_n \sqrt{\zeta^2 - 1}\; t + \frac{\zeta}{\sqrt{\zeta^2 - 1}} \sinh \omega_n \sqrt{\zeta^2 - 1}\; t \right]$$

$$\zeta = 1, \qquad \phi(t) = \frac{\mathcal{R}}{\omega_n^2} - \frac{\mathcal{R}}{\omega_n^2} e^{-\omega_n t} (1 + \omega_n t) \tag{5.13}$$

$$\zeta < 1, \qquad \phi(t) = \frac{\mathcal{R}}{\omega_n^2} - \frac{\mathcal{R}}{\omega_n^2} e^{-\zeta \omega_n t}$$

$$\times \left[\cos \omega_n \sqrt{1 - \zeta^2}\; t + \frac{\zeta}{\sqrt{1 - \zeta^2}} \sin \omega_n \sqrt{1 - \zeta^2}\; t \right]$$

The curves representing $(\omega_n^2 / \mathcal{R}) \phi(t)$ are given in Fig. 5.7. There is a steady-state phase error of $\mathcal{R} / \omega_n^2$. To get a null phase error in this case, a supplementary integration would be required in the loop. On the other hand, unlike the previous case and the two following cases, there is no phase error proportional to t. Consequently, if $\mathcal{R} / \omega_n^2$ is small enough and if the damping factor ζ is large enough to limit the overshoot amplitude, the loop stays in lock. In other words, the VCO remains in synchronism with the input signal: its instantaneous frequency becomes a linear function of time of slope R.

5.3.3 SECOND-ORDER LOOP WITH $F(s) = 1/(1 + \tau_1 s)$

From the error function (Eq. 3.51) and Eq. 5.11 we get

$$\Phi(s) = \frac{\mathcal{R}}{s(s^2 + 2\zeta \omega_n s + \omega_n^2)} + \frac{2\zeta \omega_n \mathcal{R}}{s^2(s^2 + 2\zeta \omega_n s + \omega_n^2)}$$

The reciprocal transform of the first term of the right-hand member is identical with the result given in the above paragraph. For the second term, we have to find the reciprocal transform of an expression of the type

$$\frac{1}{s^2(s + \alpha)(s + \beta)}$$

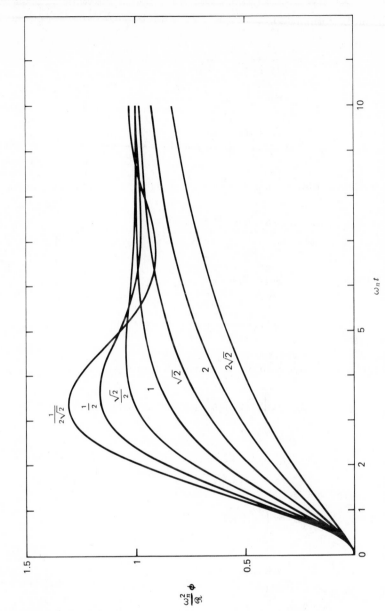

FIGURE 5.7. Phase error of a second-order loop $[F(s) = (1 + \tau_2 s)/(\tau_1 s)]$ for a linearly varying frequency of the input signal.

which can be formulated

$$\frac{\alpha+\beta}{\alpha^2\beta^2}\frac{s+k}{(s+\alpha)(s+\beta)}-\frac{\alpha+\beta}{\alpha^2\beta^2}\frac{s-k'}{s^2}$$

with

$$k=\frac{(\alpha+\beta)^2-\alpha\beta}{\alpha+\beta}\quad\text{and}\quad k'=\frac{\alpha\beta}{\alpha+\beta}$$

The reciprocal transform of $1/s^2(s+\alpha)(s+\beta)$ then becomes, in general,

$$\frac{\alpha+\beta}{\alpha^2\beta^2}\left[\frac{(k-\alpha)e^{-\alpha t}-(k-\beta)e^{-\beta t}}{\beta-\alpha}+k't-1\right]$$

and, for the particular case where $\alpha=\beta$ ($\zeta=1$),

$$\frac{2}{\alpha^3}\left\{\left[(k-\alpha)t+1\right]e^{-\alpha t}+k't-1\right\}$$

The reciprocal transform of the second term of $\Phi(s)$ is

$$\zeta>1,\qquad\frac{2\zeta}{\omega_n^3}e^{-\zeta\omega_n t}\left[\cosh\omega_n\sqrt{\zeta^2-1}\;t\right.$$

$$\left.+\frac{2\zeta^2-1}{2\zeta\sqrt{\zeta^2-1}}\sinh\omega_n\sqrt{\zeta^2-1}\;t\right]+\frac{t}{\omega_n^2}-\frac{2\zeta}{\omega_n^3}$$

$$\zeta=1,\qquad\frac{2}{\omega_n^3}e^{-\omega_n t}\left(\frac{\omega_n}{2}t+1\right)+\frac{t}{\omega_n^2}-\frac{2}{\omega_n^3}$$

$$\zeta<1,\qquad\frac{2\zeta}{\omega_n^3}e^{-\zeta\omega_n t}\left[\cos\omega_n\sqrt{1-\zeta^2}\;t\right.$$

$$\left.+\frac{2\zeta^2-1}{2\zeta\sqrt{1-\zeta^2}}\sin\omega_n\sqrt{1-\zeta^2}\;t\right]+\frac{t}{\omega_n^2}-\frac{2\zeta}{\omega_n^3}$$

If we combine this additional term with that corresponding to the reciprocal transform of the $\Phi(s)$ first term, we obtain Eqs. 5.14:

$$\zeta > 1, \quad \phi(t) = \frac{\Re}{\omega_n^2}(1 - 4\zeta^2) + \frac{2\zeta \Re t}{\omega_n}$$

$$-\frac{\Re}{\omega_n^2} e^{-\zeta\omega_n t}\left[(1 - 4\zeta^2)\cosh\omega_n\sqrt{\zeta^2 - 1}\ t \right.$$

$$\left. + \frac{\zeta}{\sqrt{\zeta^2 - 1}}\left[1 - 2(2\zeta^2 - 1)\right]\sinh\omega_n\sqrt{\zeta^2 - 1}\ t \right]$$

$$\zeta = 1, \quad \phi(t) = -\frac{3\Re}{\omega_n^2} + \frac{2\Re t}{\omega_n} + \frac{\Re}{\omega_n^2} e^{-\omega_n t}(3 + \omega_n t) \qquad (5.14)$$

$$\zeta < 1, \quad \phi(t) = \frac{\Re}{\omega_n^2}(1 - 4\zeta^2) + \frac{2\zeta \Re t}{\omega_n}$$

$$-\frac{\Re}{\omega_n^2} e^{-\zeta\omega_n t}\left[(1 - 4\zeta^2)\cos\omega_n\sqrt{1 - \zeta^2}\ t \right.$$

$$\left. + \frac{\zeta}{\sqrt{1 - \zeta^2}}\left[1 - 2(2\zeta^2 - 1)\right]\sin\omega_n\sqrt{1 - \zeta^2}\ t \right]$$

The steady-state error consists of a constant term, the value of which depends on the damping factor ζ, and a term increasing linearly versus time, similar to the $\Re t/K$ term of the first-order loop response, since $\omega_n/2\zeta = K$ for this type of loop. But to obtain a sufficiently small value for the natural angular frequency ω_n, without deteriorating the damping factor, the loop gain has to be kept very small. In these circumstances, the ramp $\Re$ will not long be applicable before the phase error $\Re t/K$ grows too large and the loop falls out of lock.

5.3.4 SECOND-ORDER LOOP WITH $F(s) = (1 + \tau_2 s)/(1 + \tau_1 s)$

The error function is an expression similar to that in Section 5.3.3 above. We simply substitute ω_n^2/K for $2\zeta\omega_n$ in the second term numerator. The reciprocal transform becomes

$$\zeta > 1, \qquad \phi(t) = \frac{\mathcal{R}}{\omega_n^2}\left(1 - 2\zeta\frac{\omega_n}{K}\right) + \frac{\mathcal{R}\,t}{K}$$

$$-\frac{\mathcal{R}}{\omega_n^2}e^{-\zeta\omega_n t}\left[\left(1 - 2\zeta\frac{\omega_n}{K}\right)\cosh\omega_n\sqrt{\zeta^2 - 1}\,t\right.$$

$$\left. + \frac{\zeta - (\omega_n/K)(2\zeta^2 - 1)}{\sqrt{\zeta^2 - 1}}\sinh\omega_n\sqrt{\zeta^2 - 1}\,t\right]$$

$$\zeta = 1, \qquad \phi(t) = \frac{\mathcal{R}}{\omega_n^2}\left(1 - 2\frac{\omega_n}{K}\right) + \frac{\mathcal{R}\,t}{K}$$

$$-\frac{\mathcal{R}}{\omega_n^2}e^{-\omega_n t}\left[1 - 2\frac{\omega_n}{K} + \left(\omega_n - \frac{\omega_n^2}{K}\right)t\right] \qquad (5.15)$$

$$\zeta < 1, \qquad \phi(t) = \frac{\mathcal{R}}{\omega_n^2}\left(1 - 2\zeta\frac{\omega_n}{K}\right) + \frac{\mathcal{R}\,t}{K}$$

$$-\frac{\mathcal{R}}{\omega_n^2}e^{-\zeta\omega_n t}\left[\left(1 - 2\zeta\frac{\omega_n}{K}\right)\cos\omega_n\sqrt{1 - \zeta^2}\,t\right.$$

$$\left. + \frac{\zeta - (\omega_n/K)(2\zeta^2 - 1)}{\sqrt{1 - \zeta^2}}\sin\omega_n\sqrt{1 - \zeta^2}\,t\right]$$

If the loop gain K is very high, the above $\phi(t)$ expressions are identical with those obtained in Section 5.3.2. Consequently, when K is very high, the curves of Fig. 5.7 can be used. However, the term $\mathcal{R}t/K$ should not be overlooked, since, even in the case of a high K, it can cause the phase detector to operate outside the linear zone and, ultimately, cause the loop to fall out of lock. As compared with the case described in Section 5.3.3., for given ω_n and ζ, we can use a much larger loop gain K. This means that, for a given frequency ramp R, the VCO will stay in synchronism with the input signal (within the frequency error R/K) for much longer.

NUMERICAL EXAMPLE.   The parameters of a phase-locked loop using a loop filter $F(s) = (1 + \tau_2 s)/(1 + \tau_1 s)$ are as follows:

$$K = 2\pi \times 10^5 \text{ rad/s}, \qquad \omega_n = 10^2 \text{ rad/s} \quad \text{and} \quad \zeta = \tfrac{1}{2}.$$

The VCO central frequency is 10 MHz.

The loop is locked on a frequency-modulated generator output signal, of frequency 10 MHz for $t < 0$. At instant $t = 0$, the frequency of this signal is made to vary linearly versus time, with a slope $R = 1$ kHz/s.

Since the quantity ω_n/K is much smaller than 1, the phase error is closely approximated by Eq. 5.13 (case $\zeta < 1$) increased by the quantity $\mathscr{R}t/K = 2\pi R t/K$ (see Eq. 5.15).

For $t \leqslant 10/\omega_n = 0.1$ s, the term $\mathscr{R}t/K$ can be disregarded, since it is equal to a maximum of 10^{-3} rad. The phase error is then given by the curve of Fig. 5.7, corresponding to $\zeta = \tfrac{1}{2}$.

The maximum phase error occurs at $t = 3.7/\omega_n = 37$ ms and is equal to $\phi_M = 1.16 \ \mathscr{R}/\omega_n^2 = 1.16 \ 2\pi \times 10^3/10^4 = 0.73$ rad.

For $t > 0.1$ s, the transient response vanishes and the phase error becomes

$$\phi(t) \cong \frac{\mathscr{R}}{\omega_n^2} + \frac{\mathscr{R}t}{K}$$

It reaches $\pi/2$ for the time t, so that

$$\frac{\pi}{2} = \frac{2\pi \times 10^3}{10^4} + \frac{2\pi \times 10^3}{2\pi \times 10^5} t$$

that is $t = 94.2$ s.

In other words, the VCO will follow the input signal frequency variation, with a frequency error of

$$\frac{1}{2\pi} \frac{\mathscr{R}}{K} = \frac{10^{-2}}{2\pi} \text{ Hz}$$

up to instant $t = 94.2$ s, when the loop drops out of lock.

CHAPTER 6
SINUSOIDAL
OPERATING CONDITIONS

The phase-locked loop response to a sinusoidal phase modulation of the input signal is directly derived from the transfer and error function expressions given in Chapter 3. We shall first analyze phase modulation and then frequency modulation. In this chapter, we shall designate the transfer and error functions, respectively, $H(j\Omega)$ and $[1 - H(j\Omega)]$. The notation $\omega = 2\pi f$ will be used to indicate the input signal angular frequency and the VCO central frequency and $\Omega = 2\pi F$ the modulation angular frequency.

6.1 SINUSOIDAL PHASE MODULATION

The signal y_i applied to the loop input is expressed

$$y_i(t) = A \sin \left[\omega t + m_i \sin (\Omega t + \Theta_i) \right]$$

The modulation index m_i is assumed to be small enough for the phase error, resulting from the input signal phase modulation $\phi_i(t) = m_i \sin(\Omega t + \Theta_i)$, to be such that only the linear part of the phase detector characteristic is used. We shall base our computation of the phase error on this a priori assumption, availing ourselves of the possibility to use the formula thus obtained to check the validity of the hypothesis.

Under these conditions, the steady-state VCO output signal phase ϕ_0 is also a sinusoidal signal, of angular frequency Ω, which can be expressed

$$\phi_0(t) = m_0 \sin (\Omega t + \Theta_0)$$

In the same way, the steady-state error signal can be expressed

$$\phi(t) = m \sin (\Omega t + \Theta)$$

The quantities m_0 and Θ_0, on the one hand, and m and Θ, on the other hand, can be derived from m_i and Θ_i using the transfer function (Eq. 3.6) and the error function (Eq. 3.7):

$$\frac{\Phi_0}{\Phi_i}(j\Omega) = H(j\Omega), \qquad \frac{\Phi}{\Phi_i}(j\Omega) = 1 - H(j\Omega)$$

that is,

$$m_0 = m_i \operatorname{Mod}\left[H(j\Omega)\right]$$

$$\Theta_0 = \Theta_i + \operatorname{Arg}\left[H(j\Omega)\right]$$

$$\text{(6.1)}$$

$$m = m_i \operatorname{Mod}\left[1 - H(j\Omega)\right]$$

$$\Theta = \Theta_i + \operatorname{Arg}\left[1 - H(j\Omega)\right]$$

$$\text{(6.2)}$$

We shall introduce successively the transfer $H(j\Omega)$ and error $[1 - H(j\Omega)]$ function expressions corresponding to the different loops to get the expressions for the VCO modulation signal and the phase error signal.

6.1.1 FIRST-ORDER LOOP

The transfer and error functions are given, respectively, by Eqs. 3.10 and 3.11:

$$H(j\Omega) = \frac{K}{K + j\Omega}, \qquad 1 - H(j\Omega) = \frac{j\Omega}{K + j\Omega}$$

Substituting the first of these equations in Eqs. 6.1, we get

$$m_0 = \frac{K}{\sqrt{K^2 + \Omega^2}} m_i$$

$$\Theta_0 = \Theta_i - \operatorname{Arctan}\frac{\Omega}{K}$$

$$\text{(6.3)}$$

The VCO phase modulating signal is thus expressed:

$$\phi_0(t) = \frac{Km_i}{\sqrt{K^2 + \Omega^2}} \sin\left(\Omega t + \Theta_i - \operatorname{Arctan}\frac{\Omega}{K}\right)$$

As regards the error signal, Eq. 3.11 has simply to be substituted in Eqs. 6.2, giving

$$m = \frac{\Omega}{\sqrt{K^2 + \Omega^2}} m_i$$

$$\Theta = \Theta_i + \frac{\pi}{2} - \operatorname{Arctan}\frac{\Omega}{K}$$

$$\text{(6.4)}$$

The error signal can then be given in the form:

$$\phi(t) = \frac{\Omega m_i}{\sqrt{K^2 + \Omega^2}} \sin\left(\Omega t + \Theta_i + \frac{\pi}{2} - \text{Arc}\tan\frac{\Omega}{K}\right)$$

The first-order loop acts like a one-pole low-pass filter, of time constant $\tau = 1/K$, with respect to the modulation $\phi_i(t)$ of the input signal $y_i(t)$ and the modulation $\phi_0(t)$ of the VCO output signal $y_0(t)$.

As regards the error signal, a first-order phase-locked loop looks like a phase demodulator followed by a one-pole high-pass filter of time constant $\tau = 1/K$.

These two functions are illustrated in the block diagrams of Figs. 6.1 and 6.2.

Concerning the diagram of Fig. 6.1, we draw attention to the fact that it is simply a model and that no part of the real loop actually provides access to the modulation signals ϕ_i and ϕ_0. The diagram of Fig. 6.2 is also a model but, if there is still no direct access to $\phi_i(t)$ in the loop, there is, on the other hand, a signal directly proportional to ϕ: the phase detector output signal u_1, since $u_1 = K_1\phi$.

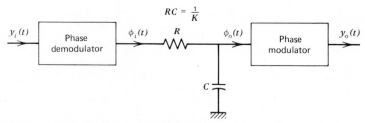

FIGURE 6.1. Model of a first-order loop for input signal modulation and VCO signal modulation.

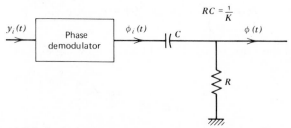

FIGURE 6.2. Model of a first-order loop for input signal modulation and phase error in the loop.

From the analysis of the loop response to transient conditions recounted in Chapter 5, it is clear that to reduce static errors and avoid transient disturbances, the loop gain K has to be relatively high; the filtering of the modulation transferred from the input signal to the VCO signal thus remains very approximate. The phase demodulator application is not much better since all the frequency modulation components below $K/2\pi$ are cut off by the high-pass filter.

6.1.2 SECOND-ORDER LOOP WITH $F(s) = (1 + \tau_2 s)/\tau_1 s$

The transfer and error functions applicable are those given by Eqs. 3.43 and 3.44:

$$H(j\Omega) = \frac{\omega_n^2 + j2\zeta\omega_n\Omega}{\omega_n^2 - \Omega^2 + j2\zeta\omega_n\Omega}$$

$$1 - H(j\Omega) = \frac{-\Omega^2}{\omega_n^2 - \Omega^2 + j2\zeta\omega_n\Omega}$$

Substituting the expression for $H(j\Omega)$ in Eqs. 6.1, we get for the VCO modulation signal

$$m_0 = m_i \sqrt{\frac{\omega_n^4 + 4\zeta^2\omega_n^2\Omega^2}{\left(\omega_n^2 - \Omega^2\right)^2 + 4\zeta^2\omega_n^2\Omega^2}}$$

and

$$\Theta_0 = \Theta_i + \operatorname{Arc\,tan} 2\zeta\frac{\Omega}{\omega_n} - \operatorname{Arc\,tan}\frac{2\zeta\omega_n\Omega}{\omega_n^2 - \Omega^2}$$

If we use the reduced variable $x = \Omega/\omega_n$, these expressions become

$$m_0 = m_i \sqrt{\frac{1 + 4\zeta^2 x^2}{\left(1 - x^2\right)^2 + 4\zeta^2 x^2}}$$

and   (6.5)

$$\Theta_0 = \Theta_i + \operatorname{Arc\,tan} 2\zeta x - \operatorname{Arc\,tan}\frac{2\zeta x}{1 - x^2}$$

The curves representing $20\log_{10}(m_0/m_i)$ and $\Theta_0 - \Theta_i$ versus $x = \Omega/\omega_n$, for different values of the damping factor ζ, are given in Figs. 6.3 and 6.4.

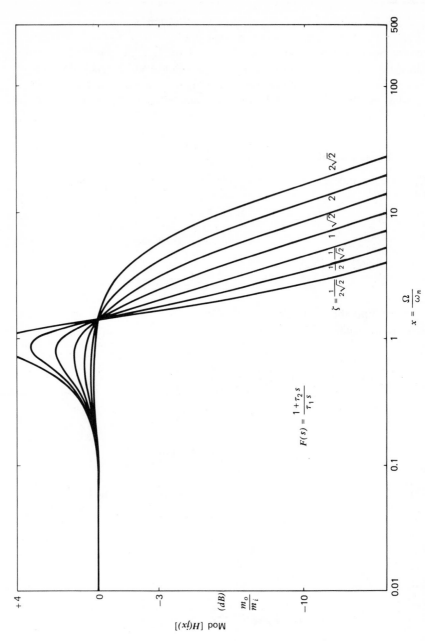

FIGURE 6.3. Transfer function of a second-order loop with $F(s) = (1 + \tau_2 s)/(\tau_1 s)$: amplitude-frequency characteristic.

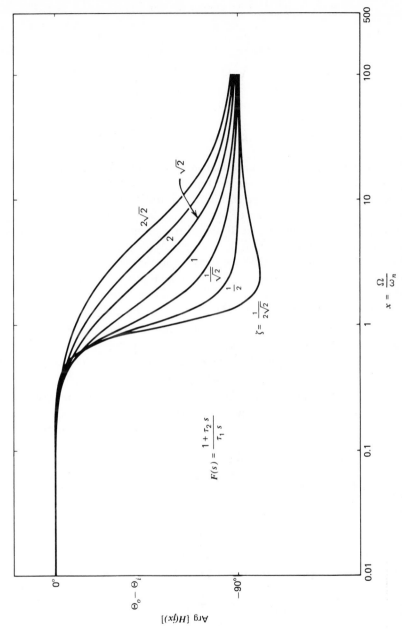

FIGURE 6.4. Transfer function of a second-order loop with $F(s) = (1 + \tau_2 s)/(\tau_1 s)$: phase-frequency characteristic.

107

As regards the error signal, if we substitute the expression for $[1-H(j\Omega)]$ in Eqs. 6.2, we get

$$m = m_i \frac{x^2}{\sqrt{(1-x^2)^2 + 4\zeta^2 x^2}}$$

$$\Theta = \Theta_i + \pi - \text{Arc tan} \frac{2\zeta x}{1-x^2}$$

(6.6)

The curves representing $20\log_{10}(m/m_i)$ and $\Theta - \Theta_i$ versus x are given in Figs. 6.5 and 6.6.

The loop gain value K has no influence on the above expressions (strictly speaking, its value is infinite; see Section 2.3, Remark 2). This leaves two possibilities: either narrow low-pass filtering of the modulation or demodulation of the input signal so as to lose only a fraction of the modulation spectrum, while maintaining a small phase error in the case of input signal or VCO frequency drift.

On the curves concerning the modulation transferred from the input signal to the VCO signal (Fig. 6.3), it will be noted that the amplitude characteristics all exceed 0 dB for $x < 1$, the maximum value being larger for lower damping factor values. All the curves cut the 0-dB axis at a point $x = \sqrt{2}$ with a negative slope increasing as the damping factor decreases. The asymptotic slope (for $x \to \infty$) is -1 (or -6 dB/octave).

The curves in Fig. 6.5 show that the error signal is a reliable image of the input signal modulation for $x > 1$, particularly when the damping factor ζ remains in the vicinity of $\sqrt{2}/2$ or 1. The very low frequencies are attenuated: the loop acts like a phase demodulator followed by a high-pass filter with an asymptotic slope of $+2$ ($+12$ dB/octave) when x approaches 0.

As for the first-order loop, we give in Figs. 6.7 and 6.8, models of the two functions: transfer of modulation from the input signal to the VCO signal and demodulation of the input signal.

6.1.3 SECOND-ORDER LOOP WITH $F(s) = 1/(1+\tau_1 s)$

The transfer and error functions corresponding to the use of a low-pass filter, of time constant τ_1, as loop filter, are given by Eqs. 3.48 and 3.49:

$$H(j\Omega) = \frac{\omega_n^2}{\omega_n^2 - \Omega^2 + j2\zeta\omega_n\Omega}$$

$$1 - H(j\Omega) = \frac{-\Omega^2 + j2\zeta\omega_n\Omega}{\omega_n^2 - \Omega^2 + j2\zeta\omega_n\Omega}$$

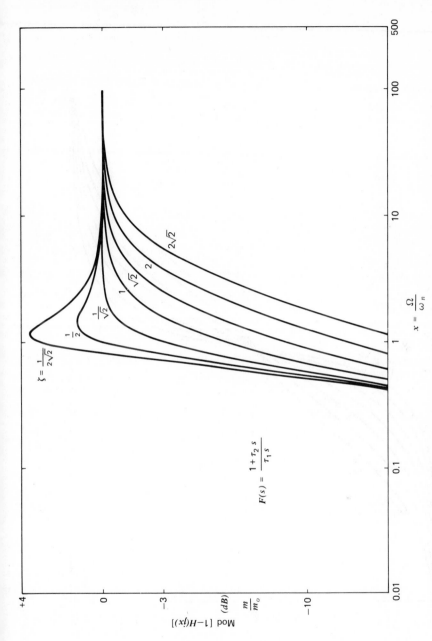

FIGURE 6.5. Error function of a second-order loop with $F(s) = (1 + \tau_2 s)/(\tau_1 s)$: amplitude-frequency characteristic.

109

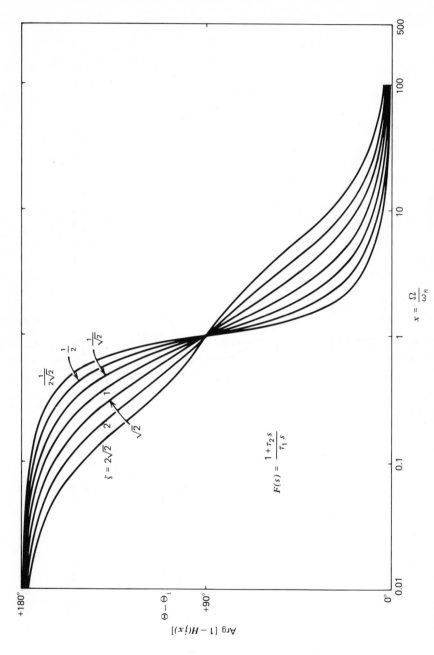

FIGURE 6.6. Error function of a second-order loop with $F(s) = (1 + \tau_2 s)/(\tau_1 s)$: phase-frequency characteristic.

110

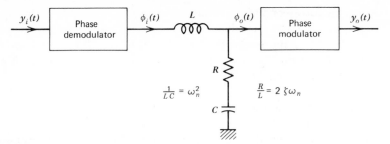

FIGURE 6.7. Model of a second-order loop with $F(s)=(1+\tau_2 s)/(\tau_1 s)$ for input signal and VCO signal modulations.

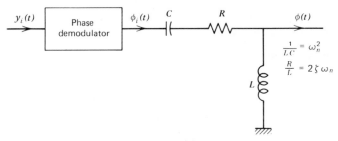

FIGURE 6.8. Model of a second-order loop with $F(s)=(1+\tau_2 s)/(\tau_1 s)$ working as an input signal phase demodulator.

The modulation index m_0 and the modulation argument Θ_0 of the VCO output signal are obtained by substituting $H(j\Omega)$ in Eqs. 6.1:

$$m_0 = m_i \frac{1}{\sqrt{\left(1-x^2\right)^2 + 4\zeta^2 x^2}}$$

$$\Theta_0 = \Theta_i - \operatorname{Arc\,tan} \frac{2\zeta x}{1-x^2} \tag{6.7}$$

The amplitude and argument of the error signal are obtained, respectively, in substituting $1-H(j\Omega)$ in Eqs. 6.2:

$$m = m_i \sqrt{\frac{x^4 + 4\zeta^2 x^2}{\left(1-x^2\right)^2 + 4\zeta^2 x^2}}$$

$$\Theta = \Theta_i + \pi - \operatorname{Arc\,tan} \frac{2\zeta}{x} - \operatorname{Arc\,tan} \frac{2\zeta x}{1-x^2} \tag{6.8}$$

The various corresponding curves are given in Figs. 6.9–6.12. With respect to the previous case, the results are reversed: the asymptotic slope of the low-pass filtering of the modulation transferred to the VCO is -2 on Fig. 6.9, instead of -1 on Fig. 6.3, while the asymptotic slope of the high-pass filtering of the error signal is $+1$ in Fig. 6.11, instead of $+2$ in Fig. 6.5.

The models shown in Figs. 6.7 and 6.8 can be used to represent these filtering functions, but the filter output take-off has then to be located between capacitor C and resistor R.

It should also be remembered, when comparing this case and the following one, that the loop gain K cannot be chosen freely: for given ζ and ω_n, the value of K is determined such that the loop performances under transient conditions are fairly disappointing.

6.1.4   SECOND-ORDER LOOP WITH $F(s) = (1 + \tau_2 s)/(1 + \tau_1 s)$

The transfer and error functions are given by Eqs. 3.53 and 3.54:

$$H(j\Omega) = \frac{\omega_n^2 + j\left(2\zeta\omega_n - \omega_n^2/K\right)\Omega}{\omega_n^2 - \Omega^2 + j 2\zeta\omega_n\Omega}$$

$$1 - H(j\Omega) = \frac{-\Omega^2 + j\left(\omega_n^2/K\right)\Omega}{\omega_n^2 - \Omega^2 + j 2\zeta\omega_n\Omega}$$

As regards the signal modulating the VCO signal and changing directly the variable $x = \Omega/\omega_n$, we get

$$m_0 = m_i \sqrt{\frac{1 + \left(2\zeta - \omega_n/K\right)^2 x^2}{\left(1 - x^2\right)^2 + 4\zeta^2 x^2}}$$

$$\Theta_0 = \Theta_i + \operatorname{Arc tan}\left(2\zeta - \frac{\omega_n}{K}\right)x - \operatorname{Arc tan}\frac{2\zeta x}{1 - x^2}$$

(6.9)

As for the error signal or demodulated signal,

$$m = m_i \sqrt{\frac{x^4 + \left(\omega_n^2/K^2\right)x^2}{\left(1 - x^2\right)^2 + 4\zeta^2 x^2}}$$

$$\Theta = \Theta_i + \pi - \operatorname{Arc tan}\frac{\omega_n}{Kx} - \operatorname{Arc tan}\frac{2\zeta x}{1 - x^2}$$

(6.10)

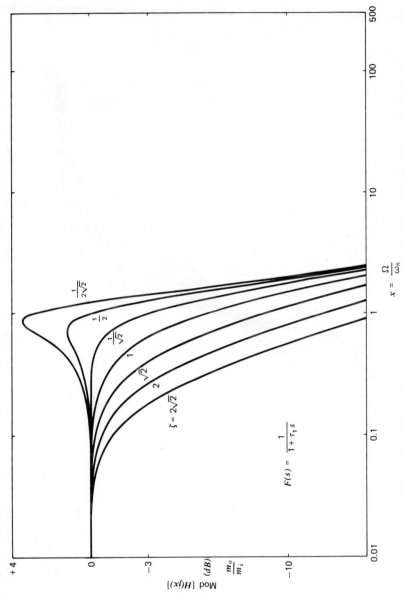

FIGURE 6.9. Transfer function of a second-order loop with $F(s) = 1/(1 + \tau_1 s)$: amplitude-frequency characteristic.

113

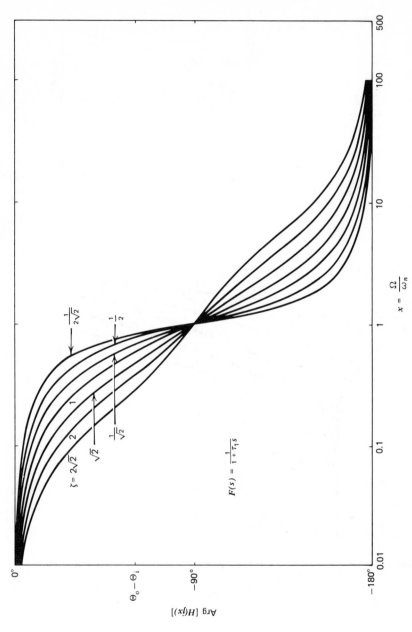

FIGURE 6.10. Transfer function of a second-order loop with $F(s) = 1/(1 + \tau_1 s)$: phase-frequency characteristic.

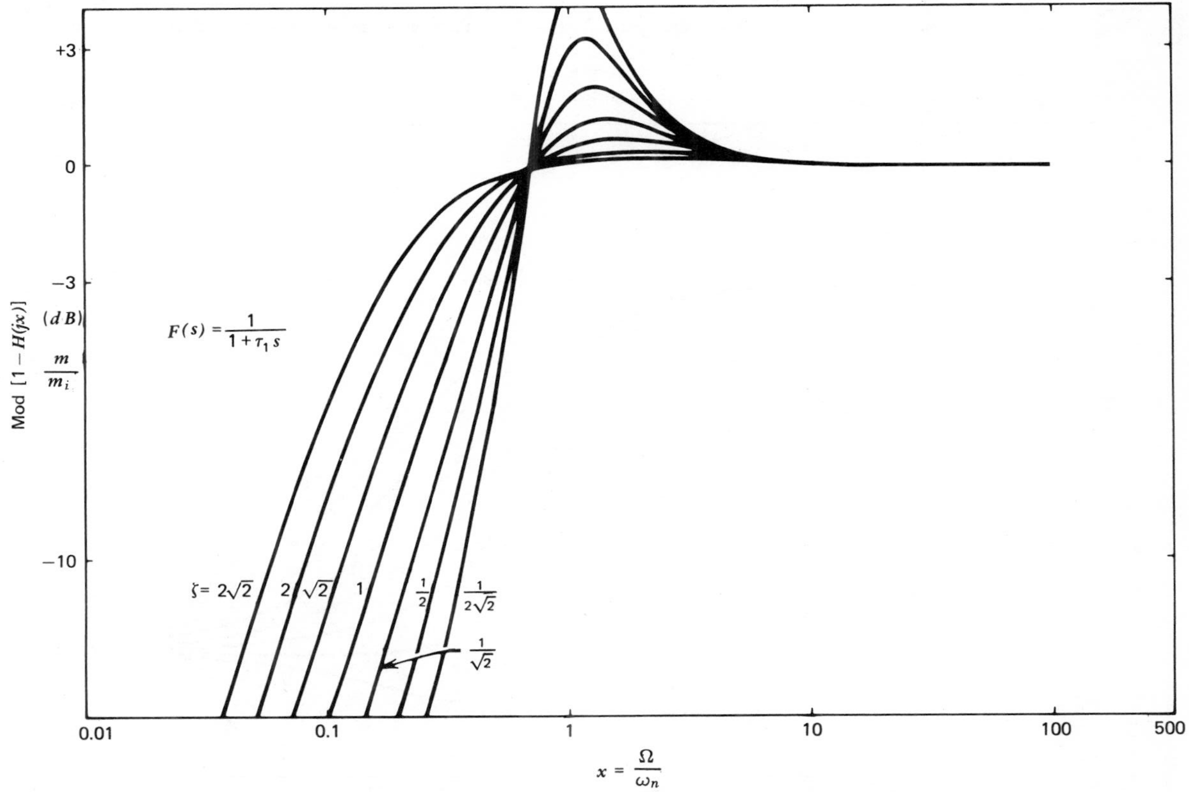

+3
0
−3
−10
Mod [1 − H(jx)]
(dB)
m / m_i
F(s) = 1 / (1 + τ₁s)
ζ = 2√2
2
√2
1
1/2
1/(2√2)
1/√2
0.01
0.1
1
10
100
500
x = Ω / ω_n

FIGURE 6.11. Error function of a second-order loop with $F(s)=1/(1+\tau_1 s)$: amplitude-frequency characteristic.

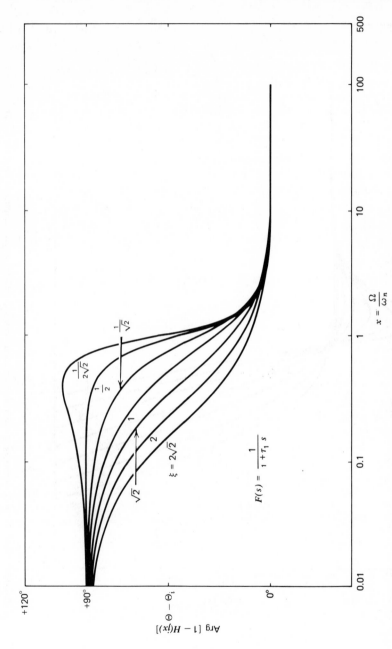

FIGURE 6.12. Error function of a second-order loop with $F(s) = 1/(1 + \tau_1 s)$: phase-frequency characteristic.

Generally speaking, the loop gain K is much larger than the natural angular frequency ω_n and consequently the corresponding curves are practically identical with those in Figs. 6.3–6.6.

But, for the error signal, it is worth noting that the asymptotic slope for $x \to 0$ is only $+1$ and not $+2$. Correlatively, the phase shift representation is not a monotonic decreasing curve from π to 0: it starts at $\pi/2$ for $x = 0$, increases to around π for a low value of x, then decreases from this value to 0 as x increases. In the "useful" bandwidth, when this loop is used as an input signal phase demodulator, above $x = 1$ for example, the phase-shift of the demodulated signal with respect to the input signal modulation is very close to that given by the curve in Fig. 6.6.

The filtering accomplished by the transfer and error functions is shown in the models given in Figs. 6.13 and 6.14. These diagrams are practically identical with those in Figs. 6.7 and 6.8. The only difference is the potentiometer take-off on the resistor at μR (on the capacitor C side) with

$$\mu = \frac{2\zeta\omega_n - \omega_n^2/K}{2\zeta\omega_n}$$

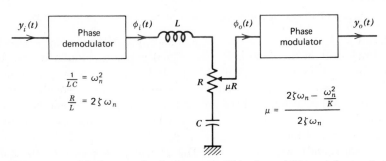

FIGURE 6.13. Model of a second-order loop with $F(s) = (1 + \tau_2 s)/(1 + \tau_1 s)$ for input signal and VCO signal modulations.

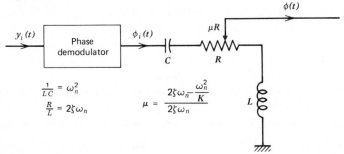

FIGURE 6.14. Model of a second-order loop with $F(s) = (1 + \tau_2 s)/(1 + \tau_1 s)$ working as a phase demodulator.

For the second-order loops (Sections 6.1.2, 6.1.3, and 6.1.4), when the input signal is phase modulated by a sinusoidal signal of angular frequency Ω, we can draw the following conclusions from the various formulas and curves:

- If the angular frequency Ω is lower than the loop natural angular frequency ω_n, the input signal modulation is transferred to the VCO signal, with practically no attenuation or phase shift. The phase error in the loop, which is the instantaneous phase difference between the input signal and the VCO signal, stays low under steady-state conditions, even if the input signal modulation index goes beyond the phase detector linear zone.

- If the angular frequency Ω is higher than ω_n, the input signal modulation is not transferred to the VCO output signal. The loop phase error is here a very good duplicate of the input signal modulation. This case corresponds to the use of the phase-locked loop as a phase demodulator. However, if the loop phase error is to be kept low enough for the phase comparator characteristic to remain within its linear zone, the modulation index m_i of the input signal has to remain below $\pi/6$ (or $\pi/4$, according to the degree of distortion tolerated) for a sinusoidal detector, or below $\pi/2$ or π for the two types of linear detector discussed in Sections 2.1.5 and 2.1.6.

NUMERICAL EXAMPLE. We receive a signal at 2 MHz, having a possible frequency drift of up to 20 kHz. This signal is phase modulated, at a low index, by a low-frequency signal with spectral components ranging from 50 Hz to 10 kHz. A phase-locked loop is used for the demodulation. A first-order loop is unsuitable because the loop gain K value would have to be around $2\pi \times 50$ rad/s, which is much too low to account for the possible frequency drift of the 2-MHz signal. A second-order loop will be preferred, with a sufficiently high loop gain. The choice of $K = 2\pi \times 2 \times 10^5$ rad/s guarantees a maximum phase error, induced by the frequency drift, of 0.1 rad.

The loop natural angular frequency ω_n should be chosen in the vicinity of $2\pi \times 50$ rad/s. This precludes the use of a loop filter of the type $F(s) = 1/(1 + \tau_1 s)$, since the damping factor $\zeta = \omega_n / 2K$ would be around 1.25×10^{-4}, which produces a very high peak in the demodulation bandwidth (Fig. 6.11), with considerable risk of instability (Section 4.4).

We shall use a loop filter of the type $F(s) = (1 + \tau_2 s)/(1 + \tau_1 s)$ [approximated as $F(s) = (1 + \tau_2 s)/\tau_1 s$, bearing in mind the value of ω_n/K]. The corresponding amplitude response curve is given in Fig. 6.5. The optimum damping factor is $\zeta = \sqrt{2}/2$ (maximally flat response). With $\omega_n = 2\pi \times 50$ rad/s, a 50-Hz component of the modulation signal will be weakened by -3 dB.

6.2 ESTABLISHMENT OF STEADY-STATE CONDITIONS

The expressions given in Section 6.1 are only valid for steady-state conditions, that is to say, assuming that the input signal modulation has existed for a long time. It can be of interest, for certain particular modulations, to know the loop response when modulation $\phi_i(t) = m_i \sin \Omega t$ is applied to the input signal at instant $t = 0$. Two methods can be used. The first consists in solving a differential equation of type 3.18, 3.24, or 3.30, making allowances for $\phi_i(t)$ and initial conditions.

The second method, more direct, consists in using the operational computation. The Laplace transform of the expression

$$\phi_i(t) = m_i \sin \Omega t \cdot \Upsilon(t)$$

is

$$\Phi_i(s) = m_i \frac{\Omega}{s^2 + \Omega^2}$$

The loop phase error, $\Phi(s)$ in operational notation, is then given by Eq. 3.9, which leads to

$$\Phi(s) = \Phi_i(s)\left[1 - H(s)\right] = m_i \frac{\Omega}{s^2 + \Omega^2}\left[1 - H(s)\right]$$

If we use the error function expression $[1 - H(s)]$ of equations 3.46, 3.51, or 3.56 and then take the reciprocal transform of $\Phi(s)$, we obtain the expression of $\phi(t)$.

The procedure described is used below in the case of a loop with integrator and correction $[F(s) = (1 + \tau_2 s)/\tau_1 s]$, that is, taking Eq. 3.46 as the expression of the error function,

$$\Phi(s) = m_i \frac{\Omega s^2}{\left(s^2 + \Omega^2\right)\left(s^2 + 2\zeta \omega_n s + \omega_n^2\right)}$$

which can be broken down into

$$\Phi(s) = \frac{m_i \Omega}{s^2 + 2\zeta \omega_n s + \omega_n^2} - \frac{m_i \Omega^3}{\left(s^2 + \Omega^2\right)\left(s^2 + 2\zeta \omega_n s + \omega_n^2\right)}$$

The reciprocal transform of the first term of the right-hand member is similar to the loop response to a frequency step $\Delta \omega = m_i \Omega$, as given in Section 5.2.2. If we consider the case where the damping factor $\zeta > 1$, the reciprocal transform is

$$m_i\Omega \frac{e^{-\omega_n\left(\zeta - \sqrt{\zeta^2-1}\,\right)t} - e^{-\omega_n\left(\zeta + \sqrt{\zeta^2-1}\,\right)t}}{2\omega_n\sqrt{\zeta^2-1}}$$

If we break down the second term of the right-hand member into simple elements and regroup the reciprocal transforms, we get the term

$$\frac{m_i\Omega^3}{\omega_n^4 + 2(2\zeta^2-1)\omega_n^2\Omega^2 + \Omega^4}\left(2\zeta\omega_n\cos\Omega t - \frac{\omega_n^2-\Omega^2}{\Omega}\sin\Omega t\right)$$

and the term

$$\frac{m_i\Omega^3}{2\omega_n\sqrt{\zeta^2-1}}\left[\frac{e^{-\omega_n\left(\zeta + \sqrt{\zeta^2-1}\,\right)t}}{\omega_n^2\left(\zeta + \sqrt{\zeta^2-1}\,\right)^2 + \Omega^2} - \frac{e^{-\omega_n\left(\zeta - \sqrt{\zeta^2-1}\,\right)t}}{\omega_n^2\left(\zeta - \sqrt{\zeta^2-1}\,\right)^2 + \Omega^2}\right]$$

The overall response $\phi(t)$ can be expressed in the form of the sum of two terms, one representing the steady-state response $\phi_P(t)$ and the other the transient response $\phi_T(t)$:

$$\phi(t) = \phi_P(t) + \phi_T(t)$$

with

$$\phi_P(t) = \frac{m_i\Omega^3}{\omega_n^4 + 2(2\zeta^2-1)\omega_n^2\Omega^2 + \Omega^4}$$

$$\times\left(2\zeta\omega_n\cos\Omega t - \frac{\omega_n^2-\Omega^2}{\Omega}\sin\Omega t\right)$$

and

$$\phi_T(t) = -\frac{m_i\Omega}{2\omega_n\sqrt{\zeta^2-1}}\left[e^{-\omega_n\left(\zeta + \sqrt{\zeta^2-1}\,\right)t} - e^{-\omega_n\left(\zeta - \sqrt{\zeta^2-1}\,\right)t}\right]$$

$$+ \frac{m_i\Omega^3}{2\omega_n\sqrt{\zeta^2-1}}\left[\frac{e^{-\omega_n\left(\zeta + \sqrt{\zeta^2-1}\,\right)t}}{\omega_n^2\left(\zeta + \sqrt{\zeta^2-1}\,\right)^2 + \Omega^2} - \frac{e^{-\omega_n\left(\zeta - \sqrt{\zeta^2-1}\,\right)t}}{\omega_n^2\left(\zeta - \sqrt{\zeta^2-1}\,\right)^2 + \Omega^2}\right]$$

By changing the variable $x = \Omega/\omega_n$, we can cross check that the response $\phi_P(t)$ corresponds to the solution given in Section 6.1.2.

The transient term comprises two terms. The first corresponds to the loop response to a frequency step $m_i\Omega$. But, when the modulation $\phi_i(t)$ $= m_i \sin \Omega t \cdot \Upsilon(t)$ is applied to the input signal, the instantaneous frequency does undergo a frequency step $(d\phi_i/dt)_{t=0} = m_i\Omega$.

The second transient component is negligible if Ω is much smaller than ω_n or, on the contrary, lessens the influence of the first term, if Ω is much larger than ω_n. In other words, the transient term represents the small signal required to change $\phi(t)$ from its null value at $t=0$ to the steady-state value, as shown in Fig. 6.15. Even though the steady-state demodulated signal $\phi_p(t)$ phase-leads the modulation signal $\phi_i(t)$, it will be observed that the establishment of signal $\phi(t)$ is always delayed with respect to $\phi_i(t)$ [$\phi(t)$ increases more slowly than $\phi_i(t)$], as would be expected, since in any physically constructable system, the output signal cannot precede the input signal.

NUMERICAL EXAMPLE. We use as a phase demodulator a second-order loop defined by $K = 2\pi \ 10^5$ rad/s, $\omega_n = 2\pi \ 50$ rad/s, $\zeta = \sqrt{2}$. At instant $t=0$, a sinusoidal modulation is applied to the input signal:

$$\phi_i(t) = m_i \sin \Omega t \quad \text{with} \quad \Omega = 2\pi \times 2.5 \times 10^2 \text{ rad/s}$$

The steady state output signal from the phase detector phase-leads by 30.5° the input signal modulation (Fig. 6.6); its amplitude is smaller than m_i by -0.95 dB (Fig. 6.5).

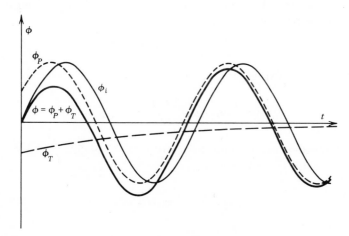

FIGURE 6.15. Transient response of a second-order loop to a sinusoidal phase modulation of the input signal.

The transient response term $\phi_T(t)$ above, is quickly damped. For $t=0$, it equals

$$\phi_T(0) = m_i \frac{250^3}{2 \times 50} \left[\frac{1}{50^2(\sqrt{2}+1)^2 + 250^2} \right.$$

$$\left. - \frac{1}{50^2(\sqrt{2}-1)^2 + 250^2} \right]$$

that is,

$$\phi_T(0) = -0.456 \, m_i$$

We cross check that $\phi(0) = \phi_T(0) + \phi_P(0) = 0$ since

$$\phi_P(0) = \frac{m_i}{1.11} \sin{(30.5°)}$$

$$= m_i \frac{0.508}{1.11} = 0.456 \, m_i$$

6.3 SINUSOIDAL FREQUENCY MODULATION

We now assume that the signal $y_i(t)$ applied to the loop input is frequency modulated by a sinusoidal signal of angular frequency Ω, with a peak frequency deviation Δf_i. This means that the instantaneous frequency of signal $y_i(t)$ is expressed

$$f_{i_{\text{inst}}} = f + \Delta f_i \cos{(\Omega t + \Theta_i)} \tag{6.11}$$

Signal $y_i(t)$ can be expressed in the form

$$y_i(t) = A \sin \left[\omega t + \frac{2\pi \Delta f_i}{\Omega} \sin{(\Omega t + \Theta_i)} \right]$$

This modulation can be interpreted as a phase modulation of the input signal by a signal, of angular frequency Ω, expressed as

$$\phi_i(t) = \frac{2\pi \Delta f_i}{\Omega} \sin{(\Omega t + \Theta_i)}$$

$$= \frac{\Delta \omega_i}{\Omega} \sin{(\Omega t + \Theta_i)}$$

The only difference with the phase modulation case is that the equivalent modulation index m_i is not independent of the modulation angular frequency Ω, but follows the law

$$m_i = \frac{\Delta\omega_i}{\Omega} \tag{6.12}$$

The VCO modulation $\phi_0(t)$ and error $\phi(t)$ signals are then derived, as for the phase modulation, from the modulation $\phi_i(t)$ defined above, by means of the transfer and error functions.

Providing we remain in the linear domain, the VCO output signal $y_0(t)$ can be expressed in the form

$$y_0(t) = B \cos \left[\omega t + m_0 \sin (\Omega t + \Theta_0) \right]$$

where m_0 and Θ_0 are related to m_i and Θ_i by Eqs. 6.1 and m_i is given by Eq. 6.12. Consequently,

$$y_0(t) = B \cos \left\{ \omega t + \frac{\Delta\omega_i}{\Omega} \operatorname{Mod} \left[H\left(j\Omega \right) \right] \right.$$

$$\left. \times \sin \left(\Omega t + \Theta_i + \operatorname{Arg} \left[H\left(j\Omega \right) \right] \right) \right\}$$

that is, the instantaneous frequency of the VCO output signal is written

$$f_{0_{\text{inst}}} = f + \Delta f_i \operatorname{Mod} \left[H\left(j\Omega \right) \right] \cos \left\{ \Omega t + \Theta_i + \operatorname{Arg} \left[H\left(j\Omega \right) \right] \right\}$$

If this expression is compared with that of the input signal instantaneous frequency (Eq. 6.11), we observe that the frequency modulation transferred from the input signal to the VCO signal is modified in amplitude by $\operatorname{Mod}[H(j\Omega)]$ and in phase by $\operatorname{Arg}[H(j\Omega)]$:

$$f_{0_{\text{inst}}} = f + \Delta f_0 \cos (\Omega t + \Theta_0) \tag{6.13}$$

with

$$\begin{aligned} \Delta f_0 &= \Delta f_i \operatorname{Mod} \left[H\left(j\Omega \right) \right] \\ \Theta_0 &= \Theta_i + \operatorname{Arg} \left[H\left(j\Omega \right) \right] \end{aligned} \tag{6.14}$$

The curves $\Delta f_0 / \Delta f_i$ and the curves $(\Theta_0 - \Theta_i)$ versus Ω or versus the reduced variable $x = \Omega/\omega_n$, are thus identical with the curves representing m_0 / m_i and $(\Theta_0 - \Theta_i)$ obtained for the phase modulation investigation. The curves in Figs. 6.3 and 6.4 or those in Figs. 6.9 and 6.10 are to be used, depending on the loop filter employed. In the same way, as regards the frequency modulation transferred from the input signal to the VCO signal, the models shown in Figs. 6.1, 6.7 and 6.13 are applicable, if we replace the phase demodulator and phase modulator by frequency demodulator and frequency modulator.

As regards the error signal, which, in any case, corresponds to an instantaneous phase, and not frequency, the phenomenon is slightly modified, for the steady-state expression is

$$\phi(t) = \Delta\phi \sin(\Omega t + \Theta) \tag{6.15}$$

with

$$\Delta\phi = m_i \operatorname{Mod}[1 - H(j\Omega)]$$

$$\Theta = \Theta_i + \operatorname{Arg}[1 - H(j\Omega)]$$

The curves giving the phase difference between the error signal and the input signal modulation, interpreted as a phase modulation, remain unchanged (Figs. 6.6 and 6.12). On the other hand, the error signal amplitude $\Delta\phi$ is related to the input magnitude $\Delta\omega_i$ by the expression

$$\Delta\phi = \Delta\omega_i \frac{\operatorname{Mod}[1 - H(j\Omega)]}{\Omega} \tag{6.16}$$

We shall analyze this formula for each of the loop types considered previously.

6.3.1   FIRST-ORDER LOOP

Taking into account the expression $\operatorname{Mod}[1 - H(j\Omega)]$ we get

$$\Delta\phi = \frac{\Delta\omega_i}{\sqrt{K^2 + \Omega^2}} \tag{6.17}$$

If the modulation frequency $\Omega \ll K$, $\Delta\phi \cong \Delta\omega_i/K$ and the linearity condition will require the input signal angular frequency deviation to remain below the loop gain.

If the modulation frequency $\Omega \gg K$, $\Delta\phi \cong \Delta\omega_i/\Omega = m_i$ and the linearity condition will require the input signal phase modulation index to remain low.

6.3.2   SECOND-ORDER LOOP WITH $F(s) = (1 + \tau_2 s)/\tau_1 s$.

Taking into account Eq. 3.44, Eq. 6.16 becomes

$$\Delta\phi = \Delta\omega_i \frac{\Omega}{\sqrt{(\omega_n^2 - \omega^2)^2 + 4\zeta^2 \omega_n^2 \Omega^2}}$$

that is, using the reduced variable $x = \Omega/\omega_n$,

$$\Delta\phi = \frac{\Delta\omega_i}{\omega_n} \frac{x}{\sqrt{(1-x^2)^2 + 4\zeta^2 x^2}} \tag{6.18}$$

The curves representing $20\log_{10}[(\omega_n/\Delta\omega_i)\Delta\phi]$ versus x, for different values of the damping factor ζ, are given in Fig. 6.16. It will be noted that these curves are symmetrical with respect to the axis $x = 1$ when a logarithmic scale is used for the variable x. We can check in the above expression that if we replace x by $1/x$, $\Delta\phi$ remains unchanged.

The error signal maximum amplitude corresponds to $\Omega = \omega_n$ ($x = 1$) and the value of this maximum is inversely proportional to the damping factor ζ:

$$\Delta\phi_M = \frac{\Delta\omega_i}{2\zeta\omega_n}$$

Let us calculate for what values of x we get a $\Delta\phi$ attenuated by 3 dB with respect to the maximum value $\Delta\phi_M$. In other words, we are trying to find the values of x that solve

$$\frac{x^2}{(1-x^2)^2 + 4\zeta^2 x^2} = \frac{1}{2}\frac{1}{4\zeta^2}$$

If we state that $x^2 = u$, we get

$$8\zeta^2 u = (1-u)^2 + 4\zeta^2 u$$

that is,

$$u^2 - 2(2\zeta^2 + 1)u + 1 = 0$$

the solutions of which are

$$u' = 2\zeta^2 + 1 + 2\zeta\sqrt{\zeta^2 + 1} = \left(\sqrt{\zeta^2 + 1} + \zeta\right)^2$$

$$u'' = 2\zeta^2 + 1 - 2\zeta\sqrt{\zeta^2 + 1} = \left(\sqrt{\zeta^2 + 1} - \zeta\right)^2$$

Since the x solutions are the positive determinations of the u-solution square roots,

$$x' = \sqrt{\zeta^2 + 1} + \zeta \quad \text{and} \quad x'' = \sqrt{\zeta^2 + 1} - \zeta$$

On the one hand, we confirm that $x'x'' = 1$ and on the other hand, $x' - x'' = 2\zeta$.

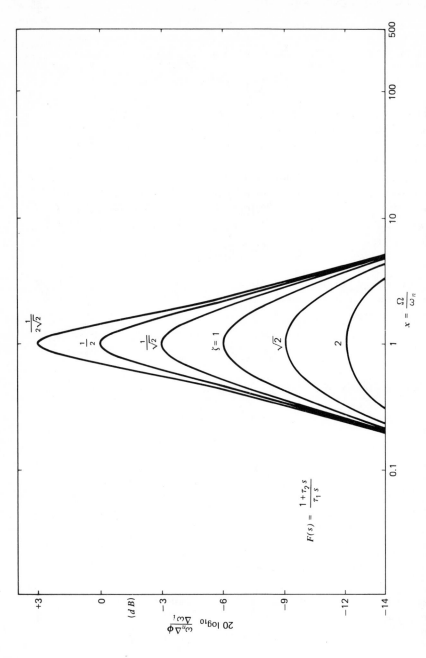

FIGURE 6.16. Amplitude-frequency characteristic of the phase error of a second-order loop $[F(s) = (1 + \tau_2 s)/(\tau_1 s)]$ for a sinusoidal frequency modulation of the input signal.

This result is very interesting. It means that if we want to determine, or check, the natural angular frequency and damping factor values of a second-order loop, with a filter $F(s) = (1 + \tau_2 s)/\tau_1 s$, we simply have to excite the loop by means of a signal, frequency modulated by a sinusoidal signal of angular frequency Ω. The frequency deviation Δf_i is set at a sufficiently low value, such as

$$\frac{2\pi \Delta f_i}{2\hat{\zeta}\hat{\omega}_n} < \frac{1}{2} \text{ rad}$$

$\hat{\zeta}$ and $\hat{\omega}_n$ are the estimated values of ζ and ω_n (we can always proceed by successive approximations). By means of an ac voltmeter connected at the phase detector output, we can read the two angular frequencies Ω' and Ω'' corresponding to -3 dB of the maximum value. Then

$$\omega_n^2 = \Omega'\Omega'', \qquad 2\zeta\omega_n = \Omega' - \Omega''$$

NUMERICAL EXAMPLE. With the frequency deviation set at 2×10^3 Hz, the loop falls out of lock during testing. We then set the frequency deviation at 10^3 Hz and pin-point roughly $F' = 1200$ Hz and $F'' = 300$ Hz. From this, we derive

$$\hat{\omega}_n = 2\pi\sqrt{300 \times 1200} = 3770 \text{ rad/s}$$

$$\hat{\zeta} = \frac{2\pi(1200 - 300)}{2 \times 3770} = 0.75$$

and

$$\Delta\phi_M = \frac{2\pi \times 10^3}{2 \times 0.75 \times 3770} = 1.11 \text{ rad}$$

Since the phase detector is sinusoidal, this phase error is unacceptable and the test is repeated with the frequency deviation set at 400 Hz. We then note

$$F' = 1225 \text{ Hz}, \qquad F'' = 330 \text{ Hz}$$

From which we derive

$$\omega_n = 2\pi\sqrt{330 \times 1225} = 3995 \text{ rad/s}$$

$$\zeta = \frac{2\pi(1225 - 330)}{2 \times 3995} = 0.704$$

6.3.3 SECOND-ORDER LOOP WITH FILTER $F(s) = 1/(1 + \tau_1 s)$

In this case we get, using Eqs. 3.49 and 6.16,

$$\Delta\phi = \frac{\Delta\omega_i}{\omega_n} \sqrt{\frac{x^2 + 4\zeta^2}{\left(1 - x^2\right)^2 + 4\zeta^2 x^2}} \tag{6.19}$$

The curves representing $20\log_{10}[(\omega_n/\Delta\omega_i)\Delta\phi]$ versus x are given in Fig. 6.17. This time, the angular frequency value Ω corresponding to the maximum phase error depends on the value of the damping factor ζ. All the curves cross the axis $+3$ dB at $x = \sqrt{2}\,/2$. When the damping factor ζ is very low, the curves tend to become symmetrical around $x = 1$ (the above expression for $\Delta\phi$ is equivalent to that given in Section 6.3.4) and the phase error is maximum for $\Omega \cong \omega_n$. Contrarily, when the damping factor ζ greatly exceeds 1, the phase error is maximum for the low values of the modulation angular frequency Ω and increases as the damping factor increases.

6.3.4 SECOND-ORDER LOOP WITH FILTER $F(s) = (1 + \tau_2 s)/(1 + \tau_1 s)$

In this case

$$\Delta\phi = \frac{\Delta\omega_i}{\omega_n} \sqrt{\frac{x^2 + \omega_n^2/K^2}{\left(1 - x^2\right)^2 + 4\zeta^2 x^2}} \tag{6.20}$$

As stated previously, when the loop gain K is much larger than the natural angular frequency ω_n, the results are approximately the same as for the loop with filter $F(s) = (1 + \tau_2 s)/\tau_1 s$. The curves of Fig. 6.16 can be used; their essential properties remain valid, in particular, the possibility of computing ω_n and ζ from Ω' and Ω'' if K is large. Strictly speaking, the symmetry of the response curves round the axis $x = 1$ is here only local since, when Ω tends towards 0, $\Delta\phi$ no longer approaches 0 but $\Delta\omega_i/K$. It would, in fact, be interesting to show that this result is obtained by two completely different methods.

When the input signal is frequency modulated according to a law $\omega_i = \omega + \Delta\omega_i \sin\Omega t$ and the angular frequency value Ω is very low, so that the input signal angular frequency only varies very slowly, at a given instant, the difference between ω_i and the central angular frequency ω is $\Delta\omega = \Delta\omega_i \sin\Omega t$. As we saw in Section 5.2.4, the corresponding phase error is

$$\phi(t) = \frac{\Delta\omega}{K} = \frac{\Delta\omega_i}{K} \sin\Omega t$$

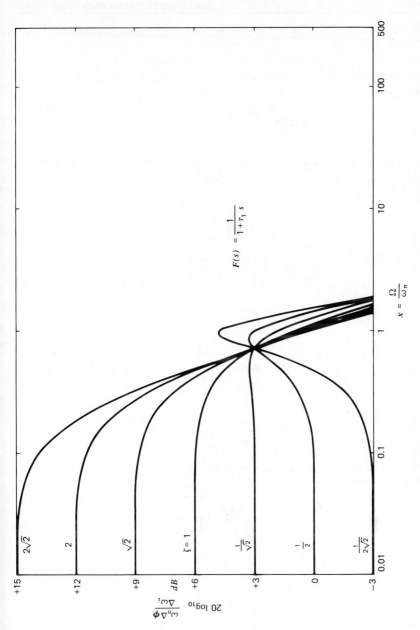

FIGURE 6.17. Amplitude-frequency characteristic of the phase error of a second-order loop with $[F(s) = 1/(1 + \tau_1 s)]$ for a sinusoidal frequency modulation of the input signal.

129

(this is, of course, the steady-state phase error since $\sin \Omega t$ is assumed to vary far more slowly than the transient responses in the servo device).

However, as mentioned at the beginning of Section 6.3, the law $\omega_i = \omega + \Delta\omega_i \sin \Omega t$ can be interpreted as an input signal phase modulation, according to the law

$$\phi_i(t) = -\frac{\Delta\omega_i}{\Omega} \cos \Omega t = m_i \cos \Omega t$$

The error signal expression is $\phi(t) = m \cos(\Omega t + \Theta)$ where m and Θ are given by Eqs. 6.10, from which we derive

$$m = m_i \sqrt{\frac{\dfrac{\Omega^4}{\omega_n^4} + \dfrac{\omega_n^2}{K^2} \dfrac{\Omega^2}{\omega_n^2}}{\left(1 - \dfrac{\Omega^2}{\omega_n^2}\right)^2 + 4\zeta^2 \dfrac{\Omega^2}{\omega_n^2}}}$$

$$= -\Delta\omega_i \sqrt{\frac{\dfrac{\Omega^2}{\omega_n^4} + \dfrac{1}{K^2}}{\left(1 - \dfrac{\Omega^2}{\omega_n^2}\right)^2 + 4\zeta^2 \dfrac{\Omega^2}{\omega_n^2}}}$$

$$\Theta = \pi - \operatorname{Arc\,tan} \frac{\omega_n^2}{K\Omega} - \operatorname{Arc\,tan} \frac{2\zeta\Omega/\omega_n}{1 - \Omega^2/\omega_n^2}$$

When $\Omega \to 0$, $m \to -\Delta\omega_i/K$ and $\Theta \to \pi/2$.

Consequently, for a very low Ω,

$$\phi(t) = -\frac{\Delta\omega_i}{K} \cos\left(\Omega t + \frac{\pi}{2}\right)$$

that is,

$$\phi(t) = \frac{\Delta\omega_i}{K} \sin \Omega t$$

6.4   USE OF A PHASE-LOCKED LOOP AS A DISCRIMINATOR

In Section 6.3, we discussed the phase error in the loop when the input signal is frequency modulated. This is, in fact, the first step when we wish to check the linearity hypothesis. We shall now see that we can pick up in the loop a signal which, under certain conditions, is a good duplicate of the input signal modulation.

If we go back to the first-order loop error signal expression and multiply it by K_1, the phase detector sensitivity, we get a signal $u_1(t)$ such that

$$u_1(t) = K_1 \frac{\Delta \omega_i}{\sqrt{K^2 + \Omega^2}} \sin\left(\Omega t + \Theta_i + \frac{\pi}{2} - \text{Arc} \tan \frac{\Omega}{K} \right)$$

that is,

$$u_1(t) = \Delta u_1 \cos\left(\Omega t + \Theta_i - \text{Arc} \tan \frac{\Omega}{K} \right)$$

with

$$\Delta u_1 = K_1 \frac{\Delta \omega_i}{\sqrt{K^2 + \Omega^2}}$$

If we compare this expression with Eq. 6.11, we observe that $u_1(t)$ is a duplicate of the modulation signal, differing only by a multiplication constant and a low-pass filtering, of time constant $1/K$.

In other words, the first-order loop behaves like a discriminator of slope K_D, followed by an RC low-pass filter such that $RC = 1/K$, with

$$K_D = \lim_{\Omega \to 0} \frac{\Delta u_1}{\Delta f_i} = 2\pi \frac{K_1}{K} = \frac{2\pi}{K_3}$$

in V/Hz if K_3 is expressed in rad/s/V, or

$$K_D = \frac{1}{K_3}$$

in V/Hz if K_3 is expressed in Hz/V.* This function is shown in diagram form in Fig. 6.18.

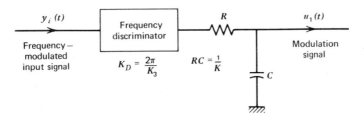

FIGURE 6.18. Model of a first-order loop working as a frequency demodulator.

*We suppose here $K_2 = 1$. If $K_2 \neq 1$, K_3 must be replaced by $K_2 K_3$.

It will be noted that the discriminator slope is independent of K_1, the phase detector sensitivity, but inversely proportional to K_3, the VCO modulation sensitivity. Furthermore, the demodulation bandwidth is proportional to the loop gain K. At a given VCO modulation sensitivity K_3, the bandwidth widens as the phase detector sensitivity K_1 increases. At a given phase detector sensitivity K_1, the bandwidth widens as the VCO modulation sensitivity increases, but the discriminator slope diminishes as K_3 is increased.

NUMERICAL EXAMPLE.   We have a VCO where $K_3 = 2 \times 10^5$ rad/s/V and two phase detectors where $K_1 = 1$ V/rad and $K_1' = 0.1$ V/rad. In both cases the discriminator slope will be

$$K_D = \frac{2\pi}{2 \times 10^5} = 31.4 \ \mu V/Hz$$

In the first case, the -3-dB bandwidth will be

$$\frac{2 \times 10^5}{2\pi} = 31.8 \ kHz$$

and in the second case, it will be

$$\frac{2 \times 10^4}{2\pi} = 3.18 \ kHz$$

If, in the second case ($K_1' = 0.1$ V/rad) we use another VCO where $K_3' = 2 \times 10^6$ rad/s/V, the -3-dB bandwidth will be 31.8 kHz, but the discriminator slope will be $K_D' = 3.14 \ \mu V/Hz$.

For the second-order loop with low-pass filter $F(s) = 1/(1 + \tau_1 s)$, we get a comparable result if the time constant τ_1 is small (for example, when τ_1 corresponds to an unavoidable parasitic time constant in a case where construction of a first-order loop was intended).

The error signal amplitude is given by Eq. 6.19:

$$\Delta\phi = \Delta\omega_i \frac{2\zeta}{\omega_n} \sqrt{\frac{1 + x^2/4\zeta^2}{\left(1 - x^2\right)^2 + 4\zeta^2 x^2}}$$

For this type of loop, $\omega_n/2\zeta = K$ (see Eq. 3.47) and the phase detector output signal $\Delta u_1 = K_1 \Delta\phi$ tends toward $\Delta\omega_i/K_3$ for low modulation frequencies. The discriminator slope is, as in the case of the first-order loop, $K_D = 2\pi/K_3$ in V/Hz when K_3 is expressed in rad/s/V. But, the loop effect is equivalent to that of a low-pass filter of asymptotic slope $-1(-6$ dB/

octave) connected to the discriminator, and producing a filtering curve which depends on the damping factor ζ, as can be seen on the curves in Fig. 6.17.

The disadvantage of these curves is that when the time constant τ_1 varies, ζ and ω_n vary simultaneously so that we have to change curves and at the same time change the $\Delta\phi$ multiplier. Since the new parameter, as compared with the first-order loop, is the time constant τ_1, it may prove more convenient to analyze the different curves with respect to τ_1.

Let us suppose that $\tau_1 = \alpha/K$, that is $\omega_n = \sqrt{K/\tau_1} = K/\sqrt{\alpha}$ and, since $2\zeta\omega_n = 1/\tau_1$, $\zeta = 1/2\sqrt{\alpha}$. The error signal amplitude is expressed

$$\Delta\phi = \frac{\Delta\omega_i}{K} \sqrt{\frac{1 + \dfrac{\Omega^2}{4\zeta^2\omega_n^2}}{\left(1 - \dfrac{\Omega^2}{\omega_n^2}\right)^2 + 4\zeta^2\dfrac{\Omega^2}{\omega_n^2}}}$$

that is,

$$\Delta\phi = \frac{\Delta\omega_i}{K} \sqrt{\frac{1 + \alpha^2\dfrac{\Omega^2}{K^2}}{\left(1 - \alpha\dfrac{\Omega^2}{K^2}\right)^2 + \dfrac{\Omega^2}{K^2}}} \tag{6.21}$$

The filtering curve variations are clearly shown in Fig. 6.19. Instead of plotting $20\log_{10}(\Delta\phi\,\omega_n/\Delta\omega_i)$ versus $x = \Omega/\omega_n$, the parameter being ζ, we have plotted $20\log_{10}(\Delta\phi K/\Delta\omega_i)$ versus $x' = \Omega/K$, the parameter being α, so that $\tau_1 = \alpha/K$. By this method, all the curves have the same ordinate when Ω approaches 0, which is more satisfactory in that the discriminator slope is independent of the value of τ_1.

When α is null, that is, when there is no time constant τ_1, the filtering curve corresponds to that of a first-order loop.

As regards the second-order loops with integrator and correction [$F(s) = (1 + \tau_2 s)/\tau_1 s$] or with low-pass filter and correction [$F(s) = (1 + \tau_2 s)/(1 + \tau_1 s)$], the curves in Fig. 6.16 show that the phase detector output voltage $u_1(t)$, proportional to the error signal, cannot be used as a discriminator output signal. On the other hand, we noted at the beginning of Section 6.3 that the frequency modulation of the VCO is a duplicate of the input signal frequency modulation, taking into account a filter $H(j\Omega)$ representing the loop effect. But the VCO frequency modulation is directly proportional to

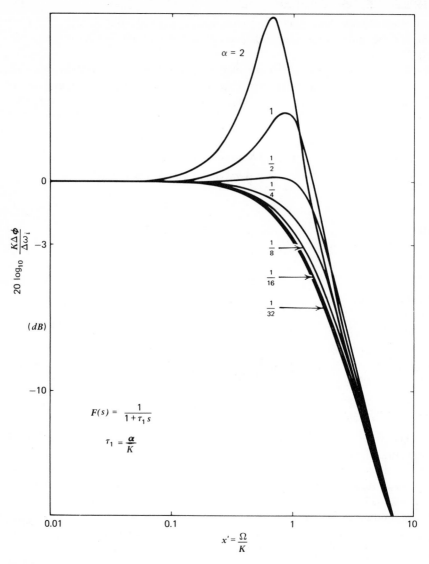

FIGURE 6.19. Amplitude-frequency characteristic of a second-order loop with $F(s) = 1/(1 + \tau_1 s)$ working as a frequency demodulator (the error signal being the output signal).

the command signal u_2; the VCO command signal can therefore be used as a discriminator output signal. If the input modulation signal is formulated as $\Delta f_i \cos(\Omega t + \Theta_i)$, then the steady-state VCO modulation signal is formulated as $\Delta f_0 \cos(\Omega t + \Theta_0)$, Δf_0 and Θ_0 being related to Δf_i and Θ_i by Eqs. 6.14.

But, $\Delta f_0 = (K_3/2\pi)\Delta u_2$ (K_3 being expressed in rad/s/V), so signal $u_2(t)$ is expressed

$$u_2(t) = \Delta u_2 \cos(\Omega t + \Theta_0)$$

with

$$\Delta u_2 = \frac{2\pi \Delta f_i}{K_3} \text{Mod}\left[H(j\Omega) \right]$$

When signal $u_2(t)$ is used as output signal, a second-order loop behaves like a discriminator of slope

$$K_D = \lim_{\Omega \to 0} \frac{\Delta u_2}{\Delta f_i} = \frac{2\pi}{K_3}$$

in V/Hz (when K_3 is in rad/s/V), followed by a low-pass filter, the characteristics of which, given in Section 6.1, are indicated on the model shown in Fig. 6.20. This model only applies in the case of a loop with integrator and correction. For a loop with low-pass filter and correction, as we saw in Section 6.1.4, a potentiometer has to be connected in place of the resistor R, the output being picked up at μR (on the C side) with

$$\mu = \frac{2\zeta\omega_n - \omega_n^2/K}{2\zeta\omega_n}$$

For the loop with a filter $F(s) = (1 + \tau_2 s)/\tau_1 s$ [or for a filter $F(s) = (1 + \tau_2 s)/(1 + \tau_1 s)$ if K is very large], the response curves (Fig. 6.3) all reflect an overshoot in the discriminator bandwidth and the asymptotic slope is -6 dB/octave.

If we want to avoid this overshoot in the bandwidth and at the same time obtain a stronger filtering for the higher frequencies (for instance, up to -12 dB/octave), we have to include a supplementary filtering of the output signal $u_2(t)$. This is often accomplished by a low-pass filter of time constant τ_2 (15). Another method, when the loop filter used is of the passive type (low-pass and correction), consists in using as output the signal $u_2'(t)$ picked-up between R_2 and C (16). In both cases, the signal $u_2'(t)$ thus obtained is related to $u_2(t)$ by the transfer function

$$\frac{U_2'(j\Omega)}{U_2(j\Omega)} = \frac{1}{1 + j\Omega\tau_2}$$

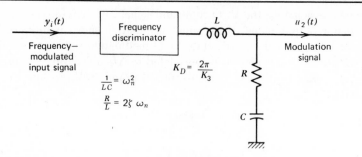

FIGURE 6.20. Model of a second-order loop $[F(s)=(1+\tau_2 s)/(\tau_1 s)]$ working as a frequency demodulator.

that is,

$$\text{Mod}\left[\frac{U_2'}{U_2}(j\Omega)\right] = \frac{1}{\sqrt{1+\Omega^2\tau_2^2}}$$

$$= \frac{1}{\sqrt{1+\omega_n^2\tau_2^2(\Omega^2/\omega_n^2)}}$$

But for a second-order loop with filter $F(s)=(1+\tau_2 s)/\tau_1 s$, or with filter $F(s)=(1+\tau_2 s)/(1+\tau_1 s)$ if K is large, $\omega_n\tau_2=2\zeta$ and consequently

$$\text{Mod}\left[\frac{U_2'}{U_2}(j\Omega)\right] = \frac{1}{\sqrt{1+4\zeta^2 x^2}}$$

which means that

$$\frac{\Delta u_2'}{\Delta f_i} = \frac{2\pi}{K_3}\frac{1}{\sqrt{(1-x^2)^2+4\zeta^2 x^2}}$$

This expression is identical, within the multiplication constant representing the discriminator slope, with the transfer function amplitude of a second-order loop with filter $F(s)=1/(1+\tau_1 s)$. The curves in Fig. 6.9 indicate the filtering function to be used. The asymptotic slope is -12 dB/octave and there is no bandwidth overshoot if the damping factor ζ is above or equal to $\sqrt{2}/2$.

Remark. The cause of the systematic overshoot in the bandwidth and the asymptotic slope to -6 dB/octave in the curves of Fig. 6.3 is the correction network $(1 + \tau_2 s)$. If this correction network can be avoided, we return to the case of the second-order loop with low-pass filter $F(s) = 1/(1 + \tau_1 s)$. But, as in this case we are using signal u_2 as output signal (instead of signal u_1, at the beginning of this section) the filtering function is also given by the curves in Fig. 6.9 instead of Fig. 6.17 or 6.19. This solution, as already mentioned, has the disadvantage of precluding the independent choice of ω_n and ζ when the loop gain K is fixed. We shall discuss further the qualities and defects of the different types of discriminator in the section dealing with operating performances in the presence of noise (Section 7.6).

FIRST NUMERICAL EXAMPLE. For the construction of a frequency discriminator we have a phase detector where $K_1 = 1$ V/rad and a VCO where $K_3 = 2 \times 10^5$ rad/s/V. The discriminator slope will be

$$K_D = \frac{2\pi}{K_3} = 31.4 \ \mu V/Hz$$

1. The circuit is constructed without a loop filter. The discriminator -3-dB bandwidth is $K/2\pi = 31.8$ kHz and the asymptotic slope is -6 dB/octave.
2. We use a loop filter having a transfer function

$$F(s) = \frac{1}{1 + \tau_1 s}$$

with

$$\tau_1 = \frac{1}{2K} = 2.5 \ \mu s \quad \left(\omega_n = 2.83 \times 10^5 \ rad/s, \quad \zeta = 0.707 \right)$$

(a) The output signal is picked up at the phase detector output: the -3 dB-bandwidth is $1.8K/2\pi = 57.2$ kHz and the asymptotic slope remains -6 dB/octave (Fig. 6.19).
(b) The output signal is picked up at the VCO modulation input: the -3-dB bandwidth is $\omega_n/2\pi = 45$ kHz and the asymptotic slope is -12 dB/octave (Fig. 6.9).

SECOND NUMERICAL EXAMPLE. We use a high-gain $(K = 2 \times 10^7 \ \text{rad/s})$ phase-locked loop with a loop filter $F(s) = (1 + \tau_2 s)/(1 + \tau_1 s)$ to perform a frequency demodulation. The time constants τ_1 and τ_2 are so chosen that $\omega_n = 10^5$ rad/s and $\zeta = 1$ $(\tau_1 = 2$ ms, $\tau_2 = 20 \ \mu s)$.

1. The output signal is picked up at the VCO modulation input: the -3-dB bandwidth is 2.45 $\omega_n/2\pi = 39$ kHz, the overshoot of the response curve is $+1.25$ dB, and the asymptotic slope is -6 dB/octave (Fig. 6.3).

2. The output signal is picked up between resistor R_2 and capacitor C of the loop filter: the -3-dB bandwidth is 0.65 $\omega_n/2\pi = 10.3$ kHz, but the response curve no longer overshoots and the asymptotic slope is -12 dB/octave (Fig. 6.9).

3. The output signal being picked up as in (1) above, if the loop filter resistor R_2 is short-circuited, the Fig. 6.9 response curve remains valid. But, for the values of K and τ_1 under consideration the damping factor would be very low ($\zeta = 0.25 \times 10^{-2}$) and would cause a very high overshoot of the response curve and a serious risk of instability for the servo device.

CHAPTER 7
ADDITIVE
NOISE RESPONSE

In this chapter we shall assume that the signal applied to the loop input $y_i(t) = A \sin(\omega t + \theta_i)$ is accompanied by a noise $n(t)$ that has the properties of a stationary Gaussian random process. The mean value of this process is assumed to be null and its one-sided power spectral density N_0 (W/Hz) is considered uniform within the frequency band $f - W/2$ to $f + W/2$. The sources of noise are many and various: noise from the emitter circuits that is radiated at the same time as the useful signal, galactic noise, antenna noise, receiver input noise, noise from the circuits preceding the phase-locked loop, and so on. The results obtained are valid insofar as these different sorts of noise applied to the phase detector with the useful signal correspond to the definition given above.

7.1 PHASE DETECTOR OPERATING PRINCIPLE

7.1.1 ADDITIVE NOISE ANALYSIS

Let us take a sinusoidal signal accompanied by an additive noise:

$$y(t) = A \sin(\omega t + \theta) + n(t)$$

It has been shown (17) that if the frequency band W is narrow and symmetrical with respect to the central frequency f, the process $n(t)$ can be broken down as follows:

$$n(t) = n_1(t) \sin \omega t + n_2(t) \cos \omega t$$

where $n_1(t)$ and $n_2(t)$ are two low-frequency random processes, which are Gaussian, stationary, and statistically independent, having a null mean value and a two-sided power spectral density N_0 that is uniform throughout the frequency range $-W/2$ to $+W/2$ (Fig. 7.1).

In particular, the power of these two processes is the same and the power of each process taken individually is equal to the power of process $n(t)$:

$$\overline{n_1^2(t)} = \overline{n_2^2(t)} = \overline{n^2(t)} = N_0 W$$

The restriction as regards the bandwidth W of noise $n(t)$, which has to be narrow as compared with the central frequency f, is not stringent. Viterbi (18) has in fact shown this analysis to be valid providing that the spectral density N_0 of process $n(t)$ is negligible beyond a frequency of $2f$. In other words, to return to Fig. 7.1, the spectral density N_0 of process $n(t)$ can range from 0 to $2f$. From this we infer that the spectral density of processes $n_1(t)$ and $n_2(t)$ can range from $-f$ to $+f$, but will be identically null beyond this interval.

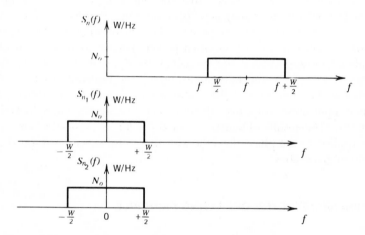

FIGURE 7.1. Power spectral density of the $n(t)$, $n_1(t)$, and $n_2(t)$ processes.

7.1.2 CASE OF THE ANALOG MULTIPLIER

Although real multipliers are, as yet, rarely used as phase detectors in phase-locked loops, this case provides a convenient model of a sinusoidal phase detector, frequently employed in specialized literature. Let us therefore consider the case where the input signal is multiplied by a reference signal having the same angular frequency, then filtered to get rid of the $2f$ term (Fig. 7.2). When the input signal is accompanied by noise $n(t)$ decomposed as described above and multiplied by the reference signal $y_R(t)$

$= B\cos(\omega t + \theta_R)$, we get

$$u(t) = AB\sin(\omega t + \theta_i)\cos(\omega t + \theta_R)$$
$$+ n_1(t)B\sin\omega t\cos(\omega t + \theta_R)$$
$$+ n_2(t)B\cos\omega t\cos(\omega t + \theta_R)$$

or again,

$$u(t) = \frac{AB}{2}\sin(2\omega t + \theta_i + \theta_R) + \frac{AB}{2}\sin(\theta_i - \theta_R) + \frac{Bn_1(t)}{2}\sin(2\omega t + \theta_R)$$
$$- \frac{Bn_1(t)}{2}\sin\theta_R + \frac{Bn_2(t)}{2}\cos(2\omega t + \theta_R) + \frac{Bn_2(t)}{2}\cos\theta_R$$

The $\sin(2\omega t + \theta_i + \theta_R)$ term is assumed to have been filtered out. The terms

$$\frac{Bn_1(t)}{2}\sin(2\omega t + \theta_R) \quad \text{and} \quad \frac{Bn_2(t)}{2}\cos(2\omega t + \theta_R)$$

represent 2ω angular frequency signals, amplitude modulated by the low-frequency random processes $Bn_1(t)/2$ and $Bn_2(t)/2$. These two signals are then also filtered out, so that we get at the phase detector output the signal $u_1(t)$

$$u_1(t) = \frac{AB}{2}\sin(\theta_i - \theta_R) + \frac{Bn_2(t)}{2}\cos\theta_R - \frac{Bn_1(t)}{2}\sin\theta_R$$

It is customary to assume that $AB/2 = K_1$, the phase detector sensitivity; hence

$$u_1(t) = K_1\sin(\theta_i - \theta_R) + K_1\frac{n_2(t)}{A}\cos\theta_R - K_1\frac{n_1(t)}{A}\sin\theta_R \qquad (7.1)$$

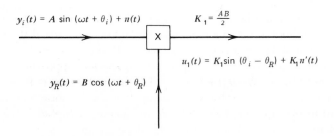

FIGURE 7.2. Model of an analog multiplier in presence of noise.

We observe that the useful signal $K_1 \sin(\theta_i - \theta_R)$ is now accompanied by a noise term $K_1 n'(t)$ with

$$n'(t) = \frac{n_2(t)}{A} \cos\theta_R - \frac{n_1(t)}{A} \sin\theta_R \qquad (7.2)$$

Since the processes $n_1(t)$ and $n_2(t)$ are statistically independent, we can write

$$\overline{n'(t)} = \frac{\cos\theta_R}{A} \overline{n_1(t)} - \frac{\sin\theta_R}{A} \overline{n_2(t)} = 0$$

$$\overline{n'^2(t)} = \frac{\cos^2\theta_R}{A^2} \overline{n_1^2(t)} + \frac{\sin^2\theta_R}{A^2} \overline{n_2^2(t)} = \frac{\overline{n^2(t)}}{A^2} = \frac{N_0 W}{A^2}$$

Finally, the power spectral density of the two processes $n_1(t)$ and $n_2(t)$ being uniform from $-W/2$ to $+W/2$, the spectral density of $n'(t)$ is uniform from $-W/2$ to $+W/2$.

In conclusion, when an analog multiplier is used as phase detector and when the input signal consists of a sinusoidal signal $A \sin(\omega t + \theta_i)$ accompanied by a noise $n(t)$, which is a Gaussian process, having a null mean value and a one-sided power spectral density N_0 from $f - W/2$ to $f + W/2$, the output signal consists of the useful signal, identical with the useful signal without noise, together with a noise term $K_1 n'(t)$. The constant K_1 is the phase detector sensitivity and $n'(t)$ is a Gaussian process, having a null mean value and a two-sided power spectral density $N_0' = N_0/A^2$ from $-W/2$ to $+W/2$.

It is apparent that the result obtained can be interpreted as a phase modulation of the input signal by the random process $n'(t)$. That is to say, we can write

$$y_i(t) = A \sin(\omega t + \theta_i) + n(t)$$

$$\cong A \sin[\omega t + \theta_i + \varphi_i(t)] \quad \text{with } \varphi_i(t) = n'(t).$$

However, it should be remembered that the modulation in this case is an equivalent and not a real modulation of the input signal, for, if it were a real input signal modulation, the output would be $K_1 \sin[\theta_i - \theta_R + n'(t)]$ and there would be a restriction on $[\theta_i - \theta_R + n'(t)]$ to permit approximation of the output signal by $K_1(\theta_i - \theta_R) + K_1 n'(t)$. But this restriction is nonexistent in the present case: whatever the value of $\varphi_i(t) = n'(t)$, the noise term is $K_1 n'(t)$. In regard to the equivalent modulation, the analog multiplier behaves like a linear-characteristic phase detector, with slope K_1 (in V/rad).

Remark. This is, of course, only valid as regards the multiplication operation by signal $y_R(t) = B \cos(\omega t + \theta_R)$, where θ_R is a constant independent of the noise term $n(t)$. We shall develop this point further in the section dealing with loop operation.

7.1.3   CASE OF THE MULTIPLIER BY $\text{Sign}[\cos(\omega t + \theta_R)]$.

A sinusoidal characteristic phase detector much used in practical applications is the four-diode configuration, where the reference signal amplitude is higher than that of the input signal. We saw in Section 2.1.3 that a device of this type works like a multiplier by the sign (polarity) of the reference signal.

Let us consider the circuit shown in Fig. 7.3, where the input signal is accompanied by a noise $n(t)$ and the reference signal is expressed as $y_R(t) = \text{Sign}[\cos(\omega t + \theta_R)]$. If we break down y_R into Fourier series, we get:

$$y_R(t) = \frac{4}{\pi} \left\{ \cos(\omega t + \theta_R) + \cdots \right.$$

$$\left. + \frac{(-1)^p}{2p+1} \cos[(2p+1)(\omega t + \theta_R)] + \cdots \right\}$$

The multiplication of $A \sin(\omega t + \theta_i)$ by $y_R(t)$ leads to

$$u_S(t) = \frac{2A}{\pi} \left[\sin(\theta_i - \theta_R) + \sin(2\omega t + \theta_i + \theta_R) + \tfrac{1}{3}\sin(2\omega t + 3\theta_R - \theta_i) \right.$$

$$\left. - \tfrac{1}{3}\sin(4\omega t + 3\theta_R + \theta_i) + \cdots \right]$$

After filtering, only the first term remains, leaving a useful output signal

$$u_{1S} = \frac{2A}{\pi} \sin(\theta_i - \theta_R).$$

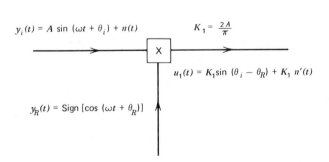

FIGURE 7.3.   Model of a four-diode phase detector (or pseudomultiplier) in presence of noise.

In the same way, the product of $y_R(t)$ by the breakdown of $n(t)$ is written

$$u_N(t) = \frac{4}{\pi} n_1(t) \sin \omega t \sum_{p=0}^{\infty} (-1)^p \frac{\cos\left[(2p+1)(\omega t + \theta_R)\right]}{2p+1}$$

$$+ \frac{4}{\pi} n_2(t) \cos \omega t \sum_{p=0}^{\infty} (-1)^p \frac{\cos\left[(2p+1)(\omega t + \theta_R)\right]}{2p+1}$$

After filtering, bearing in mind that $n_1(t)$ and $n_2(t)$ are low-frequency random processes [in fact, as previously noted, the argument holds if their power spectral density is strictly null outside the interval $(-f, +f)$], only the noise term $u_{1N}(t)$ remains:

$$u_{1N}(t) = \frac{2}{\pi} n_2(t) \cos \theta_R - \frac{2}{\pi} n_1(t) \sin \theta_R$$

The phase detector output signal $u_1(t) = u_{1S}(t) + u_{1N}(t)$ is written as follows, using $2A/\pi = K_1$ to denote the phase detector sensitivity:

$$u_1(t) = K_1 \sin(\theta_i - \theta_R) + K_1 n'(t) \tag{7.3}$$

The noise term

$$n'(t) = \frac{n_2(t)}{A} \cos \theta_R - \frac{n_1(t)}{A} \sin \theta_R$$

is thus exactly identical with that given in Eq. 7.2. The conclusions given at the end of the previous subsection are also valid in the present case: $n'(t)$ is a low-frequency Gaussian process, with a two-sided power spectral density N_0', uniform from $-W/2$ to $+W/2$, so that

$$N_0' = \frac{N_0}{A^2} = \frac{N_0}{2S}$$

[S being the power of the sinusoidal signal applied to the phase detector input, measured in a 1 Ω resistor, like the power density N_0 of the input noise $n(t)$].

The mean value of the process $\overline{n'(t)} = 0$, and its variance, is given by

$$\overline{n'^2(t)} = \frac{\overline{n^2(t)}}{A^2} = \frac{N_0 W}{A^2}$$

Remark.   If the noise $n(t)$ is a true white noise (infinite bandwidth), the noise term at the detector output can be calculated as described below. We observe, in this case, that the power spectral density increases by 1 dB, to become

$$N_0' = \frac{N_0}{A^2} \cdot \frac{\pi^2}{8}$$

7.1.4 PHASE DETECTOR LINEAR OVER AN INTERVAL $-\pi/2$ TO $+\pi/2$.

In the course of our investigation of the different types of phase detector, we observed that this characteristic can be interpreted as the result of the multiplication of a square input signal by a square reference signal of the same frequency (Section 2.1.5).

In presence of noise, we have to distinguish two different cases: a naturally square input signal, to which the noise $n(t)$ is added; and a sinusoidal input signal, accompanied by the noise $n(t)$. In the latter case, the square waveform is obtained by hard limiting of the input signal. In the first case, the signal applied to the multiplier is expressed

$$y_i(t) = A \, \mathrm{Sign} \left[\sin(\omega t + \theta_i) \right] + n(t)$$

In the second case, we have

$$y_i(t) = A' \, \mathrm{Sign} \left[A \sin(\omega t + \theta_i) + n(t) \right]$$

(a) Let us first examine the first case. The output signal results from the multiplication of y_i by the reference signal $y_R(t) = \mathrm{Sign}[\cos(\omega t + \theta_R)]$ followed by low-pass filtering designed to remove the high-angular-frequency terms. The useful signal is obtained by effecting the operation

$$A \, \mathrm{Sign} \left[\sin(\omega t + \theta_i) \right] \times \mathrm{Sign} \left[\cos(\omega t + \theta_R) \right]$$

Proceeding in the same way as in Section 2.1.5, we get

$$u_{1S} = \frac{2A}{\pi}(\theta_i - \theta_R) \qquad \text{if} \quad -\frac{\pi}{2} < \theta_i - \theta_R < +\frac{\pi}{2}$$

and

$$u_{1S} = \frac{2A}{\pi}(\pi - \theta_i + \theta_R) \qquad \text{if} \quad +\frac{\pi}{2} < \theta_i - \theta_R < +\frac{3\pi}{2}$$

Before calculating the output noise term $u_{1N}(t)$, we should bear in mind that, since the useful signal is a square signal, the noise $n(t)$ bandwidth cannot be restricted. In this case, we must therefore assume the noise $n(t)$ to be a stationary Gaussian process, with a null mean value and a one-sided power spectral density N_0 uniform from $f = 0$ to f approaching infinity.

It is always possible to assume that this noise results from the addition of noises, denoted $\nu^{(0)}(t)$, $\nu^{(1)}(t)$, $\nu^{(2)}(t),\ldots$ obtained in the following way (see Fig. 7.4):

- $\nu^{(0)}(t)$ corresponds to $n(t)$ filtered by an ideal filter with bandwidth $(0,2f)$; this is, of course, a purely theoretical filter, not physically realizable, since its attenuation must be null in the passband and infinite elsewhere and no phase shift must be introduced.
- $\nu^{(1)}(t)$ corresponds to $n(t)$ filtered in the $(2f,4f)$ band;
- $\nu^{(2)}(t)$ corresponds to $n(t)$ filtered in the $(4f,6f)$ band...

Since the noise $n(t)$ is Gaussian and the filters are nonoverlapping, the noises $\nu^{(0)}(t), \nu^{(1)}(t), \nu^{(2)}(t),\ldots$ are statistically independent Gaussian noises (19). However, each noise $\nu^{(p)}(t)$ confirms the narrow-band hypothesis, as defined at the end of Section 7.1.1. Consequently, each noise can be broken down into

$$\nu^{(p)}(t) = \nu_1^{(p)}(t)\sin\left[(2p+1)\omega t\right] + \nu_2^{(p)}(t)\cos\left[(2p+1)\omega t\right]$$

where $\nu_1^{(p)}(t)$ and $\nu_2^{(p)}(t)$ are, whatever the value of p, two statiscally independent Gaussian random processes, with a null mean value and a two-sided power spectral density N_0 uniform from $-f$ to $+f$ and null elsewhere. In particular,

$$\overline{\nu_1^{(p)}(t)^2} = \overline{\nu_2^{(p)}(t)^2} = \overline{\nu^{(p)}(t)^2} = 2fN_0$$

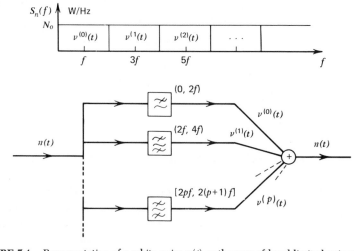

FIGURE 7.4. Representation of a white noise $n(t)$ as the sum of band-limited noises.

The noise $n(t)$ can thus be expressed

$$n(t) = \nu_1^{(0)}(t)\sin\omega t + \nu_2^{(0)}(t)\cos\omega t + \nu_1^{(1)}(t)\sin 3\omega t + \nu_2^{(1)}(t)\cos 3\omega t + \cdots$$
$$+ \nu_1^{(p)}(t)\sin\left[(2p+1)\omega t\right] + \nu_2^{(p)}(t)\cos\left[(2p+1)\omega t\right] + \cdots$$

When $n(t)$ is multiplied by $y_R(t)$ (after break down into Fourier series) and only the frequency terms below f are retained, we get

$$u_{1N}(t) = \frac{2}{\pi}\left[\nu_2^{(0)}(t)\cos\theta_R - \nu_1^{(0)}(t)\sin\theta_R + \cdots \right.$$
$$\left. + (-1)^p \frac{\nu_2^{(p)}(t)\cos\left[(2p+1)\theta_R\right] - \nu_1^{(p)}(t)\sin\left[(2p+1)\theta_R\right]}{2p+1} + \cdots \right]$$

Let $u_{1N}(t) = K_1 n'(t)$, K_1 being the phase detector sensitivity and $n'(t)$ a random process expressed as

$$n'(t) = \frac{1}{A}\sum_{p=0}^{\infty}(-1)^p \frac{\nu_2^{(p)}(t)\cos\left[(2p+1)\theta_R\right] - \nu_1^{(p)}(t)\sin\left[(2p+1)\theta_R\right]}{2p+1} \qquad (7.4)$$

The properties of $\nu_1^{(p)}(t)$ and $\nu_2^{(p)}(t)$ being as defined, this leads to $\overline{n'(t)} = 0$ and

$$\overline{n'^2(t)} = \frac{1}{A^2}\sum_{p=0}^{\infty}\frac{\overline{\nu^{(p)}(t)^2}}{(2p+1)^2} = \frac{2fN_0}{A^2}\sum_{p=0}^{\infty}\frac{1}{(2p+1)^2}$$

that is,

$$\overline{n'^2(t)} = \frac{2fN_0}{A^2}\cdot\frac{\pi^2}{8}$$

Since each process $\nu_1^{(p)}(t)$ and $\nu_2^{(p)}(t)$ has a uniform spectral density in the band $(-f, +f)$, this also applies to $n'(t)$, where the two-sided power spectral density N_0' from $-f$ to $+f$ is

$$N_0' = \frac{N_0}{A^2}\frac{\pi^2}{8} = \frac{N_0}{S}\frac{\pi^2}{8}$$

and

$$N_0' = 0 \quad \text{elsewhere}$$

If the output terms having frequency components above f are not eliminated, the above result remains valid for the spectral density of $n'(t)$

between $-f$ and $+f$ since we have kept all the terms comprising components below f (none of the rejected terms has a component below f). To consider the spectral density beyond f, we should have to include these rejected terms, which is beyond the scope of the present study.

For the same signal power S and the same spectral density N_0 for the noise $n(t)$ accompanying the signal, the spectral density N_0' of the equivalent modulation process $\varphi_i(t) = n'(t)$ of the input signal is about 4 dB higher than in the case of a sinusoidal phase detector. Three of the 4 dB stem from the fact that, for the same peak amplitude A, there is a 3-dB variation between the power of a square signal and that of a sinusoidal signal. The remaining 1 dB $(\pi^2/8)$ originates from the fact that if the signal is square, we have to consider that the input noise bandwidth is infinite. Since the reference signal spectrum itself extends to infinity, we obtain supplementary noise output components, corresponding to the multiplication of the reference signal $3f$ component, with the noise components located around $3f$, the multiplication of the reference signal $5f$ component, with the noise components located around $5f, \ldots$.

(b) The second case to be considered is that where the sinusoidal input signal accompanied by noise $n(t)$ is hard-limited before being applied to the multiplier. As the input signal is sinusoidal, it is plausible to assume, in this case, the noise $n(t)$ bandwidth to be narrow as compared with the frequency f of the signal. We must here calculate the product

$$u(t) = A' \, \text{Sign} \left[A \sin(\omega t + \theta_i) + n(t) \right] \times \text{Sign} \left[\cos(\omega t + \theta_R) \right]$$

In the absence of noise, $\text{Sign}[A \sin(\omega t + \theta_i)] = \text{Sign}[\sin(\omega t + \theta_i)]$ and the result is the same as in the case (a) above, if we substitute A' for A. The output signal is

$$u_{1S} = \frac{2A'}{\pi}(\theta_i - \theta_R) \qquad \text{if} \quad -\frac{\pi}{2} < \theta_i - \theta_R < \frac{\pi}{2}$$

$$= \frac{2A'}{\pi}(\pi - \theta_i + \theta_R) \qquad \text{if} \quad \frac{\pi}{2} < \theta_i - \theta_R < \frac{3\pi}{2}$$

The process $n(t)$ of presumably narrow bandwidth $(W \ll 2f)$ can be broken down into

$$n(t) = n_1'(t) \sin(\omega t + \theta_i) + n_2'(t) \cos(\omega t + \theta_i)$$

where the processes $n_1'(t)$ and $n_2'(t)$ have similar properties to those of processes $n_1(t)$ and $n_2(t)$ (Section 7.1.1). The quantity under the function $\text{Sign}[\cdot]$ is then expressed

$$A \sin(\omega t + \theta_i) + n_1'(t) \sin(\omega t + \theta_i) + n_2'(t) \cos(\omega t + \theta_i)$$

or again

$$\sqrt{[A + n_1'(t)]^2 + n_2'^2(t)} \; \sin\left[\omega t + \theta_i + \varphi_i(t)\right]$$

with

$$\tan\varphi_i(t) = \frac{n_2'(t)}{A + n_1'(t)} \tag{7.5}$$

The term $\varphi_i(t)$ represents the equivalent phase modulation of the input signal by the noise $n(t)$.

When there is a high signal-to-noise ratio at the output of the bandpass filter W preceding the limiter, the terms $n_1'(t)$ and $n_2'(t)$ remain smaller than A. In this case, Eq. 7.5 can be replaced by the following equation:

$$\varphi_i(t) = \frac{n_2'(t)}{A} \tag{7.6}$$

Furthermore, except in the immediate vicinity of $+\pi/2$ or $-\pi/2$, the quantity $\varphi_i(t)$ being small, the limit conditions remain unchanged, so that, taking $2A'/\pi = K_1$, the phase detector sensitivity, we can write

$$u_1(t) = K_1\left[\theta_i - \theta_R + \varphi_i(t)\right] \qquad \text{if} \quad -\frac{\pi}{2} < \theta_i - \theta_R < \frac{\pi}{2}$$

or

$$u_1(t) = K_1\left[\pi - \theta_i + \theta_R - \varphi_i(t)\right] \qquad \text{if} \quad \frac{\pi}{2} < \theta_i - \theta_R < \frac{3\pi}{2}$$

With the notation used for the input signal and the reference signal, the phase difference $(\theta_i - \theta_R)$ generally remains around zero for a phase-locked loop. Consequently, when the signal-to-noise ratio is high, the phase detector output signal can be separated into a useful term $K_1(\theta_i - \theta_R)$ and a noise term $K_1\varphi_i(t) = K_1 n_2'(t)/A$. In the same way as for a sinusoidal detector, the system behaves as if the input signal were modulated by a low-frequency, Gaussian, random process, having a null mean value and a two-sided power spectral density $N_0' = N_0/A^2 = N_0/2S$ in the frequency band $(-W/2, +W/2)$.

But, unlike the case of the sinusoidal detector, the present case involves a restriction: the quantity $[\theta_i - \theta_R + \varphi_i(t)]$ has to be small, which is only confirmed if the signal-to-noise ratio is high.

When the signal-to-noise ratio is not high enough for $n_1'(t)$ and $n_2'(t)$ to be smaller than A, Eq. 7.6 is no longer valid and the quantity $\varphi_i(t)$, given by Eq. 7.5, cannot be disregarded in the limit conditions, even if $\theta_i - \theta_R \cong 0$. The phase detector output signal is a random process, not easy to separate into a term related to the useful signal, and a noise term. The investigation increases in complexity: since the limiter is a nonlinear device, appropriate methods have to be applied. It can be shown (20) that the output signal of the perfect limiter, producing a signal $\pm A'$, can be expressed in the form

$$s(t) = \frac{4A'}{\pi} \sum_{p=0}^{\infty} \frac{1}{2p+1} \sin\left\{(2p+1)\left[\omega t + \theta_i + \varphi_i(t)\right]\right\}$$

where $\varphi_i(t)$ still represents the input signal equivalent modulation given in Eq. 7.5.

If we multiply the signal $s(t)$ by the signal $\text{Sign}\left[\cos(\omega t + \theta_R)\right]$ split up into Fourier series,

$$y_R(t) = \frac{4}{\pi} \sum_{q=0}^{\infty} \frac{(-1)^q}{2q+1} \cos\left[(2q+1)(\omega t + \theta_R)\right]$$

and only keep the terms comprising frequency components below f, that is to say, if we write $p = q$, we get for the signal $u(t)$

$$u(t) = \frac{16A'}{\pi^2} \sum_{0}^{\infty} \frac{(-1)^p}{(2p+1)^2} \sin\left\{(2p+1)\left[\omega t + \theta_i + \varphi_i(t)\right]\right\}$$

$$\times \cos\left[(2p+1)(\omega t + \theta_R)\right]$$

The phase detector output signal $u_1(t)$ is obtained by low-pass filtering to remove the high-angular-frequency terms:

$$u_1(t) = \frac{8A'}{\pi^2} \sum_{p=0}^{\infty} \frac{(-1)^p}{(2p+1)^2} \sin\left\{(2p+1)\left[\theta_i - \theta_R + \varphi_i(t)\right]\right\} \qquad (7.7)$$

This expression can also be formulated

$$u_1(t) = \frac{2A'}{\pi}\left[\theta_i - \theta_R + \varphi_i(t)\right] \qquad \text{if} \quad -\frac{\pi}{2} < \theta_i - \theta_R + \varphi_i(t) < \frac{\pi}{2}$$

or

$$u_1(t) = \frac{2A'}{\pi}\left[\pi - \theta_i + \theta_R - \varphi_i(t)\right] \qquad \text{if} \quad \frac{\pi}{2} < \theta_i - \theta_R + \varphi_i(t) < \frac{3\pi}{2}$$

which is observed to be exactly the same as for the noiseless case, with the insertion of $\varphi_i(t)$ in both the output signal expression and the limit conditions.

   Basing calculations on the series expansion of $u_1(t)$ (Eq. 7.7) and introduc-
ing the probability density function $p(\varphi_i)$ of the $\varphi_i(t)$ process (21), it is
possible to determine the mean value of the output signal $\overline{u_1(t)} = f(\theta_i - \theta_R)$
representing the component proportional to the useful signal. The curves in
Fig. 7.5 (22) show that the characteristic ceases to be linear when the
signal-to-noise ratio preceding the limiter $A^2/2N_0W$ becomes small, and that
it takes on a sinusoidal appearance when the ratio $A^2/2N_0W$ approaches
zero.

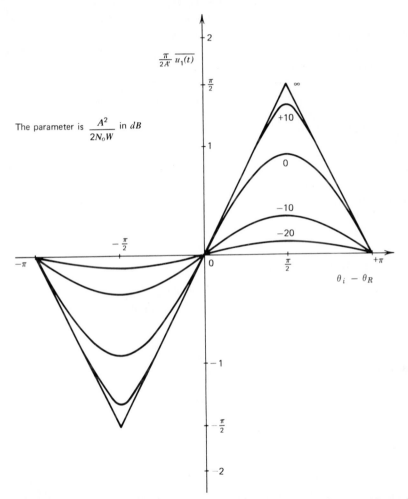

FIGURE 7.5.   Characteristic of a triangular phase detector in presence of noise, A' being the
peak amplitude of the limiter output square wave signal (by permission of A. Pouzet; from an
unpublished CNES Internal Document).

By means of rather long mathematical derivations, it is also possible to calculate the power spectral density of the noise $n'(t)$ at the detector output. In particular, the spectral density in the vicinity of the origin $f = 0$ (which is the critical zone for a phase-locked loop) increases by $+0.7$ dB if the filter preceding the limiter is an ideal rectangular filter (by only 0.3 dB if the filter is a bandpass one-pole filter) as compared with the value N_0/A^2 when the signal-to-noise ratio $A^2/2N_0W$ approaches zero (22).

To summarize, if the phase-locked loop is designed to operate in very good signal-to-noise ratio conditions, it may be interesting to insert a hard limiter in the circuit preceding the multiplier and use a square reference signal. The phase detector characteristic will then be linear over a wider range than for the sinusoidal detector and the spectral density of the equivalent modulation by the noise is identical. But, as soon as the signal-to-noise ratio preceding the phase detector drops to a few decibels (or, even to around 0 dB or below), the detector is affected by signal–noise interaction. Its performance will then be inferior to that of a sinusoidal detector, which, as we have seen, remains linear with respect to additive noise, whatever the input signal-to-noise ratio.

Remark. When the limiter is followed by a bandpass filter designed to extract the fundamental of the limited signal, we have the same problem as for a sinusoidal detector preceded by a narrow band limiter. We shall examine the performances of a phase-locked loop preceded by a limiter in Section 9.2.

7.1.5 PHASE DETECTOR LINEAR FROM $-\pi$ TO $+\pi$

The operating principle of this detector has been described in Section 2.1.6. Since the input signal circuit comprises at the input a hard limiter, followed by a differentiator (by means of which positive pulses are obtained when the input signal sign changes from negative to positive), the behavior will be modified in the presence of noise. The higher the noise level, the greater will be the disturbance observed, as in the previous example.

When the signal-to-noise ratio preceding the phase detector is high, the power spectral density of the equivalent modulation of the input signal by the noise is identical with that corresponding to the use of a sinusoidal detector. The fact that the characteristic is linear from $-\pi$ to $+\pi$ can then be an advantage in several respects (lessening of the intermodulation if the loop is used as a phase or frequency demodulator, better transient state performances, and so on).

But, as soon as the signal-to-noise ratio at the detector input weakens, the interaction between signal and noise has adverse effects on performance: the useful signal (that is, the term proportional to the phase difference $\theta_i - \theta_R$) is reduced, and the spectral density N_0' of the equivalent modulation by the noise $n(t)$ increases faster than the N_0/A^2 law applicable in the case of high signal-to-noise ratio. The deterioration of the spectral density in the vicinity of $f=0$ is greater as the signal-to-noise ratio at the input of the limiter $A^2/2N_0W$ lessens and equals about 2.9 dB when $A^2/2N_0W \to 0$ (22). The useful signal decreases as $|\theta_i - \theta_R|$ approaches π and the characteristic tends to become sinusoidal when $A^2/2N_0W \to 0$, as can be seen in Fig. 7.6 (22). This phenomenon is experimentally verified: if we plot the characteristic of this type of phase detector for variable conditions of the input signal-to-noise ratio, we observe that the linearity over the interval $(-\pi, +\pi)$ disappears and the curves obtained are very close to those given in Fig. 7.6.

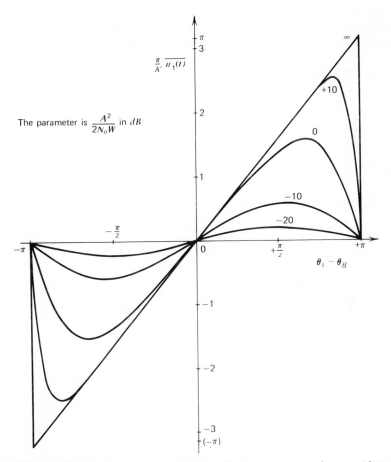

FIGURE 7.6. Characteristic of a sawtooth phase detector in presence of noise, A' being the peak amplitude of the flip-flop output signal (by permission of A. Pouzet; from an unpublished CNES Internal Document).

7.2 ADDITIVE NOISE RESPONSE OF THE LOOP

We saw in the previous section that when the signal applied at the input of a phase detector is accompanied by an additive noise $n(t)$, we can define an equivalent phase modulation of the input signal by a low-frequency noise process $n'(t)$. When the input signal is sinusoidal, the properties of this modulation are identical, whether the detector be of the analog multiplier type, or of the multiplier by the sign of the reference signal type. In both cases, the detector characteristic is sinusoidal. For both types of linear phase detector [over the interval $(-\pi/2, +\pi/2)$ or over the interval $(-\pi, +\pi)$], the sinusoidal signal accompanied by the noise $n(t)$ has to pass through a hard limiter. If the signal-to-noise ratio preceding the limiter is high enough, we can also define an equivalent modulation $n'(t)$ that has the same properties as for the sinusoidal detectors. But, when the signal-to-noise ratio is not high enough, performance deteriorates: the useful signal decreases and there is a relative increase in the spectral density of the process $n'(t)$ related to the fact that the limiter is not a linear device. Since in the present section we are dealing with linear operation, we shall restrict our investigations to the case of the sinusoidal phase detector (linear with respect to noise). The results obtained will be applicable to "linear" detectors, bearing in mind the restrictions accompanying their use.

Let us consider the loop diagram of Fig. 7.7. The signal applied at the loop input consists of a useful signal $y_u(t) = A \sin(\omega t + \theta_i)$ and noise $n(t)$. The reference signal applied at the other input of the phase detector is now the signal $y_0(t)$ from the VCO, with

$$y_0(t) = B \cos\left[\omega t + \varphi_0(t)\right]$$

In the absence of noise, the loop is in lock with a null phase error: $\varphi_0 = \theta_i$ (using our notation, for the input signal and VCO signal are, in fact, in phase-quadrature).

In the presense of noise, the signal u_1 becomes a noise signal $u_1(t)$. The noise signal $u_2(t)$, obtained by filtering $u_1(t)$ in the loop filter, is applied to

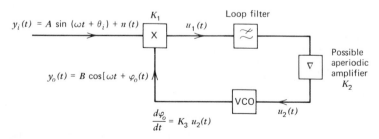

FIGURE 7.7. Phase-locked loop in presence of noise.

the VCO modulation input, producing a VCO phase modulation by a random process $\varphi_0(t)$, which we shall now further define.

First of all, it is important to note that the VCO random phase variations referred to are variations with respect to a noiseless phase value. Since the loop in lock is assumed to have a null phase error in the absence of noise, these VCO random phase variations can be defined with respect to the phase of signal $y_u(t) = A \sin(\omega t + \theta_i)$ and not with respect to signal $y_i(t)$ actually applied to the loop input and formulated (Section 7.1.2):

$$y_i(t) = A \sin(\omega t + \theta_i) + n(t)$$

$$\cong A \sin\left[\omega t + \theta_i + n'(t)\right]$$

For $n'(t)$, in this expression, is only an equivalent modulation, an abstract representation of the influence of noise $n(t)$ on signal $y_u(t)$. In fact, what we are really trying to do is to lock the VCO phase to the $y_u(t)$ phase, despite the noise $n(t)$. The phase error is really $-\varphi_0(t)$ and not $n'(t) - \varphi_0(t)$. This aspect differs fundamentally from that dealt with in Chapter 8, where the input signal $y_i(t)$ is really phase modulated by a random process. Since $-\varphi_0(t)$ represents the instantaneous phase error between the sinusoidal input signal $y_u(t) = A \sin(\omega t + \theta_i)$ and the VCO signal, it must therefore remain small enough for the linear hypothesis to apply, whereby the sine is replaced by the corresponding angle. In Chapter 12, we shall discuss the loop behavior when this hypothesis is no longer valid.

A second important point to note is that the representation of the equivalent modulation $\varphi_i(t) = n'(t)$ obtained in Section 7.1, is only valid if the reference signal phase θ_R is assumed to be a constant, or a very slowly varying quantity versus time, but independent of the noise processes $n_1(t)$ and $n_2(t)$. The results obtained will thus only be valid if the VCO random phase variations $\varphi_0(t)$ are assumed to be much slower than the variations of $n_1(t)$ and $n_2(t)$. This is confirmed when the bandwidth W preceding the loop is much larger than the loop equivalent noise bandwidth defined further on.

Finally, in certain applications, it is customary to consider the phase-locked loop as if it were a narrow-band filter with respect to the noisy input signal. But here again, this is only an analogy. If the loop really behaved like a filter, the output signal would consist of a sinusoidal signal (from the VCO) accompanied by an additive noise. In reality, this is not the case: the VCO signal is not accompanied by an additive noise but is really phase modulated by a random process. We shall come back to this point in Section 7.4.

With these remarks in mind, it is now quite simple to derive the properties of $\varphi_0(t)$ from those of $n'(t)$. We determined in Chapter 3 the equation relating the output phase (the VCO phase) to the input phase [the equivalent modulation $n'(t)$]. This relationship is the loop transfer function given by Eq. 3.6:

$$\frac{\Phi_0(j\omega)}{\Phi_i(j\omega)} = H(j\omega) = \frac{KF(j\omega)}{j\omega + KF(j\omega)}$$

It follows that the process $\varphi_0(t)$ is the output signal of a filter, having a transfer function $H(j\omega)$, driven by the process $n'(t)$. As we have seen, $n'(t)$ is a stationary Gaussian noise, having a null mean value and a uniform spectral density N_0/A^2 (in rad^2/Hz) in the frequency band $(-W/2, +W/2)$.

The process $\varphi_0(t)$ is thus a stationary Gaussian noise, having a null mean value and a power spectral density $S_{\varphi_0}(f)$ (in rad^2/Hz) from $-W/2$ to $+W/2$, given by the expression

$$S_{\varphi_0}(f) = \frac{N_0}{A^2}|H(j2\pi f)|^2 \tag{7.8}$$

In particular, the standard deviation σ_{φ_0} (in rad) is derived from

$$\sigma_{\varphi_0}^2 = \frac{N_0}{A^2}\int_{-W/2}^{+W/2}|H(j2\pi f)|^2\,df$$

Since, in most cases, $|H(j2\pi f)|^2 \cong 0$ for frequencies below $W/2$, we can simplify this equation as follows:

$$\sigma_{\varphi_0}^2 = \frac{N_0}{A^2}\int_{-\infty}^{+\infty}|H(j2\pi f)|^2\,df \tag{7.9}$$

The $H(j\omega)$ expressions are given in Chapter 3 for different types of loop filter. The expressions given below are easily derived from integral tables.

First-Order Loop $[F(j\omega) = 1]$

$$H(j\omega) = \frac{K}{j\omega + K} \qquad \text{(Equation 3.10)}$$

$$\sigma_{\varphi_0}^2 = \frac{N_0}{A^2}\frac{1}{2\pi}\int_{-\infty}^{+\infty}\frac{K^2}{K^2 + \omega^2}\,d\omega = \frac{N_0}{A^2}\frac{K}{2} \tag{7.10}$$

Second-Order Loop with Integrator and Correction $[F(j\omega) = (1 + j\omega\tau_2)/j\omega\tau_1]$

$$H(j\omega) = \frac{\omega_n^2 + j2\zeta\omega_n\omega}{\omega_n^2 - \omega^2 + j2\zeta\omega_n\omega} \qquad \text{(Equation 3.43)}$$

$$\sigma_{\varphi_0}^2 = \frac{N_0}{A^2}\frac{1}{2\pi}\int_{-\infty}^{+\infty}\frac{\omega_n^4 + 4\zeta^2\omega_n^2\omega^2}{\left(\omega_n^2 - \omega^2\right)^2 + 4\zeta^2\omega_n^2\omega^2}\,d\omega$$

$$= \frac{N_0}{A^2}\frac{\omega_n}{4\zeta}(1 + 4\zeta^2) \tag{7.11}$$

Second-Order Loop with Low-Pass Filter $[F(j\omega)=1/(1+j\omega\tau_1)]$

$$H(j\omega)=\frac{\omega_n^2}{\omega_n^2-\omega^2+j2\zeta\omega_n\omega} \qquad \text{(Equation 3.48)}$$

$$\sigma_{\varphi_0}^2=\frac{N_0}{A^2}\frac{1}{2\pi}\int_{-\infty}^{+\infty}\frac{\omega_n^4}{\left(\omega_n^2-\omega^2\right)^2+4\zeta^2\omega_n^2\omega^2}\,d\omega=\frac{N_0}{A^2}\frac{\omega_n}{4\zeta} \qquad (7.12)$$

Second-Order Loop with Low-Pass Filter and Correction $[F(j\omega)=(1+j\omega\tau_2)/(1+j\omega\tau_1)]$

$$H(j\omega)=\frac{\omega_n^2+j\left(2\zeta\omega_n-\omega_n^2/K\right)\omega}{\omega_n^2-\omega^2+j2\zeta\omega_n\omega} \qquad \text{(Equation 3.53)}$$

$$\sigma_{\varphi_0}^2=\frac{N_0}{A^2}\frac{1}{2\pi}\int_{-\infty}^{+\infty}\frac{\omega_n^4+\left(2\zeta\omega_n-\omega_n^2/K\right)^2\omega^2}{\left(\omega_n^2-\omega^2\right)^2+4\zeta^2\omega_n^2\omega^2}\,d\omega$$

$$=\frac{N_0}{A^2}\frac{\omega_n}{4\zeta}\left[1+\left(2\zeta-\frac{\omega_n}{K}\right)^2\right] \qquad (7.13)$$

We also mentioned in Section 4.6 the case of a third-order loop corresponding to the use of a loop filter having a transfer function

$$F(j\omega)=\left(\frac{1+j\omega\tau_2}{j\omega\tau_1}\right)^2$$

The transfer function $H(j\omega)$ of the loop is then

$$H(j\omega)=\frac{K\left(\dfrac{\tau_2}{\tau_1}\right)^2\left(\dfrac{1}{\tau_2^2}-\omega^2+j\dfrac{2\omega}{\tau_2}\right)}{K\left(\dfrac{\tau_2}{\tau_1}\right)^2\left(\dfrac{1}{\tau_2^2}-\omega^2\right)+j\left[K\left(\dfrac{\tau_2}{\tau_1}\right)^2\dfrac{2\omega}{\tau_2}-\omega^3\right]}$$

With this notation, the standard deviation of the VCO phase modulation can be calculated using the formula (23)

$$\sigma_{\varphi_0}^2=\frac{N_0}{A^2}\frac{K}{2}\left(\frac{\tau_2}{\tau_1}\right)^2\frac{\dfrac{2K\tau_2}{\tau_1^2}+\dfrac{3}{\tau_2^2}}{\dfrac{2K\tau_2}{\tau_1^2}-\dfrac{1}{\tau_2^2}}$$

We note that if $2K\tau_2^3 = \tau_1^2$, $\sigma_{\varphi_0}^2$ approaches infinity. We saw in Section 4.6 that this relationship is also the limit condition for loop stability.

If we choose τ_1 and τ_2 so that $K\tau_2^3/\tau_1^2 = 2\sqrt{2}$ (see Section 7.3), the formula can be simplified as follows:

$$\sigma_{\varphi_0}^2 = \frac{N_0}{A^2} \frac{K}{2} \left(\frac{\tau_2}{\tau_1}\right)^2 \frac{4\sqrt{2}+3}{4\sqrt{2}-1}$$

The VCO phase modulation variance is obtained by multiplying the two-sided power spectral density of the input signal equivalent phase modulation by the quantity $\int_{-\infty}^{+\infty} |H(j2\pi f)|^2 df$. This quantity then represents the equivalent noise bandwidth of a loop having a transfer function $H(j\omega)$. As the function to be integrated has an even symmetry, we shall use the result of this integral, denoted $(2B_n)$ to designate the equivalent noise band, reserving the notation B_n for the quantity

$$B_n = \int_0^\infty |H(j2\pi f)|^2 df \tag{7.14}$$

The equivalent noise bandwidth is measured in hertz. We can verify that if we multiply by the spectral density of the equivalent modulation expressed in rad^2/Hz, we get $\sigma_{\varphi_0}^2$ in rad^2.

With this notation,

$$\sigma_{\varphi_0}^2 = \frac{N_0}{A^2} \int_{-\infty}^{+\infty} |H(j2\pi f)|^2 df = \frac{N_0}{A^2}(2B_n) \tag{7.15}$$

We can also write $\sigma_{\varphi_0}^2 = (2N_0/A^2)B_n$. This implies that if we take the quantity B_n as definition of the noise band, that is to say, the unilateral noise band (calculated on the basis of the positive frequencies alone), in order to calculate $\sigma_{\varphi_0}^2$ we have to use the one-sided spectral density of the equivalent phase modulation $n'(t)$, that is $2N_0/A^2$.

Another magnitude used in certain cases is what is incorrectly designated "signal-to-noise ratio in the loop" or "signal-to-noise ratio in the loop equivalent bandwidth." Some use the band B_n, others the band $(2B_n)$. Since we only require a definition, what is important is to specify clearly the term selected. For present purposes, we shall use the definition

$$\left(\frac{S}{N}\right)_L = \rho_L = \frac{A^2/2}{N_0 B_n} = \frac{A^2}{2N_0 B_n}$$

(N_0 being a one-sided density, we have chosen the unilateral noise band), that is,

$$\left(\frac{S}{N}\right)_L = \rho_L = \frac{1}{\sigma_{\varphi_0}^2} \tag{7.16}$$

We can derive the equivalent noise band expressions for the different types of loop from the formulas giving $\sigma_{\varphi_0}^2$. This gives:

First-Order Loop.

$$(2B_n) = \frac{K}{2} \tag{7.17}$$

$(2B_n)$ is expressed in hertz whereas K is expressed in radians per second. The equivalent noise band is directly proportional to the loop gain K. But we saw in Chapter 5 that the loop performance improves as K increases. Under these conditions a first-order loop has no adequate protection against additive noise. This is the essential reason why this loop is relatively little used, since phase-locked loops are mainly required for the reception of weak signals, which are generally deeply immersed in additive noise at the intermediate frequency amplifier outputs.

Second-Order Loop with Integrator and Correction.

$$(2B_n) = \frac{\omega_n}{4\zeta}\left(1 + 4\zeta^2\right) \tag{7.18}$$

$(2B_n)$ is expressed in hertz when ω_n is in radians per second. The equivalent noise bandwidth is independent of the loop gain K. But, as this loop only corresponds to a limit case for a loop with low-pass filter and correction and as it is unrealizable (since it is impossible to construct a perfect integrator), we shall discuss this example further on.

Second-Order Loop with Low-Pass Filter.

$$(2B_n) = \frac{\omega_n}{4\zeta} \tag{7.19}$$

$(2B_n)$ is in hertz; ω_n is in radians per second. We have seen that this type of loop sometimes corresponds to the case of a first-order loop affected by a spurious time constant τ_1. It is interesting to calculate the expression of $(2B_n)$ versus K and τ_1; this is easily accomplished using Eqs. 3.41 and 3.47:

$$2\zeta\omega_n = \frac{1}{\tau_1}, \qquad \omega_n^2 = \frac{K}{\tau_1}$$

Consequently,

$$(2B_n) = \frac{\omega_n^2}{4\zeta\omega_n} = \frac{\dfrac{K}{\tau_1}}{\dfrac{2}{\tau_1}} = \frac{K}{2} \qquad (K \text{ in rad/s}).$$

The noise bandwidth for this loop is then independent of the time constant

τ_1 and exactly identical with the noise bandwidth of the first-order loop. An explanation of this result, which may seem rather surprising at first, will be found in the fact that, for this loop, the parameters ζ and ω_n cannot be chosen independently. If we compare with the other two second-order loops, we can say that if K is fixed and we need a small ω_n, we have to choose a large τ_1. In this case, the damping factor ζ will be very small and such that the ratio $\omega_n / 4\zeta$ remains constant and equal to $K/2$. To ensure a good loop performance, K must be large, which implies that the additive noise filtering will be inefficient.

Second-Order Loop with Low-Pass Filter and Correction.

$$(2B_n) = \frac{\omega_n}{4\zeta}\left[1 + \left(2\zeta - \frac{\omega_n}{K}\right)^2\right] \tag{7.20}$$

$(2B_n)$ is in hertz; K is in radians per second. Generally speaking, the loop gain K is high as compared with the natural angular frequency ω_n and the noise bandwidth is practically independent of K and closely approximated by

$$(2B_n) \cong \frac{\omega_n}{4\zeta}(1 + 4\zeta^2)$$

which is the same formula as for the loop with perfect integrator and correction network (Eq. 7.18). The variations of $(2B_n)$ versus the damping factor ζ, ω_n being fixed, are shown in Fig. 7.8. We draw attention to the fact that there is an optimum value for ζ as regards loop performance when the sole disturbance arises from an additive white noise accompanying the input signal. It is also worth noting that the $(2B_n)$ variations around the value $(2B_n)_{min} = \omega_n$, corresponding to $\zeta_{opt} = \frac{1}{2}$, are fairly slight. As long as ζ remains inside the interval $(0.25, 1)$, $(2B_n)$ remains below $1.25\omega_n$, which corresponds

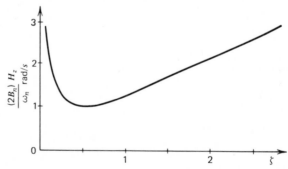

FIGURE 7.8. Variation of the equivalent noise bandwidth of a second-order loop versus damping ratio.

to a deterioration of 1 dB of the value of $\sigma_{\varphi_0}^2$ corresponding to the optimum choice.

NUMERICAL EXAMPLE. The equivalent noise temperature at a receiver input is $T_{eq} = 600°\text{K}$, and the signal power at the input is $S = -130$ dBm. The one-sided power spectral density of the noise N_0 is then, using k to designate the Boltzmann constant,

$$N_0 = kT_{eq} = 1.38 \times 10^{-23} \times 600 = 8.3 \times 10^{-21} \text{ W/Hz}$$

that is,

$$N_0 = 8.3 \times 10^{-18} \text{ mW/Hz} \quad \text{or} \quad N_0 = -170.8 \text{ dBm/Hz}$$

A second-order loop, having a loop gain $K = 2 \times 10^5$ rad/s, a natural angular frequency $\omega_n = 200$ rad/s, and damping factor $\zeta = \sqrt{2}$ is locked to the signal received. As the quantity $\omega_n/K = 10^{-3}$ is much smaller than 2ζ, the equivalent noise bandwidth $(2B_n)$, given by Eq. 7.20, is closely approximated by Eq. 7.18, which leads to

$$(2B_n) = \frac{200}{4\sqrt{2}}(1 + 4 \times 2) = 318 \text{ Hz}$$

The phase error variance due to noise (Eq. 7.15) is given by

$$10\log_{10}\sigma_{\varphi_0}^2 = 10\log_{10}\tfrac{1}{2} + 10\log_{10}N_0 - 10\log_{10}S + 10\log_{10}(2B_n)$$

$$= -3 - 170.8 + 130 + 25 = -18.8$$

from which we derive

$$\sigma_{\varphi_0}^2 = 1.32 \times 10^{-2} \text{ rad}^2 \qquad \left(\sigma_{\varphi_0} = 0.115 \text{ rad}\right)$$

and

$$\left(\frac{S}{N}\right)_L = \frac{1}{\sigma_{\varphi_0}^2} = 18.8 \text{ dB}$$

If $\zeta = 0.5$, and the other parameters remain unchanged, we get $(2B_n) = 200$ Hz, so that $(S/N)_L = 20.8$ dB and $\sigma_{\varphi_0} = 0.091$ rad.

7.3 FIRST EXAMPLE OF PARAMETER CHOICE OPTIMIZATION

Practically speaking, we rarely encounter a loop where the sole source of disturbance of the input signal is an additive white noise. For example, for a second-order loop, the optimum value for ζ will generally be different from

0.5. When the input signal is phase modulated, we have seen that a phase-locked loop behaves like a filter, having a transfer function $H(j\omega)$, both as regards this modulation and the equivalent modulation by the additive noise. We could then determine the transfer function corresponding to optimum filtering, as defined by Wiener. This work has been accomplished by Jaffe and Rechtin (24) when the input signal, in addition to the additive noise, is disturbed by the different signals discussed in Chapter 5: phase step, frequency step, and ramp. First of all, we shall briefly recall what constitutes optimum filtering. Let us take a signal $y(t)$, consisting of a "useful" signal $x(t)$ and a noise $n(t)$ applied to the input of a filter with transfer function $H(j\omega)$. Let us now suppose that we want to reduce to a minimum the mean squared deviation between the filter output signal and the "useful" part of the input signal. According to the Wiener optimum filtering theory, if we only consider filters that are physically realizable, in the case of a Gaussian, white noise $n(t)$, then $H(j\omega)$ must be chosen so that

$$H(j\omega) = 1 - \frac{\sqrt{\eta}}{\Psi(j\omega)}$$

where η is the two-sided power spectral density of noise $n(t)$ and $\Psi(j\omega)$ is defined as follows. Provided that the power spectrum $S_y(f)$ of the composite signal $y(t)$ can be expressed in the form of a rational function of f, it can always be split up into a product of two rational functions, one of which is the conjugate complex quantity of the other:

$$S_y(f) = \Psi(j2\pi f) \cdot \Psi(-j2\pi f)$$

The input signal under consideration is the signal $y(t) = \phi_i(t) + n'(t)$. The filter $H(j\omega)$, as shown in the block diagram of Fig. 7.9, has to be optimized, in order to minimize the mean square deviation $E[\phi_0^2(t) - \phi_i^2(t)]$ between $\phi_0(t)$, the VCO signal phase, and the useful part $\phi_i(t)$ of the input signal phase. The process $n'(t)$ is a Gaussian process, having a two-sided spectral density N_0/A^2 in the band $(-W/2, +W/2)$. We have to assume this band W to be larger than $H(j\omega)$ so that the noise $n'(t)$ will behave practically like a white noise.

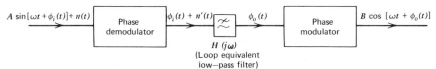

FIGURE 7.9. Model of a phase-locked loop for input signal and VCO signal modulations in presence of additive noise.

For the first example, we shall take an input signal involving a phase step of value θ:

$$\phi_i(t) = \theta \Upsilon(t) \quad \text{or} \quad \Phi_i(s) = \frac{\theta}{s}$$

As a precautionary measure, to provide against any possible increase in the phase step θ, we shall determine the optimum filter corresponding to a phase step $\lambda\theta$, λ being an adjustable multiplication coefficient. Under these conditions,

$$y(t) = \lambda\theta \Upsilon(t) + n'(t)$$

and

$$S_y(\omega) = \frac{\lambda^2\theta^2}{\omega^2} + \frac{N_0}{A^2}$$

Writing $B_0^2 = \lambda^2\theta^2 A^2 / N_0$, we can state that

$$S_y(\omega) = \frac{N_0}{A^2} \frac{B_0^2 + \omega^2}{\omega^2}$$

that is,

$$S_y(\omega) = \Psi(j\omega)\Psi(-j\omega) = \sqrt{\frac{N_0}{A^2}} \frac{B_0 + j\omega}{j\omega} \times \sqrt{\frac{N_0}{A^2}} \frac{B_0 - j\omega}{-j\omega}$$

The transfer function of the optimum filter is thus

$$H_{\text{opt}}(j\omega) = 1 - \frac{\sqrt{\dfrac{N_0}{A^2}}}{\sqrt{\dfrac{N_0}{A^2}} \dfrac{B_0 + j\omega}{j\omega}} = 1 - \frac{j\omega}{B_0 + j\omega} = \frac{B_0}{j\omega + B_0}$$

In this expression, we recognize the transfer function of the first-order loop with $K = B_0$. The first-order loop is the optimum filter, as defined by Wiener, if we take as loop gain value $K = \lambda\theta\sqrt{A^2/N_0}$, where λ is a multiplying safety coefficient with respect to the expected phase step θ. The optimum value of K depends on the signal-to-noise ratio. If this ratio is high, optimum K is large so as to reduce the phase step error. But if the signal-to-noise ratio is small, optimum K is small so as to minimize the additive noise error.

Let us now consider the case where the input signal involves an angular frequency step of value $\Delta\omega$:

$$\phi_i(t) = \Delta\omega \cdot t \cdot \Upsilon(t) \quad \text{or} \quad \Phi_i(s) = \frac{\Delta\omega}{s^2}$$

If we insert the multiplying safety coefficient λ,

$$y(t) = \lambda\Delta\omega t\Upsilon(t) + n'(t)$$

and

$$S_y(\omega) = \frac{\lambda^2\Delta\omega^2}{\omega^4} + \frac{N_0}{A^2}$$

or again

$$S_y(\omega) = \frac{N_0}{A^2} \frac{B_1^4 + \omega^4}{\omega^4} \quad \text{with } B_1^4 = \lambda^2\Delta\omega^2\frac{A^2}{N_0}$$

We can expand $S_y(\omega)$ as follows:

$$S_y(\omega) = \frac{N_0}{A^2} \frac{B_1^2 + j\omega\sqrt{2}\,B_1 + (j\omega)^2}{(j\omega)^2}$$

$$\times \frac{B_1^2 + (-j\omega)\sqrt{2}\,B_1 + (-j\omega)^2}{(-j\omega)^2}$$

which means that

$$\Psi(j\omega) = \sqrt{\frac{N_0}{A^2}} \frac{B_1^2 + j\omega\sqrt{2}\,B_1 + (j\omega)^2}{(j\omega)^2}$$

The transfer function $H_{\text{opt}}(j\omega)$ of the optimum filter is then

$$H_{\text{opt}}(j\omega) = 1 - \frac{\sqrt{\dfrac{N_0}{A^2}}}{\sqrt{\dfrac{N_0}{A^2}} \dfrac{B_1^2 + j\omega\sqrt{2}\,B_1 - \omega^2}{-\omega^2}} = 1 + \frac{\omega^2}{B_1^2 + j\omega\sqrt{2}\,B_1 - \omega^2}$$

that is,

$$H_{\text{opt}}(j\omega) = \frac{B_1^2 + j\omega\sqrt{2}\,B_1}{B_1^2 - \omega^2 + j\omega\sqrt{2}\,B_1}$$

If we compare this transfer function with that of the second-order loop with integrator and correction (Eq. 3.43), we observe that this loop corresponds to the optimum filter as defined by Wiener, taking

$$\zeta = \frac{\sqrt{2}}{2}$$

and

$$\omega_n^2 = B_1^2 = \lambda \Delta\omega \sqrt{\frac{A^2}{N_0}}$$

The multiplying coefficient λ represents the safety margin with respect to the expected angular frequency step $\Delta\omega$. The optimum damping factor in this case will be $\sqrt{2}/2$ instead of $1/2$, which would be obtained if the additive noise were considered to be the only factor of disturbance at the loop input. The choice of ω_n depends both on the angular frequency step $\lambda\Delta\omega$ and the signal-to-noise density ratio.

It is also of interest to determine the optimum Wiener filter corresponding to the case of an input signal involving a linear frequency ramp:

$$\phi_i(t) = \tfrac{1}{2}\mathcal{R}\, t^2 \Upsilon(t) \quad \text{or} \quad \Phi_i(s) = \frac{\mathcal{R}}{s^3}$$

where $\mathcal{R}$ is the input angular frequency slope expressed in rad/s^2. In this case, λ being still a safety coefficient,

$$y(t) = \tfrac{1}{2}\lambda \mathcal{R}\, t^2 \Upsilon(t) + n'(t)$$

implying that

$$S_y(\omega) = \frac{\lambda^2 \mathcal{R}^2}{\omega^6} + \frac{N_0}{A^2} = \frac{N_0}{A^2} \frac{B_2^6 + \omega^6}{\omega^6}$$

with

$$B_2^6 = \lambda^2 \mathcal{R}^2 \frac{A^2}{N_0}$$

The power spectrum of signal $y(t)$ can also be expressed

$$S_y(\omega) = \frac{N_0}{A^2} \frac{B_2^3 + j\omega 2 B_2^2 + (j\omega)^2 2 B_2 + (j\omega)^3}{(j\omega)^3}$$

$$\times \frac{B_2^3 + (-j\omega)2 B_2^2 + (-j\omega)^2 2 B_2 + (-j\omega)^3}{(-j\omega)^3}$$

consequently,

$$\Psi(j\omega) = \sqrt{\frac{N_0}{A^2}} \; \frac{B_2^3 + (j\omega)2B_2^2 + (j\omega)^2 2B_2 + (j\omega)^3}{(j\omega)^3}$$

The optimum filter transfer function

$$H_{opt}(j\omega) = 1 - \frac{\sqrt{N_0/A^2}}{\Psi(j\omega)}$$

is written

$$H_{opt}(j\omega) = \frac{B_2^3 + j\omega 2B_2^2 - 2B_2\omega^2}{B_2^3 + j\omega 2B_2^2 - 2B_2\omega^2 - j\omega^3}$$

In order to obtain this type of transfer function when using a phase-locked loop, we have to use a loop filter having a transfer function

$$F(j\omega) = \frac{B_2^3 + j\omega 2B_2^2 - 2B_2\omega^2}{-K\omega^2}$$

We can obtain a good approximation of this loop filter by the use of an active filter consisting of an inductance L_1, connected in series with an operational amplifier of gain $-G$, and a feedback network (R, L_2, C) connected in parallel to the amplifier (25). But the transfer function of a filter comprising two integrators and a double correction (roughly the equivalent of two filters like that shown in Fig. 2.24, cascade connected) as discussed in Section 4.6, is

$$F(j\omega) = \frac{1 + j\omega 2\tau_2 - \tau_2^2\omega^2}{-\tau_1^2\omega^2}$$

If we select the time constants τ_1 and τ_2 and the loop gain K so that (25)

$$\frac{K}{\tau_1^2} = B_2^3 = \lambda \mathcal{R}\sqrt{\frac{A^2}{N_0}} \quad \text{and} \quad \tau_2 = \frac{\sqrt{2}}{B_2}$$

then we obtain for the loop transfer function

$$H(j\omega) = \frac{B_2^3 + j\omega 2\sqrt{2}\, B_2^2 - 2B_2\omega^2}{B_2^3 + j\omega 2\sqrt{2}\, B_2^2 - 2B_2\omega^2 - j\omega^3}$$

which is very close to the Wiener optimum filter expression for an input signal comprising a linear frequency variation, in the presence of Gaussian white noise.

These parameter optimization examples highlight two important points. The first is that the decision as to the type of loop filter to be used to obtain optimum filtering does not depend on the value of the disturbance, but entirely on the nature of the disturbance involved (the loop filter choice does not depend on λ). The second important point is that the value of the parameters to be chosen depends on the signal-to-noise density ratio A^2/N_0. When this ratio is variable, as it is in most cases, the filter can only be optimum for a given signal-to-noise ratio value. We usually select the lowest value anticipated for the signal-to-noise ratio, since this corresponds to least favorable operating conditions. But, when the signal-to-noise ratio increases, the loop parameters can be adjusted to ensure that we keep as close as possible to the optimum filter over a wide range of signal-to-noise ratio values. Jaffe and Rechtin (24) have shown that this is effectively accomplished by adjusting the loop gain value, through the phase detector sensitivity K_1 (proportional to the input signal amplitude A), in cases where the loop is preceded by an automatic gain control device or a limiter (see Chapter 9).

NUMERICAL EXAMPLE. A signal of power $S = -127$ dBm is applied to the input of a receiver so that $N_0 = -170$ dBm/Hz. A phase-locked loop is to be used. The signal is liable to undergo angular frequency steps $\Delta\omega = 100$ rad/s. On what basis will the loop parameters be decided?

Bearing in mind the above discussion, it would be advisable to choose a second-order loop with integrator and correction network, so that $\zeta = \sqrt{2}\,/2$ and $\omega_n^2 = \lambda\Delta\omega\sqrt{A^2/N_0}$.

If we take 2 for the value of the safety coefficient λ, this leads to

$$\omega_n^2 = 2 \times 100\sqrt{4 \times 10^4} = 4 \times 10^4$$

that is,

$$\omega_n = 200 \text{ rad/s}$$

The equivalent noise bandwidth $(2B_n)$ is given by Eq. 7.18, which yields

$$(2B_n) = 200\frac{1 + 4\frac{1}{2}}{4\sqrt{2}\,/2} = 212 \text{ Hz}$$

The VCO noise-induced phase variance is then given by Eq. 7.15:

$$\sigma_{\varphi_0}^2 = \frac{N_0}{2S}(2B_n) = \frac{1}{4 \times 10^4}212 = 5.3 \times 10^{-3} \text{ rad}^2$$

The maximum phase error ϕ_M due to the angular frequency step $\Delta\omega$ is derived from the curves in Fig. 5.4:

$$\phi_M = 0.46 \frac{\Delta\omega}{\omega_n} = 0.23 \text{ rad}$$

When the input signal power increases to $S = -87$ dBm, for optimization purposes, we should have to take $\zeta = \sqrt{2}/2$ and $\omega_n = 2 \times 10^3$ rad/s, which would imply $\sigma_{\varphi_0}^2 = 5.3 \times 10^{-6}$ rad^2 and $\phi_M = 0.023$ rad. In fact, if ω_n remains at its initial value (200 rad/s), ϕ_M will keep the value of 0.23 rad but $\sigma_{\varphi_0}^2$ will only be equal to 5.3×10^{-7} rad^2.

7.4 OUTPUT SIGNAL POWER SPECTRUM

In Section 7.2, we observed that the loop does not behave like a true filter as regards additive noise, since the VCO signal is not accompanied by a filtered additive noise but is phase modulated by a filtered noise. The VCO output signal is expressed $y_0(t) = B\cos[\omega_0 t + \varphi_0(t)]$ (in this section, the VCO central frequency will be denoted $f_0 = \omega_0/2\pi$, which enables us to keep the notation $f = \omega/2\pi$ for the frequency considered as a variable).

The spectrum of signal $y_0(t)$ is that of a signal phase modulated by the process $\varphi_0(t)$, a filtered version, by the loop transfer function $H(j\omega)$, of the equivalent phase modulation $n'(t)$ of the input signal.

In other words, we are required to calculate the expression of the power spectrum of a signal, phase modulated by a Gaussian random process, the power spectrum of which is known (Eq. 7.8).

The general procedure to obtain $S_{y_0}(f)$ consists in calculating the auto-correlation function $R_{y_0}(\tau)$ and then taking the Fourier transform. A quicker method (26) is to introduce the complex function $v(t)$ defined by

$$v(t) = e^{j\varphi_0(t)}$$

The power spectrum of the real signal $y_0(t)$ is related to that of the complex signal $v(t)$ by the relation (26, p. 110)

$$S_{y_0}(f) = \frac{B^2}{4}\left[S_v(f-f_0) + S_v(-f-f_0) \right]$$

The power spectrum of signal $v(t)$ is obtained from the autocorrelation function $R_v(\tau)$ and Fourier transform. But, when $\varphi_0(t)$ is a Gaussian process, $R_v(\tau)$ is given simply by (26, p. 114)

$$R_v(\tau) = e^{-[R_{\varphi_0}(0) - R_{\varphi_0}(\tau)]}$$

where $R_{\varphi_0}(\tau)$ is the autocorrelation function of the process $\varphi_0(t)$.
Consequently,

$$S_v(f) = \int_{-\infty}^{+\infty} R_v(\tau) e^{-j2\pi f\tau} \, d\tau$$

$$= e^{-R_{\varphi_0}(0)} \int_{-\infty}^{+\infty} e^{R_{\varphi_0}(\tau)} e^{-j2\pi f\tau} \, d\tau$$

If we assume that $\varphi_0(t)$ remains small enough (remember that we are still
in the linear operating domain) for $\sigma_{\varphi_0}^2 = R_{\varphi_0}(0)$ to stay well below 1 rad^2, and
since $|R_{\varphi_0}(\tau)| \leqslant R_{\varphi_0}(0)$ (the process $\varphi_0(t)$ is a real process), we can make the
approximation

$$e^{R_{\varphi_0}(\tau)} \cong 1 + R_{\varphi_0}(\tau)$$

Under these conditions,

$$S_v(f) = e^{-\sigma_{\varphi_0}^2} \int_{-\infty}^{+\infty} \left[1 + R_{\varphi_0}(\tau) \right] e^{-j2\pi f\tau} \, d\tau$$

that is,

$$S_v(f) = e^{-\sigma_{\varphi_0}^2} \left[\delta(f) + S_{\varphi_0}(f) \right]$$

where $\delta(f)$ is the Dirac delta function and $S_{\varphi_0}(f)$ is the power spectral
density of the VCO phase $\varphi_0(t)$. Consequently,

$$S_{y_0}(f) = \frac{B^2}{4} e^{-\sigma_{\varphi_0}^2} \left[\delta(f-f_0) + \delta(-f-f_0) + S_{\varphi_0}(f-f_0) + S_{\varphi_0}(-f-f_0) \right]$$

The one-sided power spectral density (unilateral density) is then

$$S_{y_0}(f_+) = \frac{B^2}{2} e^{-\sigma_{\varphi_0}^2} \Bigg\{ \delta(f-f_0)$$

$$+ \frac{N_0}{A^2} |H[j2\pi(f-f_0)]|^2 \Bigg\} \tag{7.21}$$

If we use $P = B^2/2$ to designate the nonmodulated signal power, the effect
of the modulation by the gaussian random process $\varphi_0(t)$ on the power
spectrum of $y_0(t)$ is revealed in:

- a reduction of the f_0 discrete frequency line, since it is multiplied by
  $e^{-\sigma_{\varphi_0}^2} \cong 1 - \sigma_{\varphi_0}^2$.
- the appearance of a continuous spectrum which, to within the factor

$Pe^{-\sigma_{\varphi_0}^2}$, is the power spectrum of the modulation process $\varphi_0(t)$ translated from the frequency zero to the VCO f_0 frequency. The power of the continuous part of the spectrum is thus obtained immediately:

$$P_{\text{cont}} = Pe^{-\sigma_{\varphi_0}^2} \cdot \sigma_{\varphi_0}^2 \cong P\sigma_{\varphi_0}^2$$

The general shape of the spectrum $S_{y_0}(f)$ is given in Fig. 7.10.

If the approximation $e^{R_{\varphi_0}(\tau)} \cong 1 + R_{\varphi_0}(\tau)$ is not used, the calculation is a little more complicated. One method consists in using the complete series expansion of $e^{R_{\varphi_0}(\tau)}$. It can be shown (27) that, in addition to the two major terms defined above, we have to take into account an infinity of continuous spectra, the expression of which can be calculated step by step using a recurrence formula. The fraction of the total power represented by the unused spectra is generally very low (less than 0.8% of the total power P if $\sigma_{\varphi_0}^2 \leqslant \frac{1}{8}$ rad^2).

Finally, if we decide to avoid using the complex function $v(t)$, we have to calculate the autocorrelation function $R_{y_0}(\tau)$ of the signal $y_0(t)$. Since the phase $\varphi_0(t)$ is a Gaussian variable (assuming the operating range to be linear), it can be shown that (27)

$$R_{y_0}(\tau) = \frac{B^2}{2} \cos \omega_0 \tau \, e^{-\left[R_{\varphi_0}(0) - R_{\varphi_0}(\tau) \right]}$$

We then have to take the Fourier transform of this expression and introduce the autocorrelation function $R_{\varphi_0}(\tau)$ of the process $\varphi_0(t)$. We can use, for example, (28),

$$R_{\varphi_0}(\tau) = \frac{N_0}{A^2} \frac{\omega_n}{4} e^{-\zeta \omega_n \tau} \left[\frac{1 + 4\zeta^2}{\zeta} \cos \omega_n \sqrt{1 - \zeta^2} \, \tau \right.$$

$$\left. + \frac{1 - 4\zeta^2}{\sqrt{1 - \zeta^2}} \sin \omega_n \sqrt{1 - \zeta^2} \, \tau \right] \tag{7.22}$$

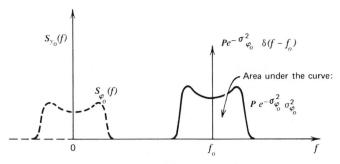

FIGURE 7.10. Power spectrum of the VCO signal.

This expression corresponds to a second-order loop with integrator and correction, when the damping factor is below 1. Otherwise, $\sqrt{1-\zeta^2}$ must be replaced by $\sqrt{\zeta^2-1}$ and the trigonometric lines by the corresponding hyperbolic lines. Reference (28) also contains the expression corresponding to a second-order loop with low-pass filter and correction.

The power spectrum obtained for $S_{y_0}(f)$, consisting of a discrete frequency line and a continuous spectrum does not allow us to conclude, despite the resemblance, that signal $y_0(t)$ consists of a pure nonmodulated sinusoidal signal (discrete frequency line) and an additive noise (continuous spectrum). An additive noise would produce a simultaneous amplitude and phase modulation of the sinusoidal signal (see Section 7.1.4b), whereas, in the present example, the VCO signal is only phase modulated and it is this phase modulation which causes the continuous part of the power spectrum of $y_0(t)$.

7.5   SIGNAL-TO-NOISE RATIO AT THE PHASE DEMODULATION OUTPUT

We discussed in Chapter 6 the possibility of using a phase-locked loop as phase demodulator or frequency demodulator and we indicated the corresponding response curves. We now calculate the noise power obtained at the output of such a demodulator, assuming, of course, that we are still in the linear operating zone. In particular, threshold problems will be dealt with in Chapter 12. When the input signal is phase modulated with a low modulation index, the error signal is a good duplicate of the modulation signal, at least when the modulation signal frequency is high enough (see Section 6.1). If we pick up the phase detector output signal, using a very high input impedance amplifier, so as not to interfere with the loop behavior, the signal obtained is a good duplicate of the modulation: we have constructed a phase demodulator.

The signal $u_1(t)$ is related to the modulation $\phi_i(t)$ of the input signal by the loop error function (Eq. 3.7) according to the equation

$$\frac{U_1(j\Omega)}{\Phi_i(j\Omega)} = K_1 \frac{\Phi(j\Omega)}{\Phi_i(j\Omega)} = K_1[1 - H(j\Omega)]$$

(In this section and in that following, we shall use the terms Ω and F for the angular frequency and the frequency of the modulation signal).

We can thus represent the phase demodulator by a perfect demodulator of slope K_1 (in V/rad) followed by a high-pass filter having a transfer function $[1 - H(j\Omega)]$ (Fig. 7.11).

When the input signal $y_i(t) = A \sin[\omega t + \phi_i(t)]$ is accompanied by a stationary Gaussian noise, having a null mean value and a power spectral

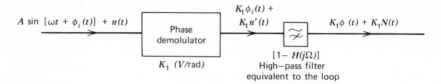

FIGURE 7.11. Model of a phase-locked loop working as a phase demodulator in presence of noise.

density N_0 from $f - W/2$ to $f + W/2$, it can be expressed in the form (Section 7.1.2)

$$y_i(t) = A \sin \left[\omega t + \phi_i(t) \right] + n(t)$$

$$\cong A \sin \left[\omega t + \phi_i(t) + n'(t) \right]$$

where $n'(t)$ is a stationary Gaussian process, having a null mean value and a power spectral density N_0/A^2 from $-W/2$ to $+W/2$. The output signal of the perfect phase demodulator is thus $K_1\phi_i(t) + K_1 n'(t)$. The output signal obtained is $K_1\phi(t) + K_1 N(t)$, where $N(t)$ represents the process $n'(t)$ filtered by $[1 - H(j\Omega)]$: $N(t) = n'(t) - \varphi_0(t)$. If we call $S_{\phi_i}(F)$ the power spectral density of the modulation $\phi_i(t)$, we get

$$\overline{\phi^2(t)} = \int_{-W/2}^{+W/2} S_{\phi_i}(F) |1 - H(j2\pi F)|^2 \, dF$$

$$\overline{N^2(t)} = \int_{-W/2}^{+W/2} \frac{N_0}{A^2} |1 - H(j2\pi F)|^2 \, dF$$

It is generally advisable to keep the part of $S_{\phi_i}(F)$ suppressed by the filter $[1 - H(j\Omega)]$ as low as possible. If we are dealing with a first-order loop, this implies that the loop gain K must be as low as possible. But, in this case, the loop performance is not very good. It is therefore preferable to use a second-order loop with integrator and correction, or low-pass filter and correction. We can then select the loop parameters so that only a small part of the spectrum of $S_{\phi_i}(F)$ will be suppressed by the demodulation, without deteriorating loop or demodulator performances. Under these conditions,

$$\overline{\phi^2(t)} \cong \int_{-W/2}^{+W/2} S_{\phi_i}(F) \, dF \cong \overline{\phi_i^2(t)}$$

$$\overline{N^2(t)} \cong \int_{-W/2}^{+W/2} \frac{N_0}{A^2} \, dF = \frac{N_0 W}{A^2}$$

(the bandwidth W preceding the loop can be chosen wide enough for $S_{\phi_i}(F)$ to be unaffected).

The output signal-to-noise ratio is then written

$$\left(\frac{S}{N}\right)_{\text{out}} = \frac{A^2}{N_0 W} \overline{\phi_i^2(t)} \tag{7.23}$$

The term "coherent" (or synchronous) phase demodulation is generally used to describe the multiplication of a low-index phase modulated sinusoidal signal, by a reference signal having the same phase (to within $\pi/2$) as the pure carrier (unmodulated). The main advantage of this type of demodulation is, at least theoretically, that it is completely unaffected by threshold phenomena: the output signal expression $K_1\phi_i(t) + K_1 n'(t)$ holds, whatever the value of $n'(t)$ (that is, at a given density N_0, for any preceding bandwidth W, providing $W \leqslant 2f$) for sinusoidal phase detectors.

A phase-locked loop with a transfer function $H(j\Omega)$, sufficiently narrow compared with $S_{\phi_i}(F)$, produces at the VCO output a signal in synchronism with the carrier, to within $\pi/2$. Since the multiplication involved is performed in the phase detector device, this loop is a coherent phase demodulator. The other restrictions are that $\phi_i(t)$ must be small enough for $K_1 \sin[\phi_i(t)]$ to be replaced by $K_1\phi_i(t)$ in the output signal expression (also to make the linear equations applicable) and that the random modulation of the VCO phase $\varphi_0(t)$ caused by input noise must also remain fairly low. This is generally easily obtained since the loop bandwidth and consequently the equivalent noise bandwidth $(2B_n)$ have to remain smaller than W. We shall consider in Chapter 12 the case where the phases $\phi_i(t)$ and $\varphi_0(t)$ drive the loop out of its linear range.

The expression obtained for the signal-to-noise ratio at the coherent demodulator output is identical to that obtained when the input signal is demodulated without the "coherent" procedure. The most usual method is to follow a standard frequency discriminator by an integrator (Fig. 7.12). Let K_D be the slope of the discriminator, of central frequency f. Since the discriminator is sensitive to instantaneous frequency variations, the output signal $u_0(t) = K_D[f_{i_{\text{inst}}}(t) - f]$, with

$$f_{i_{\text{inst}}}(t) = \frac{1}{2\pi}\frac{d}{dt}\left[\omega t + \phi_i(t)\right] = f + \frac{1}{2\pi}\frac{d\phi_i}{dt}$$

thus

$$u_0(t) = \frac{K_D}{2\pi}\frac{d\phi_i}{dt}$$

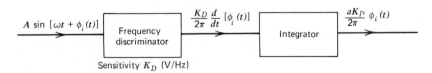

$A \sin[\omega t + \phi_i(t)]$ → Frequency discriminator → $\frac{K_D}{2\pi}\frac{d}{dt}[\phi_i(t)]$ → Integrator → $\frac{aK_D}{2\pi}\phi_i(t)$

Sensitivity K_D (V/Hz)

FIGURE 7.12. Model of a classical phase demodulator.

If the discriminator is followed by an integrator effecting

$$u(t) = a \int^t u_0(t)\, dt$$

we get

$$u(t) = \frac{aK_D}{2\pi} \phi_i(t)$$

When the input signal is accompanied by an additive noise $n(t)$ and if the noise power $\overline{n^2(t)} = N_0 W$ remains small enough as compared with the signal power $A^2/2$ (because of the discriminator threshold phenomenon) an input signal phase equivalent modulation can be defined, $\varphi_i(t) = n_2'(t)/A$ (see Section 7.1.4b), which has the same statistical properties as the process $n'(t)$. The noise term at the integrator output is then expressed as

$$N(t) = \frac{aK_D}{2\pi} \frac{n_2'(t)}{A}$$

and, consequently, the signal-to-noise ratio is

$$\frac{\overline{u(t)^2}}{\overline{N(t)^2}} = \frac{\overline{\phi_i^2(t)}}{N_0 W / A^2} = \frac{A^2}{N_0 W} \overline{\phi_i^2(t)}$$

If the two demodulators are, from this point of view, equivalent, we must not forget that the classical demodulator formula is only valid beyond the discriminator threshold. That is to say, roughly speaking, when the signal-to-noise ratio preceding the discriminator exceeds about 10 dB,

$$\frac{A^2}{2N_0 W} \geqslant 10$$

For the coherent demodulator, there is no signal-to-noise ratio restriction of this kind. The only restriction concerns the VCO random phase variations, which must remain relatively low, for example such that $\sigma_{\varphi_0}^2 \leqslant \frac{1}{10}$ rad^2, that is,

$$\frac{A^2}{N_0(2B_n)} \geqslant 10$$

As the equivalent noise band $(2B_n)$ of the loop is very small, the minimum compatible with adequate loop performances, the latter condition is more easily complied with than that imposed by the discriminator.

We also draw attention to the fact that both demodulators have the same defect in that they suppress the low-frequency components of the modulation signal. In the case of the coherent demodulator, this is because the low modulation frequencies "pass through the loop" (modulate the VCO) and are no longer available in the demodulated output. As regards the conventional demodulator, it is because the discriminator is sensitive to the instantaneous frequency and behaves like a differentiator as regards the phase modulation. This is compensated by inserting an integrator behind the discriminator. But we are unable to construct a perfect integrator having an Ω^{-1} frequency transfer function. The simplest way to construct an approximate integrator consists in using a low-pass RC filter with a very high time constant τ. Then

$$u(t) \cong \frac{1}{RC} \int^t u_0(t)\,dt$$

But this filter transfer function does not feature an asymptotic slope of -1 up to $\Omega = 0$. From $\Omega_0 = 1/RC$ the asymptotic slope becomes null and the overall transfer function for the "discriminator/pseudointegrator" unit is represented in Fig. 7.13. In order to encompass the maximum information, the time constant τ must be very large. But, as shown in Fig. 7.13, if we use a discriminator with a given slope K_D (V/Hz) and increase RC, we automatically diminish the phase demodulator sensitivity, which is equal to $K_D/2\pi RC$ (V/rad). For a coherent demodulator, on the contrary, the sensitivity remains the same as that of the phase detector K_1 (V/rad), whatever the

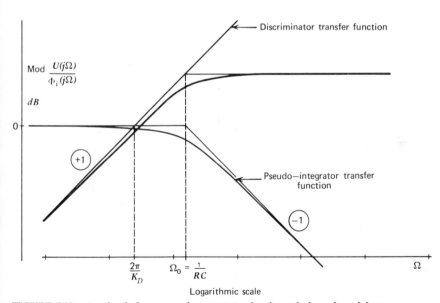

FIGURE 7.13. Amplitude-frequency characteristic of a classical phase demodulator.

low-frequency cutoff, since, in this case, the low-frequency cutoff depends solely on ω_n and ζ for a second-order loop (see Fig. 6.5).

However, the conventional demodulator features an appreciable advantage as regards the choice of modulation index. Whereas the coherent demodulator is restricted to low values, 0.5 rad or, at the limit, 1 rad, for a sinusoidal phase detector (slightly more for linear detectors), the classical demodulator can accept far higher modulation indexes. If the discriminator is perfect, the limitation depends on the acceptable distortion incurred by the bandpass filter (bandwidth W) preceding the demodulator. If this bandwidth equals, for instance, 100 kHz, and if the sinusoidal frequency modulation equals 10 kHz, we can choose, at least in the absence of noise, an index of a few radians. But, in the presence of noise, this advantage disappears and the coherent demodulator is without doubt the most sensitive demodulation device we are able to construct.

NUMERICAL EXAMPLE. A signal having a frequency stability such that the frequency can vary slowly within a range ± 4 kHz, is applied to the input of a receiver so that $N_0 = -168$ dBm/Hz. The signal power is $S = -128$ dBm. The signal is phase modulated with an index $m_i = 0.5$ rad by a sinusoidal signal, whose frequency F varies between 100 and 200 Hz.

The minimum bandwidth of the amplifier at intermediate frequency (IF) is $W = 2(4 + 0.2) = 8.4$ kHz. The signal-to-noise ratio is

$$\left(\frac{S}{N}\right)_{\mathrm{IF}} = \frac{S}{N_0 W} \Rightarrow -128 + 168 - 39.25 = +0.75 \text{ dB}$$

In this case, a conventional discriminator cannot be used to demodulate the signal. We shall use a second-order loop [with loop filter $(1 + \tau_2 s)/(1 + \tau_1 s)$] so that $K = 400$ kHz, $\omega_n = 50$ rad/s, $\zeta = \sqrt{2}/2$.

The VCO noise-induced phase variance calculated using Eqs. 7.15 and 7.18, is $\sigma_{\varphi_0}^2 = 2.65 \times 10^{-3}$ rad^2. The phase error resulting from the frequency drift of the carrier (Eq. 5.10 for t approaches infinity) is $\Delta \omega / K = 0.01$ rad. The modulation index being $m_i = 0.5$ rad, the linear operation hypothesis is verified.

Since the quantity ω_n / K is very small, the demodulator response is given by the curves in Fig. 6.5. It will be seen that for the minimum modulation frequency $F = F_{\min} = 100$ Hz (the reduced variable is $x = \Omega / \omega_n = 2\pi \times 100/50 = 12.5$), there is practically no signal diminution and the phase shift is 9°.

Since the modulation signal is $m_i \sin \Omega t$, we pick up at the phase detector (sensitivity K_1) output, the useful signal

$$K_1 m_i \sin(\Omega t + \Psi) \qquad \text{with } \Psi \leqslant 9°$$

and the noise signal $K_1 N(t)$, the power spectral density of which is $K_1^2 \times |1 - H(j2\pi F)|^2 N_0 / A^2$.

The useful signal power is $K_1^2 m_i^2 / 2$, and the noise power at the phase detector output is

$$K_1^2 \overline{N^2(t)} = K_1^2 \frac{N_0}{A^2} \int_{-W/2}^{+W/2} |1 - H(j2\pi F)|^2 \, dF$$

$$\cong K_1^2 \frac{N_0}{A^2} \int_{-W/2}^{+W/2} dF = K_1^2 \frac{N_0 W}{A^2}$$

The signal-to-noise ratio at the phase detector output is thus

$$\left(\frac{S}{N} \right)_d = m_i^2 \frac{A^2}{2 N_0 W}$$

that is,

$$\left(\frac{S}{N} \right)_d \Rightarrow -6 - 128 + 168 - 39.25 = -5.25 \text{ dB}$$

But the modulation signal frequency is such as to allow the use at the output of a bandpass filter having a central frequency of 150 Hz and a bandwith $\mathcal{B} = 100$ Hz. At this filter output, the noise power is $K_1^2 2 N_0 \mathcal{B} / A^2$ but the signal power remains as before. Consequently, the signal-to-noise ratio obtained at the filter output is

$$\left(\frac{S}{N} \right)_{out} = m_i^2 \frac{A^2}{4 N_0 \mathcal{B}}$$

that is,

$$\left(\frac{S}{N} \right)_{out} \Rightarrow -6 - 128 - 3 + 168 - 20 = +11 \text{ dB}$$

7.6 SIGNAL-TO-NOISE RATIO AT THE FREQUENCY DEMODULATION OUTPUT

We saw in Section 6.4 that if the signal applied to a phase-locked loop input is frequency modulated, the phase detector output signal $u_1(t)$ or the VCO modulation signal $u_2(t)$ (according to the type of loop under consideration) represents a low-pass filtered version of the modulation signal; the low-pass filtering function depends on the loop filter used. But we have also seen (Section 6.3) that the possible peak frequency deviation is restricted by the maximum phase error value, which must remain fairly low (if the loop is to stay in the linear operation range).

When the frequency deviation is fixed and constant as far as the highest frequency F_m of the modulation spectrum, we generally have to have the loop gain K or the natural angular frequency ω_n, according to the type of loop, much larger than $2\pi F_m$. Since the low-pass filtering cutoff frequency, according to the loop, is around K or ω_n, it is advisable to follow the loop acting as demodulator by a low-pass filter with a cutoff frequency F_m in order to reduce the noise power at the output. Under these conditions, and setting aside for the present threshold phenomena, the phase-locked loop behaves like an ideal discriminator, whatever loop filter is used. The output noise power has to be evaluated in an ideal low-pass filter $(0, F_m)$ and the formula giving the signal-to-noise ratio is identical with that corresponding to a conventional discriminator.

For certain particular modulation spectra, when the expected frequency deviation of the high-frequency components of the modulation signal is fairly limited, K or ω_n can be chosen so as to benefit from the low-pass filtering function inherent in the loop. In this case, the output signal-to-noise ratio formula depends on the type of loop filter used.

7.6.1 STANDARD DISCRIMINATOR (REMINDER)

When the signal applied to a discriminator input is accompanied by a stationary Gaussian noise $n(t)$, having a null mean value and a power spectral density N_0 from $f - W/2$ to $f + W/2$, we can, as in the case of phase modulation, define an equivalent frequency modulation by a stationary, Gaussian, low-frequency random process $n''(t)$, having a null mean value and a power spectral density $(N_0/A^2)F^2$ (in Hz^2/Hz) between $-W/2$ and $+W/2$. The noise power at the discriminator output, without a low-pass filter, is expressed as

$$\overline{u_N^2(t)} = K_D^2 \int_{-W/2}^{+W/2} \frac{N_0}{A^2} F^2 \, dF$$

$$= K_D^2 \frac{N_0}{A^2} \left[\frac{F^3}{3} \right]_{-W/2}^{+W/2}$$

that is,

$$\overline{u_N^2(t)} = K_D^2 \frac{N_0}{A^2} \frac{W^3}{12} \tag{7.24}$$

The useful signal power at the discriminator output is

$$\overline{u_S^2(t)} = K_D^2 \, \overline{f_{i_{inst}}^2(t)}$$

and, consequently, the signal-to-noise ratio at the output is

$$\left(\frac{S}{N}\right)_D = 12 \frac{A^2}{N_0} \frac{\overline{f_{i_{inst}}^2(t)}}{W^3}$$

Providing the modulation signal power spectrum is negligible beyond a certain frequency F_m, we can filter the noise at the discriminator output, using a low-pass filter to suppress the frequency components higher than F_m. Then

$$\overline{u_N^2(t)} = K_D^2 \frac{N_0}{A^2}\left[\frac{F^3}{3}\right]_{-F_m}^{+F_m} = \frac{2}{3}K_D^2\frac{N_0}{A^2}F_m^3 \qquad (7.25)$$

and

$$\left(\frac{S}{N}\right)_D = \frac{3}{2}\frac{A^2}{N_0}\frac{\overline{f_{i_{inst}}^2(t)}}{F_m^3}$$

If the modulation signal is a sinusoidal signal, we get

$$\overline{f_{i_{inst}}^2(t)} = \tfrac{1}{2}\Delta f_i^2,$$

where Δf_i represents the peak frequency deviation. Then

$$\left(\frac{S}{N}\right)_D = \frac{3}{4}\frac{A^2}{N_0}\frac{\Delta f_i^2}{F_m^3}$$

which is often given in the form

$$\left(\frac{S}{N}\right)_D = \frac{3}{2}\frac{\Delta f_i^2}{F_m^2}\left[\frac{A^2}{2N_0F_m}\right] \qquad (7.26)$$

The quantity in brackets corresponds to the signal-to-noise ratio that would be obtained by the use of an amplitude modulation at a modulation percentage of 100%. The quantity $\tfrac{3}{2}(\Delta f_i^2/F_m^2)$ is generally known as the frequency modulation improvement factor. It intervenes in cases where the input signal power exceeds the discriminator threshold, which depends on the preceding bandwidth W; the minimum bandwidth is given by the

Carson formula:

$$W = 2F_m + 2\Delta f_i$$

whereas the minimum amplitude modulation (as well as low index phase modulation) bandwidth is derived approximately from $W = 2F_m$. Thus, as a counterpart for an improved output signal-to-noise ratio, the necessary radiated signal spectrum is more cumbersome.

7.6.2 FIRST-ORDER LOOP

We saw in Section 6.4 that this loop behaves like a discriminator of slope $K_D = 2\pi/K_3$ (in V/Hz when K_3 is expressed in rad/s/V) followed by an RC low-pass filter where $RC = 1/K$. The output noise power is

$$\overline{u_N^2(t)} = K_D^2 \int_{-W/2}^{+W/2} \frac{N_0}{A^2} \frac{F^2}{1 + (2\pi F)^2/K^2} \, dF$$

from which we derive

$$\overline{u_N^2(t)} = K_D^2 \left(\frac{K}{2\pi} \right)^2 \frac{N_0}{A^2} \left(W - \frac{K}{\pi} \operatorname{Arc tan} \frac{\pi W}{K} \right) \tag{7.27}$$

Exactly as for a standard discriminator followed by a low-pass filter of time constant $1/K$, we have to make a hypothesis as to the value of K with respect to the preceding bandwidth W.

(a) $K \gg \pi W$ (that is to say $K/2\pi \gg W/2$, thus the loop gain expressed in hertz far exceeds the preceding half-bandwidth, also expressed in hertz). Then

$$\operatorname{Arc tan} \frac{\pi W}{K} \cong \frac{\pi W}{K} - \frac{\pi^3 W^3}{3K^3}$$

and

$$\overline{u_N^2(t)} = K_D^2 \frac{N_0}{A^2} \frac{W^3}{12}$$

which is Eq. 7.24 corresponding to a standard discriminator without output filtering. The maximum modulation frequency F_m is only limited by the preceding bandwidth (Carson formula). The maximum frequency deviation for the useful signal is limited by the phase detector linearity requirement, for example,

$$\Delta\phi = \frac{\Delta\omega_i}{K} < \tfrac{1}{2} \text{ rad.}$$

(b) $K \ll \pi W$; then

$$\text{Arc tan} \, \frac{\pi W}{K} \cong \frac{\pi}{2}$$

and Eq. 7.27 becomes

$$\overline{u_N^2(t)} = \left(\frac{K}{2\pi}\right)^2 K_D^2 \frac{N_0}{A^2} \left(W - \frac{K}{2}\right) \cong K_D^2 \left(\frac{K}{2\pi}\right)^2 \frac{N_0 W}{A^2}$$

The -3-dB cutoff frequency being equal to $K/2\pi$, we shall assume this value to correspond to the maximum frequency of the signal to transmit, so that

$$\overline{u_N^2(t)} = K_D^2 \frac{N_0}{A^2} F_m^2 W \qquad (7.28)$$

The output noise power in this case far exceeds that corresponding to a standard discriminator followed by a low-pass filter with a sharp F_m cutoff (Eq. 7.25). But this is only because the slope of the low-pass filter in its attenuation zone (-6 dB/octave) is insufficient to suppress the noise, whose spectral density is in F^2 over the frequency interval $(-W/2, +W/2)$. A standard discriminator followed by a low-pass filter having the same characteristics produces the same result.

It is also worth noting that it is inadvisable to choose W so that $\pi W \gg K$. In this case, the maximum modulation frequency is far smaller than $W/2$ and the maximum frequency deviation, given for instance by $2\pi\Delta f_i/K \leqslant \frac{1}{2}$ rad, is such that $\Delta f_i \ll W/4$. The bandwidth W is then far too large as compared with the frequency deviation Δf_i and the maximum frequency to be transmitted F_m. Since the lattter two magnitudes are fixed, it is better to reduce W and increase K, which will bring us back to case (a). If the demodulator is followed by a low-pass filter, with a sharp cutoff F_m, we should then get the same results as with a standard discriminator.

7.6.3 SECOND-ORDER LOOP WITH LOW-PASS FILTER

We have seen that with this type of loop filter, we can take as discriminator output either the phase detector output signal $u_1(t)$ or the VCO modulation signal $u_2(t)$. The second of these two cases (see Section 6.4) is similar to that dealt with in Section 7.6.4. When signal $u_1(t)$ is used as output signal, the loop behaves like a discriminator of slope $K_D = 2\pi/K_3$ followed by a filter, the power transfer function of which can be derived from Eq. 6.19 or Eq. 6.21:

$$|L(j\Omega)|^2 = \frac{1 + \dfrac{1}{4\zeta^2}\dfrac{\Omega^2}{\omega_n^2}}{\left(1 - \dfrac{\Omega^2}{\omega_n^2}\right)^2 + 4\zeta^2\dfrac{\Omega^2}{\omega_n^2}}$$

$$= \frac{1 + \alpha^2\dfrac{\Omega^2}{K^2}}{\left(1 - \alpha\dfrac{\Omega^2}{K^2}\right)^2 + \dfrac{\Omega^2}{K^2}}$$

The parameter α is related to the low-pass filter time constant τ_1 and the loop gain K by the relation $\tau_1 = \alpha/K$. We saw in Section 6.4 that the optimum value of α is $\frac{1}{2}$, corresponding to $\zeta = \dfrac{\sqrt{2}}{2}$ and $\omega_n = K\sqrt{2}$. Under these conditions, the amplitude transfer function is nearly flat up to $\Omega = \omega_n/\sqrt{2} = K$. The output noise power is

$$\overline{u_N^2(t)} = K_D^2 \int_{-W/2}^{+W/2} \frac{N_0}{A^2} F^2 |L(j2\pi F)|^2 \, dF$$

which can be formulated as follows, when $\pi W \gg \omega_n$:

$$\overline{u_N^2(t)} = K_D^2 \frac{N_0}{A^2} \left(\frac{\omega_n}{2\pi}\right)^3 \left(\frac{2\pi W}{4\zeta^2 \omega_n} + \frac{2\pi}{16\zeta^3}\right)$$

For this type of loop, $\omega_n/2\zeta = K$ and we can write

$$\overline{u_N^2(t)} = K_D^2 \frac{N_0}{A^2} \left(\frac{K}{2\pi}\right)^2 \left(W + \frac{K}{2}\right) \qquad (7.29)$$

Since the quantity $K/2$ is negligible as compared with W, bearing in mind the calculation hypothesis ($\pi W \gg \omega_n, \omega_n = K\sqrt{2}$), the formula obtained is comparable to Eq. 7.28. The same remarks regarding the use of this loop as a discriminator are also applicable.

7.6.4   SECOND-ORDER LOOP WITH INTEGRATOR AND CORRECTION OR WITH LOW-PASS FILTER AND CORRECTION

In this case, the discriminator slope is still $K_D = 2\pi/K_3$ and the power transfer function of the low-pass filter equivalent to the loop, when $u_2(t)$ is used as output signal, is

$$|H(j\Omega)|^2 = \frac{1 + 4\zeta^2\dfrac{\Omega^2}{\omega_n^2}}{\left(1 - \dfrac{\Omega^2}{\omega_n^2}\right)^2 + 4\zeta^2\dfrac{\Omega^2}{\omega_n^2}}$$

As this transfer function curve involves an overshoot in the useful bandwidth, a supplementary filtering is effected with a low-pass filter of time constant τ_2 (see Section 6.4) and the transfer function becomes

$$|G(j\Omega)|^2 = \left[\left(1 - \frac{\Omega^2}{\omega_n^2}\right)^2 + 4\zeta^2\frac{\Omega^2}{\omega_n^2}\right]^{-1}$$

This transfer function, which also corresponds to that of a loop with filter $F(s) = 1/(1 + \tau_1 s)$ when $u_2(t)$ is used as output signal, is shown in Fig. 6.9. The value $\zeta = \sqrt{2}/2$ corresponds to a maximally flat response curve with a -3-dB cutoff frequency of $\omega_n/2\pi$.

The noise power at the discriminator output,

$$\overline{u_N^2(t)} = K_D^2 \int_{-W/2}^{+W/2} \frac{N_0}{A^2} F^2 |G(j2\pi F)|^2 \, dF$$

can, using the hypothesis $\pi W \gg \omega_n$, be expressed as

$$\overline{u_N^2(t)} = K_D^2 \frac{N_0}{A^2} \left(\frac{\omega_n}{2\pi}\right)^3 \frac{\pi}{2\zeta}$$

If the natural angular frequency is equal to $2\pi F_m$ and if $\zeta = \sqrt{2}/2$, we get

$$\overline{u_N^2(t)} = K_D^2 \frac{N_0}{A^2} \frac{\pi}{\sqrt{2}} F_m^3 \tag{7.30}$$

If we compare this formula with Eq. 7.25, we observe that the coefficient $\frac{2}{3}$ is here replaced by $\pi/\sqrt{2}$, which implies a noise power increase of $+5.2$

dB. But, the sharp F_m cutoff value of Section 7.6.1 is optimistic in that this type of filter would certainly distort the output signal. Depending on the modulation signal spectrum (see Remark 2 below), and with the same acceptable level of distortion, we can take $\omega_n < 2\pi F_m$ and in this case the phase-locked loop is a discriminator identical in performance with a standard discriminator followed by a low-pass filter with transfer function $G(j\Omega)$.

Since the transfer function $G(j\Omega)$ is the same for both types of discriminator obtained using:

- either a loop with filter $1/(1 + \tau_1 s)$ and output $u_2(t)$;
- or a loop with filter $(1 + \tau_2 s)/\tau_1 s$ [or $(1 + \tau_2 s)/(1 + \tau_1 s)$ if K is large] and output $u_2(t)$ followed by a low-pass filter $1/(1 + \tau_2 s)$, performances are identical as regards output signal and output signal-to-noise ratio. However, another point of comparison could be the instantaneous phase error in the loop, for instance, when $\zeta = \sqrt{2}/2$ (maximally flat response curve).

Regarding the VCO phase variance due to the input noise, we have to use:

- for filter $1/(1 + \tau_1 s)$, Eq. 7.12:

$$\sigma_{\varphi_0}^2 = \frac{N_0}{A^2} \frac{\omega_n}{4\zeta} = \frac{N_0}{A^2} \frac{\omega_n}{2\sqrt{2}}$$

- for filter $(1 + \tau_2 s)/\tau_1 s$, Eq. 7.11:

$$\sigma_{\varphi_0}^2 = \frac{N_0}{A^2} \frac{\omega_n}{4\zeta}(1 + 4\zeta^2) = \frac{N_0}{A^2} \frac{3\omega_n}{2\sqrt{2}}$$

As regards the phase error due to the input signal modulation, the results in Section 6.3 and in particular the curves of Fig. 6.16 and 6.17, show that the maximum phase error (the frequency deviation being constant) occurs:

- in the case of filter $1/(1 + \tau_1 s)$, for $x = \Omega/\omega_n \cong \sqrt{2}/2$ and equals

$$\Delta\phi_M \cong \sqrt{2} \; \frac{\Delta\omega_i}{\omega_n}$$

- in the case of filter $(1 + \tau_2 s)/\tau_1 s$, for $x = \Omega/\omega_n = 1$ and equals

$$\Delta\phi_M = \frac{1}{\sqrt{2}} \frac{\Delta\omega_i}{\omega_n}$$

For a given natural angular frequency ω_n, the use of a loop filter of the type $1/(1+\tau_1 s)$ is better as regards noise ($\sigma_{\varphi_0}^2$ three times weaker) but not so good as regards modulation (maximum phase error $\Delta\phi_M$ twice as large), which makes it difficult to compare them without a more accurate definition of the modulation spectrum. We can say that for a given maximum phase error $\Delta\phi_M$ due to the modulation, even though the error due to input noise is greater, with the $(1+\tau_2 s)/\tau_1 s$ filter, we can use a frequency deviation twice as large and thus obtain a discriminator output signal-to-noise ratio four times as large. The use of this type of filter rather than the $1/(1+\tau_1 s)$ filter may thus, in certain cases, be preferable.

Remark 1. The use of these two discriminators, assuming $\omega_n \ll \pi W$, is only possible if the frequency deviation for the high-frequency modulation signals is low enough. If the natural angular frequency ω_n equals $2\pi F_m$ (point at -3 dB on the response curve $G(j\Omega)$ for $\zeta = \sqrt{2}/2$), the possible angular frequency deviation is $\sqrt{2}\,\Delta\phi_M\omega_n$ [filter $(1+\tau_2 s)/\tau_1 s$] or $(\Delta\phi_M/\sqrt{2})\omega_n$ [filter $1/(1+\tau_1 s)$]. Since the maximum acceptable phase error is generally between $\frac{1}{2}$ and 1 radian, the possible frequency deviation is relatively small. If it should be large, we have to take $\omega_n \gg 2\pi F_m$ and follow the discriminators by a low-pass filter having an F_m cutoff, more or less abrupt according to the acceptable distortion level. The signal-to-noise ratio formula applicable in this case is Eq. 7.26 and the discriminator performances are identical, apart from threshold phenomena.

Remark 2. When the modulation signal spectrum is such that the amplitude of the high-frequency modulation components is lower than that of the low-frequency modulation components, parameter optimization is feasible.

It can even be shown (29) that if the input frequency modulation spectrum is

$$S_{f_i}(\Omega) = \frac{C}{\Omega^2 + a^2}$$

(C and a being constant) a phase-locked loop with loop filter $(1+\tau s)/(1+\tau' s)$ corresponds to an optimum Wiener filter (see Section 7.3):

- as regards the input signal phase, if we consider the VCO phase as the output signal (phase-error reduction)
- and as regards the input signal frequency, if we consider the VCO command signal (or VCO frequency) as the output signal, provided that a supplementary low-pass filtering $1/(1+\tau'' s)$ is included at the output.

As in the cases reviewed in Section 7.3, the choice of constants τ, τ', and τ'' depends both on the modulation constants C and a and the signal-to-noise density ratio A^2/N_0. Strictly speaking, then, optimization is only possible for a single value of the input power (generally the lowest). The use of an automatic gain control or limiter preceding the loop enables us to keep very close to the optimum over a large variation range of input signal power (30).

NUMERICAL EXAMPLE.   A signal of power $S = A^2/2 = -108$ dBm is applied to a receiver . 'nput so that $N_0 = -168$ dBm. It is frequency modulated by a low-frequency sinusoidal signal (frequency range: 40 Hz to 10 kHz).

The signal is demodulated using a second-order phase-locked loop such that $\zeta = \sqrt{2}\,/2$. The curves of Fig. 6.9 show that the discriminator -3-dB cutoff frequency is $\omega_n/2\pi$. We shall consequently choose $\omega_n = 2\pi \times 10^4$ rad/s.

CASE A   We use a loop filter $(1 + \tau_2 s)/(1 + \tau_1 s)$. If we choose the components so that the loop gain K is greater than ω_n ($K \geqslant 2\pi \times 10^6$ rad/s) and use an additional low-pass filter $1/(1 + \tau_2 s)$, the discriminator response is given by the curves in Fig. 6.9.

The phase detector is of the sawtooth type, allowing a maximum modulation-induced phase error $\Delta\phi_M = 2$ rad. The frequency deviation Δf_i can then be chosen so that (Fig. 6.16)

$$\frac{\omega_n}{2\pi\Delta f_i}\Delta\phi_M = \frac{\sqrt{2}}{2}$$

implying that $2\pi\Delta f_i = 2\sqrt{2}\,\omega_n$, that is, $\Delta f_i = 2\sqrt{2}\,\times 10^4$ Hz.

Since the noise power is given by Eq. 7.30 and the signal power is $K_D^2 \Delta f_i^2/2$, the signal-to-noise ratio at the discriminator output is

$$\left(\frac{S}{N}\right)_D = \frac{\Delta f_i^2}{2}\frac{\sqrt{2}\,A^2}{N_0\pi F_m^3}$$

$$= \frac{A^2}{2N_0 F_m}\frac{\sqrt{2}}{\pi}\frac{\Delta f_i^2}{F_m^2}$$

that is,

$$\left(\frac{S}{N}\right)_D \Rightarrow -108 + 168 - 40 + 1.5 - 5 + 9 = +25.5 \text{ dB}$$

The VCO variance due to noise $\sigma_{\varphi_0}^2$, given by Eq. 7.11, is 3.4×10^{-2} rad^2, that is, $\sigma_{\varphi_0} = 0.184$ rad. If we choose $\Delta\phi_M = 2$ rad, we are sure that, for the type of phase detector considered, we shall stay in the linear operation zone.

CASE B.   We use a loop filter $1/(1 + \tau_1 s)$ and the VCO modulation signal as output. The discriminator response curve also corresponds to Fig. 6.9.

Taking into account ζ and ω_n chosen, K will have to be such that

$$K = \frac{\omega_n}{2\zeta} = \frac{2\pi}{\sqrt{2}} \times 10^4 \text{ rad/s}$$

The frequency deviation Δf_i is chosen so that the modulation-induced phase error will not exceed 2 rad (sawtooth phase detector). This leads to (Fig. 6.17)

$$\frac{\omega_n}{2\pi\Delta f_i} \Delta \phi_M \cong \sqrt{2}$$

implying that $2\pi\Delta f_i = (2/\sqrt{2})\omega_n$, that is, $\Delta f_i = \sqrt{2} \times 10^4$ Hz.

The discriminator output signal-to-noise ratio is given by the same formula as in case A above. The numerical calculation gives

$$\left(\frac{S}{N}\right)_D \Rightarrow -108 + 168 - 40 + 1.5 - 5 + 3 = +19.5 \text{ dB}$$

The VCO phase variance due to noise, given by Eq. 7.12, is $\sigma_{\varphi_0}^2 = 1.13 \times 10^{-2}$ rad^2, leading to $\sigma_{\varphi_0} = 0.106$ rad. Although this value is lower than in case A, it would be unrealistic to accept a higher $\Delta\phi_M$ [to increase Δf_i and $(S/N)_D$] for, owing to the value of K, this loop will be far more sensitive to an input signal frequency drift (or VCO central frequency drift).

CHAPTER 8
RESPONSE
TO RANDOM
MODULATIONS

In the previous chapter, we discussed the way in which an additive noise influenced phase-locked loop linear operation. In the course of this investigation, we calculated the VCO phase fluctuation standard deviation when the input signal is a pure sinusoid and the VCO a perfect oscillator. We also saw in Chapter 6 that the loop can be used as a phase demodulator, providing that the bandwidth of the filter equivalent to the loop is narrow and the modulation index is not too high. When the signal is frequency modulated the loop has to be chosen so that the equivalent filter bandwidth is wide enough to follow the modulation, so that the phase error will remain slight. The loop can then be used as a discriminator.

It would now be interesting to see what happens when the input signal is modulated by random processes, applied voluntarily, in the case of a modulation to be transmitted, or involuntarily, in the case of a short-term frequency stability problem. It would also be interesting to observe the influence of VCO short-term frequency fluctuations on the linear operation of the loop. This will enable us, to a certain extent, to determine whether the phase-locked loop could be effectively used for a certain type of signal and, possibly, to evaluate the loop performances.

8.1 INFLUENCE OF THE INPUT SIGNAL MODULATION

In the preceding chapter, the expression of the signal applied at the loop input was $y_i(t) = A \sin(\omega t + \theta_i) + n(t)$ where $n(t)$ represents an additive noise. When this noise has certain properties and when the phase detector is of the analog multiplier type (or multiplier by the reference signal sign), we have shown that the input signal can be expressed in an equivalent form $y_i(t)$ $= A \sin[\omega t + \theta_i + n'(t)]$ where $n'(t)$ is an equivalent phase modulation representing the effect of the additive noise. From this we were able to derive the VCO phase fluctuations $\varphi_0(t)$ due to noise $n(t)$. But we also observed that the phase detector output signal was not $K_1 \sin[\theta_i - \varphi_0(t) + n'(t)]$ but $K_1 \sin[\theta_i - \varphi_0(t)] + K_1 n'(t)$, providing, of course, that $\varphi_0(t)$ varies very slowly as compared with the variations of processes $n_1(t)$ and $n_2(t)$ (Section 7.2).

We shall now turn our attention to the case where the input signal is expressed by $y_i(t) = A \sin[\omega t + \phi_i(t)]$, where $\phi_i(t)$ is a random process. Here, the phase detector output signal is really $K_1 \sin[\phi_i(t) - \phi_0(t)]$ so that it is the quantity $[\phi_i(t) - \phi_0(t)]$ which has to stay small enough for the linearity hypothesis to hold, whereas in the previous chapter, it was only the quantity $\varphi_0(t)$ that had to stay small, independently of $n'(t)$.

Two possibilities may be encountered. If the process $\phi_i(t)$ is such that $\sigma^2_{\phi_i}$ is small enough, we can use a "narrow" loop, where the VCO does not follow the $\phi_i(t)$ variations but remains in synchronism with the pseudosignal $A \sin[\omega t + \overline{\phi_i(t)}]$.

But if the $\phi_i(t)$ variations are large, in order to stay in the linear operation zone, it is absolutely necessary for the VCO to follow insofar as possible the $\phi_i(t)$ variations. The loop bandwidth should therefore be chosen fairly "large," so that the VCO stays in synchronism with the signal $A \sin[\omega t + \phi_i(t)]$ with a phase error $\phi(t) = \phi_i(t) - \phi_0(t)$ that we shall calculate for various cases.

We shall assume that the modulation signal can in all cases be expressed as $\phi_i(t) = \alpha(t) + \theta_i$, where $\alpha(t)$ is a random process having a null mean value (if the mean value is not null, it can always be included in the constant θ_i). The VCO steady-state mean phase value is equal to the input signal mean phase value, to within $\pi/2$, and the VCO output signal can be expressed as

$$y_0(t) = B \cos\left[\omega t + \theta_i + \phi_0(t)\right]$$

where $\phi_0(t)$ is a version filtered by $H(j\omega)$ of the $\alpha(t)$ variations of the input phase. The instantaneous phase error $\phi(t) = \alpha(t) - \phi_0(t)$ is a version filtered by $[1 - H(j\omega)]$ of the $\alpha(t)$ variations (see Eqs. 3.6 and 3.7).

In particular, if we call $S_\alpha(f)$ the power spectral density of the process $\alpha(t)$, the power spectral densities $S_{\phi_0}(f)$ and $S_\phi(f)$ of the processes $\phi_0(t)$ and $\phi(t)$ are

$$S_{\phi_0}(f) = S_\alpha(f)|H(j2\pi f)|^2 \tag{8.1}$$

$$S_\phi(f) = S_\alpha(f)|1 - H(j2\pi f)|^2 \tag{8.2}$$

Since the filters $H(j\omega)$ and $[1 - H(j\omega)]$ are linear filters, if the process $\alpha(t)$ is a stationary Gaussian process, processes $\phi_0(t)$ and $\phi(t)$ are also stationary and Gaussian. Since the process $\alpha(t)$ has a null mean value, therefore

$$\overline{\phi_0(t)} = \overline{\phi(t)} = \overline{\alpha(t)} = 0$$

Finally, the process $\phi_0(t)$ and $\phi(t)$ variances are

$$\sigma^2_{\phi_0} = \int_{-\infty}^{+\infty} S_\alpha(f)|H(j2\pi f)|^2 df \qquad (8.3)$$

$$\sigma^2_\phi = \int_{-\infty}^{+\infty} S_\alpha(f)|1 - H(j2\pi f)|^2 df \qquad (8.4)$$

8.1.1 PHASE MODULATION BY A GAUSSIAN WHITE NOISE

The two-sided spectral density of process $\alpha(t)$ is shown in Fig. 8.1:

$$S_\alpha(f) = \frac{\nu}{2} \; (\text{rad}^2/\text{Hz}), \qquad -\infty < f < +\infty$$

Then

$$\sigma^2_\alpha = \overline{\alpha^2(t)} = \int_{-\infty}^{+\infty} S_\alpha(f) df$$

is infinite and a phase-locked loop cannot be locked on the signal $A \sin[\omega t + \theta_i + \alpha(t)]$. In this case, the only possible solution would be to construct a loop such that $|H(j2\pi f)|^2 \equiv 1$, so that the phase error itself would not be infinite.

But a spectral density $S_\alpha(f)$ uniform from $-\infty$ to $+\infty$ is not very realistic, since it would require infinite bandwidth to transmit the signal $y_i(t)$. We shall therefore consider the case where $S_\alpha(f) = \nu/2$ for $-\mathcal{B} < f < +\mathcal{B}$ (Fig. 8.2), so that

$$\sigma^2_\alpha = \overline{\alpha^2(t)} = \int_{-\mathcal{B}}^{+\mathcal{B}} \frac{\nu}{2} df = \nu \mathcal{B}$$

is a finite quantity. Two cases are possible:

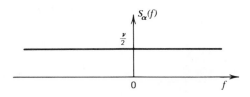

FIGURE 8.1. Power spectral density of the white-noise phase modulation process.

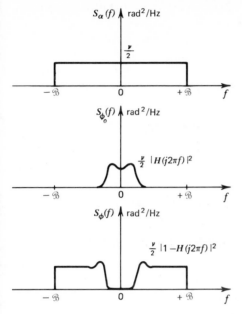

FIGURE 8.2. Power spectral densities $S_{\phi_0}(f)$ and $S_\phi(f)$ when $S_\alpha(f)$ is a bandlimited white noise.

1. If $\sigma^2_{\phi_i} = \sigma^2_\alpha$ is small enough, we can choose $H(j\omega)$ so that the VCO phase variance $\sigma^2_{\phi_0}$, given by Eq. 8.3,

$$\sigma^2_{\phi_0} = \int_{-\mathcal{B}}^{+\mathcal{B}} \frac{\nu}{2} |H(j2\pi f)|^2 \, df \cong \frac{\nu}{2}(2B_n) \qquad (8.5)$$

is as small as required [it should be remembered that we are working on the assumption that the modulation $\alpha(t)$ is the only source of error in the loop]. Consequently, the VCO output signal is expressed as

$$y_0(t) \cong B \cos(\omega t + \theta_i)$$

that is to say that the VCO is in synchronism with a pseudo input signal $A \sin(\omega t + \theta_i)$, unhampered by the modulation $\alpha(t)$.

The phase error variance σ^2_ϕ is given by Eq. 8.4:

$$\sigma^2_\phi = \int_{-\mathcal{B}}^{+\mathcal{B}} \frac{\nu}{2} |1 - H(j2\pi f)|^2 \, df \leqslant \nu \mathcal{B}$$

We then have simply to check that σ^2_α is small enough to ensure linear operation [this is necessary in particular if we also want processes $\phi_0(t)$ and $\phi(t)$ to be Gaussian].

2. However, if $\sigma^2_{\phi_i} = \sigma^2_\alpha$ is not small enough, in accordance with the same

linearity requirement, we have to choose $H(j\omega)$ so that σ_ϕ^2 remains small enough. In this case, Eq. 8.5 becomes

$$\sigma_{\phi_0}^2 \cong \frac{\nu}{2}(2B_n) \cong \frac{\nu}{2}(2\mathcal{B})$$

The VCO stays approximately in synchronism with the input signal, and the output signal is expressed as

$$y_0(t) = B\cos\left[\omega t + \theta_i + \phi_0(t)\right]$$

with

$$\phi_0(t) \cong \alpha(t)$$

The spectral densities $S_\phi(f)$ and $S_{\phi_0}(f)$ are represented in Fig. 8.2.

8.1.2   PHASE MODULATION BY A GAUSSIAN FLICKER NOISE

We call flicker noise a random process $\alpha(t)$ such that its two-sided power spectral density is

$$S_\alpha(f) = \frac{\chi}{|f|}$$

where χ is a constant expressed in rad^2. For power and bandwidth limitation reasons, an f^{-1} law cannot be applicable between $f=0$ and $f=\infty$, for

$$\sigma_\alpha^2 = 2\int_0^\infty S_\alpha(f)\,df = 2\chi[\ln f]_0^\infty \to \infty$$

It is then advisable to restrict the high-frequency spectrum, assuming that $S_\alpha(f)=0$ for $|f| > \mathcal{B}$ and also to assume that for $|f| \leqslant \varepsilon$, the spectral density remains constant and equal to χ/ε (Fig. 8.3).

Under these conditions,

$$\sigma_\alpha^2 = 2\int_0^\varepsilon \frac{\chi}{\varepsilon}\,df + 2\int_\varepsilon^\mathcal{B} \chi\frac{df}{f}$$

that is,

$$\sigma_\alpha^2 = 2\chi\left(1 + \ln\frac{\mathcal{B}}{\varepsilon}\right)$$

If the upper limitation $\mathcal{B}$ is easy to understand (natural bandwidth limitation), the lower limitation ε is necessary if we want process $\alpha(t)$ variance to be finite. The values proposed for ε generally range from one

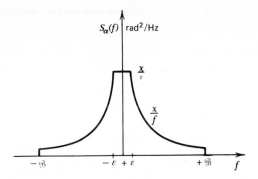

FIGURE 8.3.   Power spectral density of the flicker noise.

cycle per day to one cycle per year or even less. The model used here for the law $S_\alpha(f)$ when $|f| < \varepsilon$ is also open to criticism, but at least guarantees that the process is stationary and that the autocorrelation function $R_\alpha(\tau)$ exists (31), the latter being implicitly necessary to calculate σ_ϕ^2 and $\sigma_{\phi_0}^2$.

In any case, since ε is certainly very low, we can expect $\sigma_{\phi_i}^2 = \sigma_\alpha^2$ to be large (except if $\mathcal{B}$ is also a very low frequency). Consequently, it will probably be necessary to use a wide-bandwidth loop, so that the VCO will stay in synchronism with the modulated signal. The VCO phase variance, taking into account the $S_\alpha(f)$ representation in Fig. 8.3 and Eq. 8.3, is given by

$$\sigma_{\phi_0}^2 = 2\int_0^\varepsilon \frac{\chi}{\varepsilon}|H(j2\pi f)|^2\,df + 2\int_\varepsilon^{\mathcal{B}} \chi\frac{|H(j2\pi f)|^2}{f}\,df \qquad (8.6)$$

The phase error variance, taking account of Eq. 8.4 is expressed as

$$\sigma_\phi^2 = 2\int_0^\varepsilon \frac{\chi}{\varepsilon}|1 - H(j2\pi f)|^2\,df + 2\int_\varepsilon^{\mathcal{B}} \chi\frac{|1 - H(j2\pi f)|^2}{f}\,df \qquad (8.7)$$

We shall expand these quantities only in the case of a second-order loop with loop filter $(1 + \tau_2 s)/(1 + \tau_1 s)$. The loop transfer and error functions are then given by Eqs. 3.53 and 3.54, from which we derive

$$|H(j\omega)|^2 = \frac{\omega_n^4 + \left(2\zeta\omega_n - \omega_n^2/K\right)^2\omega^2}{\left(\omega_n^2 - \omega^2\right)^2 + 4\zeta^2\omega_n^2\omega^2}$$

$$|1 - H(j\omega)|^2 = \frac{\left(\omega_n^4/K^2\right)\omega^2 + \omega^4}{\left(\omega_n^2 - \omega^2\right)^2 + 4\zeta^2\omega_n^2\omega^2}$$

If we call $D = (\omega_n^2 - \omega^2)^2 + 4\zeta^2\omega_n^2\omega^2$ and change the variable $\omega = 2\pi f$, Eqs. 8.6 and 8.7 become

$$\sigma_{\phi_0}^2 = \frac{2\chi\omega_n^4}{2\pi\varepsilon}\int_0^{2\pi\varepsilon}\frac{d\omega}{D}$$

$$+ \frac{2\chi}{2\pi\varepsilon}\left(2\zeta\omega_n - \frac{\omega_n^2}{K}\right)^2\int_0^{2\pi\varepsilon}\frac{\omega^2\,d\omega}{D} + 2\chi\omega_n^4\int_{2\pi\varepsilon}^{2\pi\mathcal{B}}\frac{d\omega}{\omega D}$$

$$+ 2\chi\left(2\zeta\omega_n - \frac{\omega_n^2}{K}\right)^2\int_{2\pi\varepsilon}^{2\pi\mathcal{B}}\frac{\omega\,d\omega}{D}$$

and

$$\sigma_\phi^2 = \frac{2\chi}{2\pi\varepsilon}\int_0^{2\pi\varepsilon}\frac{\omega^4\,d\omega}{D}$$

$$+ \frac{2\chi}{2\pi\varepsilon}\frac{\omega_n^4}{K^2}\int_0^{2\pi\varepsilon}\frac{\omega^2\,d\omega}{D} + 2\chi\int_{2\pi\varepsilon}^{2\pi\mathcal{B}}\frac{\omega^3\,d\omega}{D}$$

$$+ 2\chi\frac{\omega_n^4}{K^2}\int_{2\pi\varepsilon}^{2\pi\mathcal{B}}\frac{\omega\,d\omega}{D}$$

The calculation can be completed with the help of an Integral Table. In particular, the expression σ_ϕ^2 can be given when $2\pi\varepsilon \to 0$ (it will be noted that for σ_ϕ^2, no calculation problem is involved if we assume $\varepsilon = 0$, since the error function $[1 - H(j2\pi f)]$ behaves like a high-pass filter of slope $+1$ as f approaches 0). The expression below is only valid if $\zeta > 1$:

$$\sigma_\phi^2 = \frac{\chi}{4\omega_n^2\zeta\sqrt{\zeta^2-1}}\left[b^2\ln\frac{(2\pi\mathcal{B})^2 + b^2}{b^2} - a^2\ln\frac{(2\pi\mathcal{B})^2 + a^2}{a^2}\right]$$

$$+ \frac{\chi(\omega_n^2/K^2)}{4\zeta\sqrt{\zeta^2-1}}\left[\ln\frac{(2\pi\mathcal{B})^2 + a^2}{a^2} - \ln\frac{(2\pi\mathcal{B})^2 + b^2}{b^2}\right]$$

with

$$b^2 = \omega_n^2\left(\zeta + \sqrt{\zeta^2-1}\right)^2 \quad \text{and} \quad a^2 = \omega_n^2\left(\zeta - \sqrt{\zeta^2-1}\right)^2$$

When $2\pi \mathcal{B} \gg b > a$, σ_ϕ^2 can be expressed in the form

$$\sigma_\phi^2 = 2\chi \ln \frac{2\pi \mathcal{B}}{\omega_n}$$

$$+2\chi \frac{\left(\zeta - \sqrt{\zeta^2 - 1}\right)^2 \ln\left(\zeta - \sqrt{\zeta^2 - 1}\right) - \left(\zeta + \sqrt{\zeta^2 - 1}\right)^2 \ln\left(\zeta + \sqrt{\zeta^2 - 1}\right)}{4\zeta\sqrt{\zeta^2 - 1}}$$

$$+ \frac{\chi\omega_n^2 \ln\left(\zeta + \sqrt{\zeta^2 - 1}\right)}{K^2\zeta\sqrt{\zeta^2 - 1}}$$

The third term is very low in the case of high loop gain K (it is null for a loop with integrator and correction); then σ_ϕ^2 consists of a term depending on the ratio $\mathcal{B}/\omega_n$ and a term depending on the damping factor ζ.

If the coefficient χ is very low and the bandwidth $\mathcal{B}$ not too large, ω_n can be chosen reasonably small without incurring a large phase error. The VCO is then practically in synchronism with the pseudosignal $A \sin(\omega t + \theta_i)$, to within a random phase error $\phi_0(t)$, whose variance can be calculated.

But if the bandwidth $\mathcal{B}$ is large, for the loop behavior to stay in the linear domain, the natural angular frequency ω_n must be large enough for σ_ϕ^2 to remain low. In this case, it is obvious that the VCO tends to be in synchronism with the real signal $A \sin[\omega t + \theta_i + \alpha(t)]$ applied at the loop input, rather than with the pseudosignal $A \sin(\omega t + \theta_i)$.

8.1.3 FREQUENCY MODULATION BY A GAUSSIAN WHITE NOISE

The signal applied to the loop input is frequency modulated by $f_i(t)$, a Gaussian noise, having a power spectral density that is uniform for $-\mathcal{B} < f < \mathcal{B}$ and null elsewhere:

$$S_{f_i}(f) = \frac{\eta}{2} \ (\mathrm{Hz^2/Hz}), \quad -\mathcal{B} < f < +\mathcal{B}$$
$$= 0 \qquad\qquad \text{elsewhere}$$

Since the phase is the integral of the instantaneous frequency, the phase modulation $\alpha(t)$ is a random process, having a spectral density

$$S_\alpha(f) = \frac{\eta}{2} \frac{1}{f^2} \ (\mathrm{rad^2/Hz}), \quad -\mathcal{B} < f < +\mathcal{B}$$
$$= 0 \qquad\qquad \text{elsewhere}$$

The integral used to calculate the variance of process $\alpha(t)$:

$$\sigma_\alpha^2 = \int_{-\mathcal{B}}^{+\mathcal{B}} \frac{\eta}{2} \frac{df}{f^2}$$

diverges for $f=0$ [but not for f approaching infinity; the upper boundary $\mathcal{B}$, only included to limit the power of the modulation process $f_i(t)$, can be omitted without complicating the calculation]. Since the $\alpha(t)$ process variance is infinite, very narrow loops cannot be locked to the input signal. The loop bandwidth has to be large enough for the VCO to be able to follow most of the variations of $\alpha(t)$ so that the phase error $\phi(t) = \alpha(t) - \phi_0(t)$ will remain small enough. Taking into account $S_\alpha(f)$ and Eq. 8.4, the phase error variance is expressed as

$$\sigma_\phi^2 = 2 \int_0^{\mathcal{B}} \frac{\eta}{2} \frac{|1 - H(j2\pi f)|^2}{f^2} df \quad \text{(in rad}^2\text{)} \tag{8.8}$$

Using the expression $|1 - H(j2\pi f)|^2$ corresponding to a second-order loop with low-pass filter and correction network, if we change the variable $\omega = 2\pi f$ and call D the denominator of $|1 - H(j\omega)|^2$:

$$\sigma_\phi^2 = 4\pi \frac{\eta}{2} \int_0^{2\pi \mathcal{B}} \frac{|1 - H(j\omega)|^2}{\omega^2} d\omega$$

that is,

$$\sigma_\phi^2 = 4\pi \frac{\eta}{2} \int_0^{2\pi \mathcal{B}} \frac{\omega^2 d\omega}{D} + 4\pi \frac{\eta}{2} \frac{\omega_n^4}{K^2} \int_0^{2\pi \mathcal{B}} \frac{d\omega}{D}$$

An Integral Table can be used to complete the calculation; assuming $2\pi \mathcal{B} \gg \omega_n$, we get

$$\sigma_\phi^2 = \frac{\eta}{2} \frac{\pi^2}{\zeta \omega_n} \left(1 + \frac{\omega_n^2}{K^2}\right)$$

If the condition $2\pi \mathcal{B} \gg \omega_n$ is not fulfilled, the phase error variance is smaller than the above quantity. If the loop filter is of the integrator plus correction type, the second term is discarded and

$$\sigma_\phi^2 = \frac{\eta}{2} \frac{\pi^2}{\zeta \omega_n} \tag{8.9}$$

This formula serves to confirm that if we widen the loop bandwidth by increasing ω_n, we can reduce the phase error.

8.1.4 FREQUENCY MODULATION BY A GAUSSIAN FLICKER NOISE

To limit the power of modulation process $f_i(t)$ and safeguard its autocorrelation function, we can reasonably state that

$$S_{f_i}(f) = \frac{\xi}{\varepsilon}, \qquad |f| < \varepsilon$$

$$= \frac{\xi}{|f|}, \qquad \varepsilon < |f| < \mathcal{B}$$

$$= 0, \qquad \text{elsewhere}$$

where ξ is a constant expressed in Hz^2. The power spectral density of process $\alpha(t)$ is then

$$S_\alpha(f) = \frac{\xi}{\varepsilon} \frac{1}{f^2}, \qquad |f| < \varepsilon$$

$$= \xi \frac{1}{|f| f^2}, \qquad \varepsilon < |f| < \mathcal{B}$$

$$= 0, \qquad \text{elsewhere}$$

The variance σ_α^2 is infinite. The input signal spectrum $S_{y_i}(f)$ is a continuous spectrum, characterizing a frequency modulation, rather than a discrete frequency line plus a continuous spectrum which characterizes a phase modulation. Consequently, as in the case discussed above, there is a minimum loop bandwidth to be observed if the VCO is to be able to follow the input phase variations $\alpha(t)$. If we set aside considerations concerning the modulation process $f_i(t)$ itself (see the observations in Section 8.1.2), as regards the calculation of σ_ϕ^2, the upper frequency limit at $\mathcal{B}$ is unnecessary, since the function to be integrated decreases fast enough when f approaches infinity. The low-frequency limit is not necessary either if the loop filter is of the integrator plus correction type, since, in this case, $|1 - H(j2\pi f)|^2$ behaves like f^4 when f approaches 0. For a low-pass filter plus correction, however, the lower limit is necessary, for $|1 - H(j2\pi f)|^2$ behaves like f^2 when f approaches 0. We shall calculate σ_ϕ^2 for the latter case, the first case result corresponding the same result with K infinite. Equation 8.4, taking into account $S_\alpha(f)$, becomes

$$\sigma_\phi^2 = 4\pi \int_0^{2\pi\varepsilon} \frac{\xi}{\varepsilon} \frac{|1 - H(j\omega)|^2}{\omega^2} d\omega$$

$$+ 8\pi^2 \int_{2\pi\varepsilon}^{\infty} \xi \frac{|1 - H(j\omega)|^2}{\omega^3} d\omega \qquad (8.10)$$

which can be split up into

$$\sigma_\phi^2 = 4\pi \int_0^{2\pi\varepsilon} \frac{\xi}{\varepsilon} \frac{\omega^2 d\omega}{D} + 4\pi \frac{\omega_n^4}{K^2} \int_0^{2\pi\varepsilon} \frac{\xi}{\varepsilon} \frac{d\omega}{D}$$

$$+ 8\pi^2 \int_{2\pi\varepsilon}^{\infty} \xi \frac{\omega d\omega}{D} + 8\pi^2 \frac{\omega_n^4}{K^2} \int_{2\pi\varepsilon}^{\infty} \xi \frac{d\omega}{\omega D}$$

If we calculate for the case where $\zeta > 1$ and assume ε to be very small (the first integral tends towards zero when $2\pi\varepsilon \to 0$),

$$\sigma_\phi^2 = 4\pi^2\xi \frac{2}{K^2} \left\{ 1 + \ln \frac{\omega_n}{2\pi\varepsilon} + \frac{1}{4\zeta\sqrt{\zeta^2 - 1}} \right.$$

$$\times \left[\frac{\ln\left(\zeta - \sqrt{\zeta^2 - 1}\right)}{\left(\zeta - \sqrt{\zeta^2 - 1}\right)^2} - \frac{\ln\left(\zeta + \sqrt{\zeta^2 - 1}\right)}{\left(\zeta + \sqrt{\zeta^2 - 1}\right)^2} \right] \right\}$$

$$+ 4\pi^2\xi \frac{2}{\omega_n^2} \frac{1}{4\zeta\sqrt{\zeta^2 - 1}} \ln \frac{\zeta + \sqrt{\zeta^2 - 1}}{\zeta - \sqrt{\zeta^2 - 1}}$$

For a loop with integrator and correction or with low-pass filter and correction, when the loop gain K is very large $[K^2 \gg \ln(\omega_n/2\pi\varepsilon)]$, the last term is the only one used:

$$\sigma_\phi^2 = \xi \frac{4\pi^2}{\omega_n^2} \frac{\ln\left(\zeta + \sqrt{\zeta^2 - 1}\right)}{\zeta\sqrt{\zeta^2 - 1}} \qquad (8.11)$$

When the damping factor ζ has any given value, the formula to be used is (32)

$$\sigma_\phi^2 = \xi \frac{4\pi^2}{\omega_n^2} f(\zeta) \tag{8.12}$$

with

$$
\begin{aligned}
f(\zeta) &= \frac{\ln\left(\zeta + \sqrt{\zeta^2 - 1}\right)}{\zeta\sqrt{\zeta^2 - 1}}, & \zeta &> 1 \\[2mm]
&= 1, & \zeta &= 1 \\[2mm]
&= \frac{\operatorname{Arc\,tan}\left(\sqrt{1 - \zeta^2}\,/\,\zeta\right)}{\zeta\sqrt{1 - \zeta^2}} & \zeta &< 1
\end{aligned}
\tag{8.13}
$$

In all cases, σ_ϕ is observed to decrease when ω_n increases, which confirms that when the input signal is frequency modulated, the loop performance can be improved by widening the bandwidth.

8.2   INFLUENCE OF VCO MODULATION

We shall now suppose that the signal applied to the loop is a nonmodulated sinusoidal signal $A \sin(\omega t + \theta_i)$ and that the signal from the VCO used as a free-running oscillator (outside the loop) is phase modulated by a random process $\theta(t)$, the power spectral density of which is known. We shall determine what happens to the VCO phase and the loop error signal when the loop is in lock (presuming linear operation). To this end, there are two ways of including the disturbance $\theta(t)$ in the loop: preceding or following the VCO. We shall see that the two methods are equivalent and produce the same results.

8.2.1   DISTURBANCE FOLLOWING THE VCO

In this case, we substitute for the VCO phase-modulated by $\theta(t)$, a perfect (nonmodulated) VCO followed by a phase modulator to which we apply the modulation $\theta(t)$ in accordance with the diagram in Fig. 8.4. If the perfect VCO output signal is expressed as $B\cos[\omega t + \phi_0(t)]$, the output from the fictitious modulator is expressed as $B\cos[\omega t + \phi_0(t) + \theta(t)]$, and this is the only physically accessible signal. Locating the operation a priori in the linear zone, we can assume θ_i to be constant. When $\theta(t) = 0$, the VCO output signal is in quadrature with the input signal and is expressed $B\cos[\omega t + \theta_i]$.

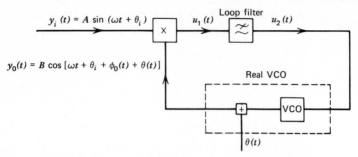

FIGURE 8.4. Model of a phase-locked loop when the VCO is parasitically phase modulated.

When the modulation $\theta(t)$ is applied, we are interested in the phase variations around θ_i. The real VCO signal is then expressed as $B\cos[\omega t + \theta_i + \phi_0(t) + \theta(t)]$, where $\theta(t)$ is the parasitic modulation and $\phi_0(t)$ the result of the loop action.

The error signal $u_1(t)$ is expressed as

$$u_1(t) = K_1 \big[\, \theta_i - \theta_i - \phi_0(t) - \theta(t)\,\big]$$

Let us suppose for a moment that the modulation $\theta(t)$ is nonrandom. This gives, in operational notation,

$$U_1(s) = -K_1 \big[\, \Phi_0(s) + \Theta(s)\,\big]$$

The VCO command signal is

$$U_2(s) = K_2 U_1(s) F(s)$$

that is,

$$U_2(s) = -K_1 K_2 F(s) \big[\, \Phi_0(s) + \Theta(s)\,\big]$$

If we assume that this signal frequency modulates the VCO, we get

$$s\Phi_0(s) = K_3 U_2(s) = -KF(s) \big[\, \Phi_0(s) + \Theta(s)\,\big]$$

which implies that

$$\frac{\Phi_0(s)}{\Theta(s)} = -\frac{KF(s)}{s + KF(s)} = -H(s) \tag{8.14}$$

But $\phi_0(t)$ represents the fictitious VCO phase modulation and the real VCO phase is*

$$\theta_0(t) = \phi_0(t) + \theta(t)$$

*It is worth noting that $\theta_0(t)$ is also the instantaneous phase error.

that is,

$$\Theta_0(s) = \Phi_0(s) + \Theta(s)$$

whence

$$\frac{\Theta_0(s)}{\Theta(s)} = 1 + \frac{\Phi_0(s)}{\Theta(s)} = 1 - H(s) \qquad (8.15)$$

When the VCO of a phase-locked loop [transfer function $H(s)$] is phase modulated by a modulation expressed $\Theta(s)$ with the loop open, the VCO closed-loop modulation is expressed $\Theta_0(s)$, so that

$$\Theta_0(s) = [1 - H(s)]\Theta(s)$$

Let us now suppose that modulation $\theta(t)$ is a random process. If we call $S_\theta(f)$ the power spectral density of this process, then the spectral density of process $\theta_0(t)$ is given by

$$S_{\theta_0}(f) = |1 - H(j2\pi f)|^2 S_\theta(f) \qquad (8.16)$$

In particular,

$$\sigma_{\theta_0}^2 = \int_{-\infty}^{+\infty} |1 - H(j2\pi f)|^2 S_\theta(f) \, df \qquad (8.17)$$

8.2.2 DISTURBANCE PRECEDING THE VCO

The VCO phase modulated by $\theta(t)$ is replaced by a perfect VCO, with the addition to the command signal $u_2(t)$ of a noise signal $n_2(t)$, producing with the loop open, modulation $\theta(t)$ (Fig. 8.5). The process $n_2(t)$ is then related to $\theta(t)$ by the equation

$$\frac{d\theta}{dt} = K_3 n_2(t)$$

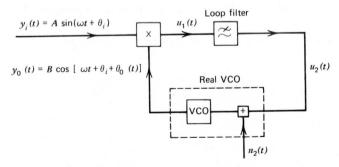

FIGURE 8.5. Model of a phase-locked loop when the VCO is parasitically frequency modulated.

The error signal $u_1(t)$ is expressed as

$$u_1(t) = -K_1\theta_0(t)$$

Let us suppose for a moment that modulation $n_2(t)$ [and therefore $\theta(t)$] is nonrandom. This gives, in operational notation,

$$U_1(s) = -K_1\Theta_0(s)$$

The loop filter output signal is written

$$U_2(s) = K_2F(s)U_1(s) = -K_1K_2F(s)\Theta_0(s)$$

and the VCO command signal expression is

$$U_2'(s) = U_2(s) + N_2(s)$$

This signal frequency modulates the VCO and consequently

$$s\Theta_0(s) = K_3[U_2(s) + N_2(s)]$$

and

$$\frac{\Theta_0(s)}{N_2(s)} = \frac{K_3}{s + KF(s)} \tag{8.18}$$

When the VCO of a phase-locked loop with a loop filter transfer function $F(s)$ is frequency modulated by a modulation $N_2(s)$, the resulting VCO closed-loop phase modulation is expressed by $\Theta_0(s)$, so that

$$\frac{\Theta_0(s)}{N_2(s)} = \frac{K_3}{s + KF(s)}$$

Let us now reconsider the case where modulation $n_2(t)$ is a random process. If we call $S_{n_2}(f)$ the spectral density of the modulation voltage $n_2(t)$ (in V^2/Hz), the spectral density of the resulting phase modulation (expressed in rad^2/Hz if K_3 is in $rad/s/V$) is expressed as

$$S_{\theta_0}(f) = \frac{K_3^2 S_{n_2}(f)}{|j2\pi f + KF(j2\pi f)|^2} \tag{8.19}$$

In particular,

$$\sigma_{\theta_0}^2 = K_3^2 \int_{-\infty}^{+\infty} \frac{S_{n_2}(f)}{|j2\pi f + KF(j2\pi f)|^2}\,df \tag{8.20}$$

Remark. When the loop is in lock, despite the fact that the VCO is a frequency-modulated oscillator, if an outside signal is applied to the VCO modulation input, the VCO is phase modulated by this outside signal. This phenomenon has a practical application in that it is a simple method of devising a phase modulator: the HF signal to be modulated is applied to the loop input, the modulation LF signal is added to the command signal $u_2(t)$, the HF modulated signal is the VCO output. The possible phase variation range is limited to the linearity zone of the phase detector used.

In the case of a first-order loop, the expressions above indicate a modulation sensitivity of $K_3/K = 1/K_1 K_2$ in rad/V, and a -3-dB bandwidth modulation of $(0, K/2\pi)$. To obtain a good modulation sensitivity, the phase detector sensitivity K_1 has to be low (we should have $K_2 = 1$). If we require a high loop gain K (if only to get a large modulation bandwidth), we need a VCO with a very high modulation sensitivity K_3.

8.2.3 EQUIVALENCE OF THE TWO METHODS

Let us state in operational notation that signal $n_2(t)$ produces modulation $\theta(t)$, with the loop open:

$$s\Theta(s) = K_3 N_2(s)$$

If we substitute this expression in Eq. 8.18, we get

$$\Theta_0(s) = \frac{K_3 N_2(s)}{s + KF(s)} = \frac{s\Theta(s)}{s + KF(s)} = [1 - H(s)]\Theta(s)$$

which is identical with Eq. 8.15.

Using another way, if we recall that

$$1 - H(j2\pi f) = \frac{j2\pi f}{j2\pi f + KF(j2\pi f)}$$

we can write

$$\frac{1}{|j2\pi f + KF(j2\pi f)|^2} = \frac{|1 - H(j2\pi f)|^2}{4\pi^2 f^2}$$

and Eq. 8.19 becomes

$$S_{\theta_0}(f) = \frac{K_3^2}{4\pi^2} \frac{S_{n_2}(f)}{f^2} |1 - H(j2\pi f)|^2$$

But, since the VCO modulation sensitivity is K_3 expressed in rad/s/V, the VCO frequency modulation spectral density $S_{f_0}(f)$ in Hz^2/Hz, is related to the modulation noise spectral density $S_{n_2}(f)$ in V^2/Hz, by the expression

$$S_{f_0}(f) = \left(\frac{K_3}{2\pi}\right)^2 S_{n_2}(f)$$

and consequently

$$S_{\theta_0}(f) = \frac{S_{f_0}(f)}{f^2} |1 - H(j2\pi f)|^2 = S_\theta(f)|1 - H(j2\pi f)|^2$$

which is identical to Eq. 8.16.

8.2.4 USE OF A PHASE-LOCKED LOOP TO CLEANUP A SIGNAL SPECTRUM

Let us consider a free-running VCO, where short-term frequency instability is such that it can be represented as a frequency modulation by a white noise, at least within a certain bandwidth $\mathcal{B}$. The spectral density of the resulting $\theta(t)$ random phase modulation is then given by

$$S_\theta(f) = \frac{\eta}{2} \frac{1}{f^2} \ (\text{rad}^2/\text{Hz}), \qquad |f| \leqslant \mathcal{B}$$
$$= 0, \qquad\qquad\qquad \text{elsewhere.}$$

The phase variance σ_θ^2 is infinite. The oscillator output signal features the continuous spectrum characteristic of frequency modulation.

Let us now suppose that the VCO is placed in a phase-locked loop where the input is a perfect sinusoidal signal (unmodulated). When the loop is in lock, the VCO phase modulation spectral density is given by Eq. 8.16:

$$S_{\theta_0}(f) = S_\theta(f)|1 - H(j2\pi f)|^2 = \frac{\eta}{2} \frac{|1 - H(j2\pi f)|^2}{f^2}$$

in particular,

$$\sigma_{\theta_0}^2 = \int_{-\mathcal{B}}^{+\mathcal{B}} \frac{\eta}{2} \frac{|1 - H(j2\pi f)|^2}{f^2} \ df$$

This expression is identical to Eq. 8.8. For a second-order loop, the calculation results in (Eq. 8.9)

$$\sigma_{\theta_0}^2 = \frac{\eta}{2} \frac{\pi^2}{\zeta\omega_n}$$

Consequently, by increasing the quantity $\zeta\omega_n$, we reduce the random variations of the VCO signal phase. The VCO phase modulation variance lessens and the VCO signal spectrum becomes a discrete frequency line, of which the power, $\frac{1}{2}B^2 e^{-\sigma_{\theta_0}^2}$, represents a considerable part of the total power, accompanied by a continuous spectrum, the power of which decreases as $\sigma_{\theta_0}^2$ (see Section 7.4).

Similarly, if the VCO frequency modulation noise, with the oscillator outside the loop, is $S_{f_0}(f) = \xi/|f|$, it is advisable to increase ω_n in order to reduce both the loop phase error and the residual phase modulation of the VCO inserted in the loop (see Eq. 8.12). We can make use of this property to obtain a high-power signal with an exceptionally pure spectrum, when the high-power signal is derived from an oscillator that can be frequency modulated. The phase of this signal can be "locked" to that of a signal of low power, but high short-term stability. The output signal will then have the power of the first signal and the spectral purity of the second.

8.3   SECOND EXAMPLE OF PARAMETER OPTIMIZATION

In Section 7.3 we gave a first example of phase-locked loop parameter optimization, in the case where the signal applied to the loop input, accompanied by a Gaussian, white additive noise, is affected by phase or frequency steps or linear frequency ramps.

Using the calculations elaborated in Sections 8.1 and 8.2 we can give other examples of parameter optimization in the case of a second-order loop. We shall first discuss an example where the input signal, accompanied by a Gaussian, white, additive noise is frequency modulated by a noise process consisting of a white noise and a flicker noise. We shall also suppose that the VCO used has the same defect. This case is often encountered when we want to use a phase-locked loop with a high-frequency input signal, where the short-term frequency instability of the signals (input and VCO) can be a relatively serious problem. We shall then examine the case where the input signal, accompanied by a Gaussian, white, additive noise, is phase modulated by a low-pass filtered noise process, for instance, by the use of another phase-locked loop. This example is typical of cases where the signal applied to the loop input comes from a coherent transponder.

8.3.1   INFLUENCE OF SHORT-TERM SIGNAL INSTABILITY

The signal applied at the loop input,

$$y_i(t) = A \sin \left[\omega t + \phi_i(t) \right] + n(t)$$

is frequency modulated by a noise process producing modulation $\phi_i(t)$, the power spectral density $S_{\phi_i}(f)$ of which is known. The useful signal is accompanied by a Gaussian, white, additive noise $n(t)$, the power spectral density of which is N_0 within the frequency band $(f - W/2, f + W/2)$.

The VCO signal expression, when the VCO is placed outside the loop, is

$$y_0(t) = B \cos \left[\omega t + \theta(t) \right]$$

In other words, it is frequency modulated by a random process producing $\theta(t)$, of power spectral density $S_\theta(f)$.

The role of the phase-locked loop is to obtain a signal the phase of which follows as closely as possible the phase variations of the input signal (otherwise the phase error will be excessive—see Sections 8.1.3 and 8.1.4—and the loop will no longer be operating in the linear zone), and this despite additive noise and VCO parasitic modulation. The overall phase error in the loop must consequently be reduced to a minimum. The sources of this error are:

- input signal frequency modulation
- additive noise
- VCO signal frequency modulation.

Since the three processes are assumed to be statistically independent, if we are to remain in the linear zone, we have to minimize the quantity σ_ϕ^2 given by

$$\sigma_\phi^2 = \int_{-\infty}^{+\infty} S_{\phi_i}(f) |1 - H(j2\pi f)|^2 \, df$$

$$+ \int_{-\infty}^{+\infty} \frac{N_0}{A^2} |H(j2\pi f)|^2 \, df$$

$$+ \int_{-\infty}^{+\infty} S_\theta(f) |1 - H(j2\pi f)|^2 \, df$$

that is,

$$\sigma_\phi^2 = \frac{N_0}{A^2} \int_{-\infty}^{+\infty} |H(j2\pi f)|^2 \, df$$

$$+ \int_{-\infty}^{+\infty} \left[S_{\phi_i}(f) + S_\theta(f) \right] |1 - H(j2\pi f)|^2 \, df$$

Let us suppose we have a second-order loop with integrator and correction (or with low-pass filter and correction if the loop gain K is much larger than ω_n). If we assume that the frequency modulations of the input signal and VCO signal can be represented by

$$S_{f_i}(f) = \frac{\eta}{2} + \frac{\xi}{|f|} \quad \text{and} \quad S_{f_0}(f) = \frac{\eta'}{2} + \frac{\xi'}{|f|}$$

implying

$$S_{\phi_i}(f) = \frac{\eta}{2} \frac{1}{f^2} + \frac{\xi}{|f|f^2} \quad \text{and} \quad S_{\theta}(f) = \frac{\eta'}{2} \frac{1}{f^2} + \frac{\xi'}{|f|f^2}$$

then, taking into account Eqs. 7.11, 8.9, and 8.12,

$$\sigma_{\phi}^2 = \frac{N_0}{A^2} \frac{\omega_n(1+4\zeta^2)}{4\zeta} + \frac{\eta+\eta'}{2} \frac{\pi^2}{\zeta\omega_n} + (\xi+\xi')\frac{4\pi^2}{\omega_n^2} f(\zeta)$$

The function $f(\zeta)$ is given by Eq. 8.13. The parameters N_0/A^2, η, η', ξ, and ξ' being given, the problem is to determine ζ and ω_n in order to minimize σ_ϕ^2.

Let us first suppose that the flicker noise is negligible as compared with the frequency-modulation white noise:

$$\sigma_{\phi}^2 = \frac{N_0}{A^2} \frac{\omega_n(1+4\zeta^2)}{4\zeta} + \frac{\eta+\eta'}{2} \frac{\pi^2}{\zeta\omega_n}$$

For a given value of ζ, σ_ϕ^2 consists of the sum of two ω_n terms, the product of which is constant. This sum is minimum when the two terms are equal, which implies that

$$\sigma_{\phi \, \text{min}}^2 = 2\frac{N_0}{A^2} \frac{\omega_{n\,\text{opt}}(1+4\zeta^2)}{4\zeta} = 2\frac{\eta+\eta'}{2} \frac{\pi^2}{\zeta\,\omega_{n\,\text{opt}}}$$

and

$$\omega_{n\,\text{opt}}^2 = \frac{\eta+\eta'}{2} \frac{A^2}{N_0} \frac{4\pi^2}{1+4\zeta^2}$$

The variations of σ_ϕ^2 versus ω_n (for a given ζ) are represented in Fig. 8.6, where the minimum corresponding to $\omega_{n\,\text{opt}}$ is perfectly clear. If we choose $\omega_n \gg \omega_{n\,\text{opt}}$, σ_ϕ^2 increases because, although the error induced by signal modulation is negligible, the error induced by the additive noise becomes considerable. On the contrary, if $\omega_n \ll \omega_{n\,\text{opt}}$, then the additive noise error is negligible as compared with the signal modulation error. If we substitute the expression $\omega_{n\,\text{opt}}$ in either of the $\sigma_{\phi\,\text{min}}^2$ expressions, we get

$$\sigma_{\phi\,\text{min}}^2 = \text{constant} \times \frac{\sqrt{1+4\zeta^2}}{\zeta}$$

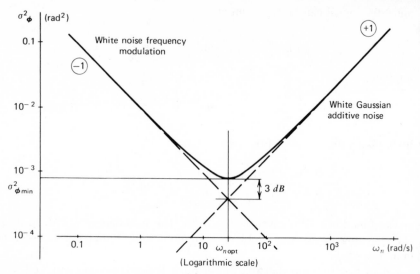

FIGURE 8.6. Phase error variance in presence of additive white noise and white noise frequency modulation (second-order loop).

The function $g(\zeta) = \sqrt{1 + 4\zeta^2}\,/\,\zeta$ is a decreasing function. It is thus advisable for ζ to be as large as possible. But when $\zeta \to \infty$, $g(\zeta) \to 2$ whereas for

$$\zeta = 1, \qquad g(\zeta) = \sqrt{5} = 2.236$$

It is obvious that there is little to be gained by having $\zeta \gg 1$, and that the most judicious choice for ζ is a value in the vicinity of 1, bearing in mind the other possible sources of error in the loop (transient phenomena, for instance).

Let us now suppose that the parasitic frequency modulation of the signals is essentially due to flicker noise. In this case, the phase error variance is given by

$$\sigma_\phi^2 = \frac{N_0}{A^2} \frac{\omega_n (1 + 4\zeta^2)}{4\zeta} + (\xi + \xi') \frac{4\pi^2}{\omega_n^2} f(\zeta)$$

The variations of σ_ϕ^2 versus ω_n, for a given ζ, are represented in Fig. 8.7. Here again, the minimum is perfectly clear and corresponds to

$$\sigma_{\phi\,\text{min}}^2 = \frac{3X}{2} \omega_{n\,\text{opt}} = \frac{3Y}{\omega_{n\,\text{opt}}^2}$$

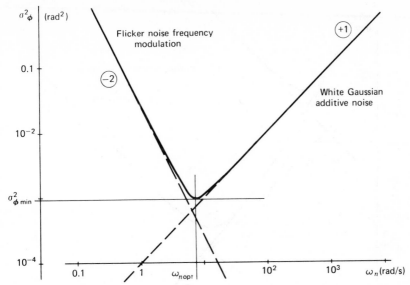

FIGURE 8.7. Phase error variance in presence of additive white noise and flicker noise frequency modulation (second-order loop).

and

$$\omega_{n\,\mathrm{opt}}^3 = \frac{2Y}{X}$$

with

$$X = \frac{N_0}{A^2}\frac{1+4\zeta^2}{\zeta} \quad \text{and} \quad Y = 4\pi^2(\xi+\xi')f(\zeta)$$

By substituting the $\omega_{n\,\mathrm{opt}}$ expression in either of the formulas giving $\sigma_{\phi\,\mathrm{min}}^2$, we can determine the optimum value for ζ, since $\sigma_{\phi\,\mathrm{min}}^2$ is minimum depending on ζ, at the same time as the function $g(\zeta)$,

$$g(\zeta) = \frac{\left(1+4\zeta^2\right)^2}{\zeta^2}f(\zeta)$$

and the latter is minimum for $\zeta = 1$. However, even though $g(\zeta)$ varies perceptibly when ζ varies around 1, since $\sigma_{\phi\,\mathrm{min}}^2$ is proportional to $[g(\zeta)]^{1/3}$, the choice of the value $\zeta = 1$ is not in fact critical and $\sigma_{\phi\,\mathrm{min}}^2$ stays very close to its minimum value while ζ remains within usual values $(0.25 < \zeta < 4)$.

More generally, when the flicker noise and the white noise have both to be taken into consideration, the simplest way to determine $\omega_{n\,\mathrm{opt}}$ is the graphical method (see Fig. 8.8).

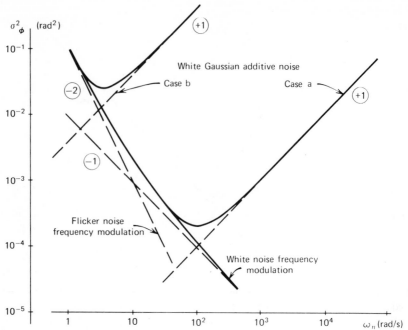

FIGURE 8.8. Phase error variance in presence of additive noise and both white and flicker noise frequency modulation (second-order loop): (a) Low-level additive noise; (b) high-level additive noise.

According to the values of parameters N_0/A^2, $(\xi+\xi')$, and $(\eta+\eta')/2$, and for a given ζ, the point where the additive white noise error curve (slope $+1$) intersects the frequency modulation noise error curve will be located in the zone where the slope is in the vicinity of -1 (case a) or -2 (case b).

In case a, we can fairly safely use the frequency modulation white noise formulas, reserving the flicker noise formulas for case b. For intermediate cases, considering the general uncertainty surrounding the value of parameters $(\xi+\xi')$ and $(\eta+\eta')/2$, the graphical method will be sufficiently accurate. Finally, the choice $\zeta=1$ (optimum flicker noise value and near-optimum white noise value) will, in any case, be practically optimum.

NUMERICAL EXAMPLE.   The signal-to-noise ratio operating conditions of a second-order phase locked loop $[F(s)=(1+\tau_2 s)/\tau_1 s$ and $\zeta=1]$ are such that $A^2/N_0=2\times10^4$. We observe that for the natural angular frequency value $\omega_n=40$ rad/s, the phase error variance is $\sigma_\phi^2=1.4\times10^{-2}$ rad^2, which is a minimum. Since the input signal (from a frequency synthesizer) is perfectly stable, we assume that the loop VCO is frequency modulated by a spurious noise.

If we suppose this noise to be a white noise, having a two-sided spectral density $\eta'/2$, we get

$$\frac{\eta'}{2} = \omega_{n\,opt}^2 \frac{N_0}{A^2} \frac{1+4\zeta^2}{4\pi^2}$$

implying that

$$\frac{\eta'}{2} = (40)^2 \times \frac{1}{2} \times 10^{-4} \times \frac{5}{40} = 10^{-2} \text{ Hz}^2/\text{Hz}$$

The minimum phase error should be

$$\sigma_{\phi\,min}^2 = 2\frac{\eta'}{2}\frac{\pi^2}{\zeta\omega_{n\,opt}}$$

that is,

$$\sigma_{\phi\,min}^2 = 2 \times 10^{-2} \times \frac{10}{40} = 5 \times 10^{-3} \text{ rad}^2$$

Considering the value measured for σ_ϕ^2, the initial hypothesis would not appear to be verified.

If we suppose the frequency modulation noise to be a flicker noise having a spectral density $\xi'/|f|$, we get

$$\xi' = \omega_{n\,opt}^3 \frac{N_0}{A^2}\frac{1+4\zeta^2}{\zeta}\frac{1}{8\pi^2}$$

implying that

$$\xi' = (40)^3 \times \frac{1}{2} \times 10^{-4} \times 5 \times \frac{1}{80} = 0.2 \text{ Hz}^2$$

The minimum phase error should be

$$\sigma_{\phi\,min}^2 = \frac{3}{2}\frac{N_0}{A^2}\frac{1+4\zeta^2}{\zeta}\omega_{n\,opt}$$

that is,

$$\sigma_{\phi\,min}^2 = \frac{3}{2} \times \frac{1}{2} \times 10^{-4} \times 5 \times 40 = 1.5 \times 10^{-2} \text{ rad}^2$$

Considering the value given for σ_ϕ^2, the hypothesis is apparently realistic: we can therefore represent the VCO spurious frequency modulation by a noise having a spectral density

$$S_{f_0}(f) = \frac{0.2}{|f|} \text{ (Hz}^2/\text{Hz)}$$

8.3.2 CASE OF CASCADED LOOPS

The signal applied to the loop input $y_i(t) = A \sin[\omega t + \phi_i(t)]$ is phase-modulated by a Gaussian random process, having a null mean value, the power spectral density $S_{\phi_i}(f)$ of which is known. This modulation could be due, for instance, to the fact that the signal in question comes from a VCO inserted in a preceding loop with transfer function $H'(j2\pi f)$. If the signal applied to the input of this loop, $A' \sin(\omega t + \theta_i)$, is accompanied by a Gaussian, white, additive noise, having a one-sided power spectral density N_0', then (Eqs. 7.8 and 7.15)

$$S_{\phi_i}(f) = \frac{N_0'}{A'^2} |H'(j2\pi f)|^2$$

in particular,

$$\sigma_{\phi_i}^2 = \frac{N_0'}{A'^2}(2B_n')$$

where $(2B_n')$ is the preceding loop equivalent noise bandwidth. Signal $y_i(t)$, itself accompanied by a Gaussian white additive noise, having a one-sided spectral density N_0, is applied to the input of a phase-locked loop having a transfer function $H(j2\pi f)$. The problem is to optimize the second loop performance.

The case here is that of a signal, phase modulated by a process of finite (and even low) power, since $\sigma_{\phi_i}^2$, representing the influence of the additive noise on the preceding loop operation, must be small.

In the absence of noise at the second loop input, it is then possible to try and synchronize the second loop VCO signal, not with the modulated input signal $A \sin[\omega t + \phi_i(t)]$ but with the pseudo (unmodulated) signal $A \sin \omega t$. But this is only possible because, as $\sigma_{\phi_i}^2$ is low enough for the first loop to stay in the linear domain, the phase error variance in the second loop, due to the input signal modulation(Eq. 8.4), is necessarily low enough to ensure that the second loop also stays in the linear domain.

If we want to make allowances for the second loop input additive noise, we have to be more careful, since it is the quantity (Eqs. 7.9 and 8.4)

$$\sigma_{\phi}^2 = \int_{-\infty}^{+\infty} S_{\phi_i}(f)|1 - H(j2\pi f)|^2 df$$

$$+ \int_{-\infty}^{+\infty} \frac{N_0}{A^2} |H(j2\pi f)|^2 df$$

which has to stay small to guarantee linear operation. Providing this is verified, the second loop VCO signal phase variance is expressed (Eqs. 7.9 and 8.3) as

$$\sigma_{\phi_0}^2 = \int_{-\infty}^{+\infty} S_{\phi_i}(f)|H(j2\pi f)|^2 df$$

$$+ \int_{-\infty}^{+\infty} \frac{N_0}{A^2}|H(j2\pi f)|^2 df$$

Substituting for $S_{\phi_i}(f)$, we get

$$\sigma_{\phi}^2 = \frac{N_0'}{A'^2} \int_{-\infty}^{+\infty} |H'(j2\pi f)|^2 |1 - H(j2\pi f)|^2 df$$

$$+ \frac{N_0}{A^2} \int_{-\infty}^{+\infty} |H(j2\pi f)|^2 df$$

$$\sigma_{\phi_0}^2 = \frac{N_0'}{A'^2} \int_{-\infty}^{+\infty} |H'(j2\pi f)|^2 |H(j2\pi f)|^2 df$$

$$+ \frac{N_0}{A^2} \int_{-\infty}^{+\infty} |H(j2\pi f)|^2 df$$

Generally speaking, the corresponding integral calculations are rather tedious, but two particular cases may be encountered:

(a) Loop $H'(j2\pi f)$ is assumed to be larger than loop $H(j2\pi f)$; then $|H'(j2\pi f)|^2 \cong 1$ in the frequency range where $H(j2\pi f)$ is significant. We can then state that

$$\sigma_{\phi_0}^2 = \left(\frac{N_0'}{A'^2} + \frac{N_0}{A^2}\right) \int_{-\infty}^{+\infty} |H(j2\pi f)|^2 df$$

$$= \left(\frac{N_0'}{A'^2} + \frac{N_0}{A^2}\right)(2B_n)$$

As regards the second loop VCO signal phase variance, it is as if the spectral density of the equivalent phase modulation by the first loop noise N_0'/A'^2 were added to the spectral density of the equivalent phase modulation of the second loop N_0/A^2.

Furthermore, if $H'(j2\pi f)$ is much larger than $H(j2\pi f)$, the error function $[1 - H(j2\pi f)]$ practically represents an all-pass filter with respect to

$H'(j2\pi f)$, and we can write

$$\sigma_\phi^2 \cong \frac{N_0'}{A'^2} \int_{-\infty}^{+\infty} |H'(j2\pi f)|^2 \, df$$

$$+ \frac{N_0}{A^2} \int_{-\infty}^{+\infty} |H(j2\pi f)|^2 \, df$$

$$= \sigma_{\phi_i}^2 + \frac{N_0}{A^2}(2B_n)$$

This means that the second loop is too narrow to follow the input phase variations, which reappear practically in entirety in the phase error. Under these conditions, without trying to minimize σ_ϕ^2, it is advisable to have $(2B_n)$ fairly small, so that $\sigma_{\phi_0}^2$ will be as small as possible and $\sigma_\phi^2 \cong \sigma_{\phi_i}^2$. When $\sigma_{\phi_i}^2$ is small enough to vouch for linear operation, $(2B_n)$ must be as narrow as possible so that the VCO distortion with respect to the pseudoinput signal $A \sin \omega t$, despite the second loop additive noise and the input signal phase modulation (which itself results from the first loop additive noise), will be minimum. This second loop pseudoinput signal is in fact the real input signal of the first loop. It is this signal that we generally try and pickup, despite the various disturbances.

(b)   Let us suppose, on the contrary, that the $H(j2\pi f)$ loop is much larger than the $H'(j2\pi f)$ loop; then $|1 - H(j2\pi f)|^2 \cong 0$ in the frequency range where $H'(j2\pi f)$ is significant; consequently,

$$\sigma_\phi^2 \cong \frac{N_0}{A^2} \int_{-\infty}^{+\infty} |H(j2\pi f)|^2 \, df = \frac{N_0}{A^2}(2B_n)$$

This means that the second loop bandwidth is large enough to follow entirely the input signal phase variations. Consequently, the only error term in the loop is that resulting from the additive noise.

However, if $H(j2\pi f)$ is larger than $H'(j2\pi f)$, then

$$|H(j2\pi f)|^2 |H'(j2\pi f)|^2 \cong |H'(j2\pi f)|^2$$

and consequently:

$$\sigma_{\phi_0}^2 \cong \frac{N_0'}{A'^2} \int_{-\infty}^{+\infty} |H'(j2\pi f)|^2 \, df$$

$$+ \frac{N_0}{A^2} \int_{-\infty}^{+\infty} |H(j2\pi f)|^2 \, df$$

$$= \sigma_{\phi_i}^2 + \frac{N_0}{A^2}(2B_n)$$

We easily recognize in the VCO signal phase variance expression the entire input signal phase fluctuation together with a term representing the additive noise. In this case, despite the additive noise, the VCO follows as well as possible the input signal $A\sin[\omega t+\phi_i(t)]$, that is, it provides no means of picking up the signal applied at the first loop input, $A\sin\omega t$. In particular, it should be noted that we are obliged to have $H(j2\pi f)$ fairly large with respect to $H'(j2\pi f)$ if $\sigma^2_{\phi_i}$ is large: in this case, the second loop does not restore the signal $A\sin\omega t$, but this inadequacy is in fact to be attributed to the first loop filtering, which fails to reduce sufficiently the variance $\sigma^2_{\phi_i}$.

NUMERICAL EXAMPLE. The frequency of the output signal of a coherent transponder is obtained by multiplying by a rational number p/q the VCO frequency of a loop, phase locked to the transponder input signal. Let A', N_0', ζ', and ω_n', be the loop parameters, with

$$\frac{A'^2}{2N_0'}=30\text{ dB-Hz},\qquad \zeta'=0.5,\qquad \omega_n'=120\text{ rad/s}$$

The phase variance of the transponder loop VCO is then

$$\sigma^2_{\phi_0'}=\frac{N_0'}{A'^2}(2B_n')=\frac{1}{2}10^{-3}\times120=6\times10^{-2}\text{ rad}^2$$

If $p/q=1.1$, the phase variance of the signal emitted is

$$\sigma^2_{\phi_i}=\left(\frac{p}{q}\right)^2\sigma^2_{\phi_0'}=1.21\times6\times10^{-2}=7.26\times10^{-2}\text{ rad}^2$$

The receiver uses a phase-locked loop characterized by

$$\frac{A^2}{2N_0}=40\text{ dB-Hz},\qquad \zeta=\frac{\sqrt{2}}{2},\qquad \omega_n=30\text{ rad/s}$$

For a rough estimate, we can assume this loop to be narrower than the transponder loop. The VCO phase variance expression will then be

$$\sigma^2_{\phi_0}\cong\left(\frac{p^2}{q^2}\frac{N_0'}{A'^2}+\frac{N_0}{A^2}\right)(2B_n)$$

that is,

$$\sigma^2_{\phi_0}\cong\frac{1}{2}(1.21\times10^{-3}+10^{-4})\times31.8=2.09\times10^{-2}\text{ rad}^2$$

The loop phase error variance is given by

$$\sigma^2_{\phi}\cong\frac{N_0}{A^2}(2B_n)+\sigma^2_{\phi_i}$$

that is,

$$\sigma_\phi^2 = \frac{1}{2} 10^{-4} \times 31.8 + \sigma_{\phi_i}^2 = 0.16 \ 10^{-2} + 7.26 \times 10^{-2}$$

$$= 7.42 \times 10^{-2} \ \text{rad}^2$$

If this quantity is considered to be excessive, the loop has to be enlarged so that it can follow the greater part of the input signal modulation.

For example, if we take $\omega_n = 500$ rad/s (keeping $\zeta = \sqrt{2}/2$), we get

$$\sigma_\phi^2 \cong \frac{N_0}{A^2}(2B_n)$$

that is,

$$\sigma_\phi^2 = \tfrac{1}{2} 10^{-4} \times 530 = 2.65 \times 10^{-2} \ \text{rad}^2$$

The disadvantage of this arrangement is that the second loop VCO is far more sensitive to the first loop noise, since we get

$$\sigma_{\phi_0}^2 \cong \sigma_{\phi_i}^2 + \frac{N_0}{A^2}(2B_n)$$

that is,

$$\sigma_{\phi_0}^2 = 7.26 \times 10^{-2} + 2.65 \times 10^{-2} = 9.91 \times 10^{-2} \ \text{rad}^2$$

CHAPTER 9
PARAMETER VARIATION

We saw in Chapter 2 that the phase detector sensitivity K_1 is proportional to the amplitude A of the signal applied at the phase detector input, at least for sinusoidal detectors. In this case, the open-loop gain or "loop gain" K, equal to the product K_1, K_2, K_3 ($K_2 = 1$ when there is no operational amplifier), is also proportional to the input signal amplitude A. We then observed, in Section 3.6, that the parameters of a second-order loop, natural angular frequency ω_n, and damping factor ζ, are related to the value of K.

These aspects should be borne in mind when we consider the loop response to various transients or when we intend using it as a phase or frequency demodulator. Similarly, performance in the presence of additive noise, or when signals are modulated by noise, depends on A, through parameters K, ω_n, and ζ. We also discussed certain cases of multiple, simultaneous disturbances.

For instance, if we consider simultaneously the effect of transients and of a Gaussian, white, additive noise, we can determine (24) the optimum Wiener filter to be used. If the type of filter depends essentially on the nature of the transient, the value of the parameters depends on the signal-to-noise ratio. For instance, when the transient is an angular frequency step $\Delta\omega$, as we saw in Section 7.3, the optimum filter would be a second-order loop with integration and correction, so that

$$\zeta = \frac{\sqrt{2}}{2}$$

$$\omega_n^2 = \lambda \Delta\omega \sqrt{\frac{A^2}{N_0}}$$

where λ is a multiplying safety coefficient, A the input signal amplitude, and N_0 the additive noise one-sided power spectral density. When the amplitude A varies, if we want the filter to remain optimum, we have to keep ζ at the value $\sqrt{2}/2$ and adjust ω_n proportionately to A. But we saw in Section 3.6

that when A varies, ω_n varies, but not proportionately, and ζ also varies. In other words, if the filter is optimum for a given value of A, it cannot be optimum for other values.

Similarly, when the signal received, disturbed by the additive noise, is modulated by a spurious signal, as we saw in Section 8.3, the optimum natural angular frequency $\omega_{n\,\text{opt}}$ for a second-order loop depends on the input signal amplitude A, the influence of the value of ζ being slight. In the case, for instance, of frequency modulation by a white noise, having a two-sided spectral density $\eta/2$, the parameters ζ and ω_n have to be such that

$$\zeta \geqslant 1$$

$$\omega_{n\,\text{opt}}^2 = \frac{\eta}{2}\frac{A^2}{N_0}\frac{4\pi^2}{1+4\zeta^2}$$

When A varies, if the variations of ζ are not very important as regards the phase error variance value σ_ϕ^2, they have a certain bearing on the expression of $\omega_{n\,\text{opt}}$.

Generally speaking, the performances required of a phase-locked loop should correspond to a wide range of input signal amplitude variations. This implies that realization of a permanently optimum filter is unrealistic. If we have an input signal with a 60-dB dynamic, A varying, for instance, between 10 mV and 10 V, in order to conserve the proportionality between ω_n and A, the natural angular frequency of the loop would have to vary, for example, from 100 Hz to 100 KHz. When the input signal is very low, the loop will be "narrowband," and when the input signal is high, the loop will tend to be "wideband." In other words, such a loop will be unsuitable for use as a filter or a demodulator.

It is for this reason that the loop parameters are determined in such a way as to be optimum in the most difficult case, which generally corresponds to the lowest input signal (that is, the worst signal-to-noise ratio conditions). Then, according to circumstances, we either ensure that the loop parameter value remains constant or that it varies within acceptable limits using an automatic-gain-control device (AGC) or a limiter. When the input signal increases, the parameter value is no longer optimum, but this is compensated by the fact that, with a better signal-to-noise ratio, one of the cases of phase error has diminished.

It is thus interesting to derive the expression corresponding to the parameters of a phase-locked loop, preceded by a regulation device when A varies. These expressions can then be used to verify that the required performances, generally specified for a low input signal, are not deteriorated, if the input signal is high, to the point of incompatibility with the expected application.

9.1   USE OF AN AUTOMATIC-GAIN-CONTROL DEVICE

An automatic gain control is a regulating chain system that adjusts the gain of an amplifier, preceding the phase-locked loop so as to keep the amplifier output power constant and thereby, to a certain extent (when the signal-to-noise ratio is high) the amplitude A_0 of the signal applied to the phase detector.

Let $\mathcal{S}_0 = A_0^2/2$ be the amplifier output power (as evaluated in a 1 Ω resistor). We call S_i and N_i the respective powers of signal and noise at the amplifier input and S and N the same variables evaluated at the output. If the AGC maintains the output power constant, then

$$S + N = \mathcal{S}_0$$

Let us first suppose that $S_i = \mathcal{S}_0$; the amplifier power gain will be $G \cong 1$, then $S \cong S_i$ and $N \cong N_i$ (generally a very low quantity). When S_i decreases, the power gain G becomes

$$G = \frac{S + N}{S_i + N_i} \qquad \text{with } S + N = \mathcal{S}_0$$

Then

$$S = GS_i = \frac{\mathcal{S}_0}{S_i + N_i} S_i$$

$$N = GN_i = \frac{\mathcal{S}_0}{S_i + N_i} N_i$$

These expressions hold whatever the value of S_i providing the amplifier gain is able to reach a maximum value G_{max} defined by

$$G_{max} = \frac{\mathcal{S}_0}{N_i}$$

If the maximum amplifier gain is insufficient, the maximum value is reached for a value of the input signal power $S_{i\,min}$ so that

$$G_{max} = \frac{\mathcal{S}_0}{S_{i\,min} + N_i}$$

and, when S_i is below $S_{i\,min}$, the gain remains at its maximum value, and we get

$$S = G_{max} S_i$$

$$N = G_{max} N_i$$

The respective variations of S and N when S_i varies, in accordance with these two maximum power gain possibilities ($G_{max} \geqslant S_0/N_i$ or $G_{max} < S_0/N_i$) are shown in Fig. 9.1.

Let us now suppose that we are in the more favorable case as regards gain control ($G_{max} \geqslant S_0/N_i$) and state that

$$S = \alpha^2 S_0$$

When the control works perfectly, that is when the entire output power is kept constant, then $S + N = S_0$. This implies that

$$N = S_0 - S = (1 - \alpha^2) S_0$$

We assume that the amplifier remains linear and thus conserves the signal-to-noise ratio:

$$\frac{S}{N} = \frac{\alpha^2 S_0}{(1 - \alpha^2) S_0} = \frac{\alpha^2}{1 - \alpha^2} = \frac{S_i}{N_i}$$

which implies that

$$\alpha^2 = \frac{S_i / N_i}{1 + S_i / N_i} \tag{9.1}$$

The variations of α versus the signal-to-noise ratio at the input of the amplifier are shown in Fig. 9.2. We note that the control is only effective when the signal-to-noise ratio exceeds about $+10$ dB. Below this value, and despite the control device, the signal power S at the amplifier output decreases, which means that the amplitude of the signal applied to the phase detector decreases according to

$$A = \alpha A_0$$

The coefficient α is sometimes called the AGC "reduction factor" or "suppression factor."

We shall call K_{10} the nominal value of the phase detector sensitivity (when the input signal amplitude is A_0). When the input signal amplitude is A, the phase detector sensitivity becomes K_1:

$$K_1 = \frac{A}{A_0} K_{10} = \alpha K_{10}$$

In the same way, keeping the index 0 for nominal values, the loop gain becomes

$$K = \alpha K_0$$

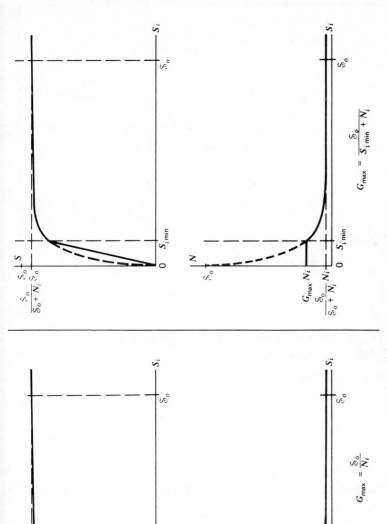

FIGURE 9.1. Signal and noise powers at the output of a constant output power amplifier (AGC).

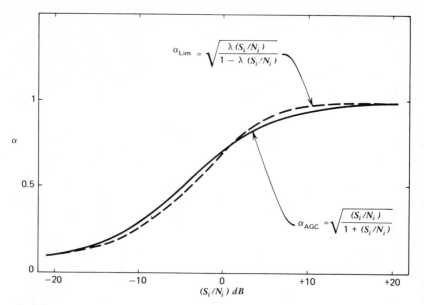

$$\alpha_{\text{Lim}} = \sqrt{\frac{\lambda \, (S_i / N_i)}{1 - \lambda \, (S_i / N_i)}}$$

$$\alpha_{\text{AGC}} = \sqrt{\frac{(S_i / N_i)}{1 + (S_i / N_i)}}$$

$(S_i / N_i) \, dB$

FIGURE 9.2. AGC and limiter suppression factors.

9.1.1 SECOND-ORDER LOOP WITH LOW-PASS FILTER AND CORRECTION NETWORK

The natural angular frequency is written (Eq. 3.41) as

$$\omega_n^2 = \frac{K}{\tau_1} = \frac{\alpha K_0}{\tau_1} = \alpha \omega_{n0}^2$$

that is

$$\omega_n = \sqrt{\alpha} \; \omega_{n0}$$

and the damping factor, given by Eq. 3.52, becomes (Section 3.6, Remark 2)

$$\zeta = \frac{1 + \alpha K_0 \tau_2}{2 \tau_1 \sqrt{\alpha} \; \omega_{n0}}$$

If K_0 is large and if the input signal variation range of the amplifier is such that the quantity $\alpha K_0 \tau_2$ is larger than 1, then

$$\zeta \cong \frac{\alpha K_0 \tau_2}{2 \tau_1 \sqrt{\alpha} \; \omega_{n0}} = \frac{K_0 \tau_2}{2 \tau_1 \omega_{n0}} \sqrt{\alpha} = \zeta_0 \sqrt{\alpha}$$

When the input signal-to-noise ratio decreases, the loop parameter value decreases, with the result that the loop properties are modified. As an example, let us examine the variations of the equivalent noise bandwidth $(2B_n)$ versus α. Equation 7.20 can be written

$$(2B_n) = \frac{\omega_n^2}{4\zeta\omega_n}\left[1 + \frac{1}{\omega_n^2}\left(2\zeta\omega_n - \frac{\omega_n^2}{K}\right)^2\right]$$

or

$$(2B_n) = \frac{K/\tau_1}{2(1+K\tau_2)/\tau_1}\left[1 + \frac{\tau_1}{K}\left(\frac{1+K\tau_2}{\tau_1} - \frac{1}{\tau_1}\right)^2\right]$$

from which we derive

$$(2B_n) = \frac{1}{2\tau_2}\frac{1}{1+1/K\tau_2}\left(1 + K\frac{\tau_2^2}{\tau_1}\right)$$

and, replacing K by αK_0, we get

$$(2B_n) = \frac{1}{2\tau_2}\frac{1}{1+1/\alpha K_0\tau_2}\left(1 + \alpha K_0\frac{\tau_2^2}{\tau_1}\right)$$

The variations of $(2B_n)$ versus α are shown in Fig. 9.3. The main variation is observed to be a linear variation versus α according to the law

$$\frac{1}{2\tau_2} + \alpha K_0\frac{\tau_2}{2\tau_1}$$

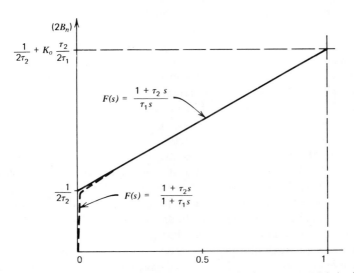

FIGURE 9.3. Equivalent noise bandwidth of a second-order loop versus AGC (or limiter) suppression factor.

It is only when α becomes very small that the $1/\alpha K_0 \tau_2$ term becomes very large and $(2B_n)$ approaches zero.

9.1.2 SECOND-ORDER LOOP WITH INTEGRATOR AND CORRECTION

The results are similar, except that in this case we have to take into account that, since $2\zeta\omega_n = K\tau_2/\tau_1$, we have exactly $\zeta = \sqrt{\alpha}\ \zeta_0$. As regards the equivalent noise bandwidth $(2B_n)$ (Eq. 7.18), the variation is strictly linear versus α since

$$(2B_n) = \frac{1}{2\tau_2}\left(1 + \alpha K_0 \frac{\tau_2^2}{\tau_1}\right)$$

The variations are shown in Fig. 9.3. We observe that this is yet another case where, as the open-loop gain K_0 is very large, a loop with filter $F(s) = (1 + \tau_2 s)/\tau_1 s$ is a very good approximation of a loop with filter $F(s) = (1 + \tau_2 s)/(1 + \tau_1 s)$.

The fact that the equivalent noise bandwidth is reduced when the input signal-to-noise ratio decreases does not correspond to optimization as defined by the Wiener theory, but is nevertheless an extremely useful factor much appreciated in various phase-locked loop applications. The drawback lies in the corresponding variations in loop properties, which can be serious.

Remark 1 The variations of α versus the input signal-to-noise ratio, shown in Fig. 9.2, are only valid if the control system is perfect, that is, capable, under any conditions, of maintaining a constant value for the total output power of the amplifier preceding the phase detector. We have already noted that, if this is to be accomplished, the amplifier must be capable of a maximum gain $G_{max} = S_0/N_i$. Furthermore, the quality of the control (under steady-state conditions, disregarding the response to fast input signal power fluctuations) depends on the detector providing the dc signal proportional to the total power and on the law governing the amplifier gain variation versus the command signal (amplified, filtered version of the difference between the detector output and a reference dc signal).

If the detector characteristic is strictly quadratic, as $y = ax^2$ for $x \geqslant 0$ (and $y = 0$ for $x < 0$), the dc component at the detector output is exactly proportional to the total power $S_0 + N_0$. But, for another characteristic, this would not be so. In particular, if the detector is linear ($y = ax$ for $x \geqslant 0$; $y = 0$ for $x < 0$), the detector output dc signal is proportional to the signal amplitude, when the signal-to-noise ratio is high, and proportional to the noise power, when the signal-to-noise ratio is very low (20).

In connection with the gain variation versus command signal law, we should mention two types: the "voltage gain linear variation," encountered when the device used is a linear amplitude modulator, and the "voltage gain logarithmic variation," which is particularly useful when we want the dc signal detected to be a usable replica over a wide variation range of the input signal power.

Remark 2. Contrary to what we have seen so far, there are certain cases where the loop parameters are required to remain strictly constant when the signal amplitude varies. When the signal-to-noise ratio preceding the phase detector is smaller than 1, we can no longer use a standard AGC device (see Fig. 9.2). In this case we use a device called a coherent AGC where the signal amplitude A detection is performed coherently by multiplication, using a reference signal that has the same phase and frequency as the input signal. We shall see in Chapter 11 how to construct a coherent amplitude detector using a phase-locked loop. When the loop phase error is low enough, the control device works correctly. When the signal-to-noise ratio in the loop bandwidth deteriorates, the control device becomes ineffective, but, as we saw in Section 7.5, the equivalent noise bandwidth $(2B_n)$ is usually very small in comparison with the bandwidth W of the filter preceding the loop. The gain control is thus far more efficient than would be obtained using a standard AGC.

If we need to consider the control device dynamic response, we have to bear in mind the law underlying the amplifier gain versus the command signal and the transfer function of the low-pass filter that extracts the dc component at the coherent amplitude detector output. Even when this filter is simply a one-pole low-pass filter, of time constant $\tau = RC$, the calculations involved are fairly tiresome and few results are available. However, it would seem feasible by choosing a relatively small time constant τ to take advantage of the rapid AGC response to a sudden variation in the input signal level, without being unduly troubled by noise disturbances in the control network (33).

NUMERICAL EXAMPLE. When the signal received at a receiver input is high, the parameters of a second-order loop with $F(s) = (1 + \tau_2 s)/(1 + \tau_1 s)$ locked to the signal received are

$$K_0 = 2 \times 10^6 \text{ rad/s}, \quad \omega_{n0} = 100 \text{ rad/s}, \quad \zeta_0 = \frac{\sqrt{2}}{2}$$

The equivalent noise bandwidth $(2B_n)_0$ is then (Eq. 7.18)

$$(2B_n)_0 = \omega_{n0} \frac{1 + 4\zeta_0^2}{4\zeta_0} = 100 \frac{3}{2\sqrt{2}} = 106 \text{ Hz}$$

The loop filter time constants τ_1 and τ_2 are respectively equal to (using Eqs. 3.41 and 3.52)

$$\tau_1 = 200 \text{ s}, \qquad \tau_2 = 14 \text{ ms}$$

When the signal received is very low, the signal-to-noise ratio in the intermediate frequency passband preceding the loop measures -13 dB. The AGC no longer suffices to keep the loop parameters constant since the suppression factor is, in this case, $\alpha = 0.21$ (Fig. 9.2).

The loop parameters become

$$K = \alpha K_0 = 0.21 \times 2 \times 10^6 = 4.2 \times 10^5 \text{ rad/s}$$

$$\omega_n = \sqrt{\alpha}\ \omega_{n0} = 0.46 \times 100 = 46 \text{ rad/s}$$

$$\zeta \cong \sqrt{\alpha}\ \zeta_0 = 0.46 \frac{\sqrt{2}}{2} = 0.325$$

$$(2B_n) = 46 \frac{1.424}{1.3} = 50.5 \text{ Hz}$$

As the equivalent noise bandwidth has decreased, the loop is better protected against additive noise. But the damping factor is rather low and the transient response will not be very good.

9.2 USE OF A LIMITER

Another way of keeping at a constant value the total power (signal plus noise) applied at the phase detector input is to use a limiter. In most cases, this limiter will consist of a hard limiter, followed by a bandpass filter centered on the fundamental of the sinusoidal input signal (see diagram in Fig. 9.4). The hard limiter output signal is $+A'$ if the sign of the composite signal applied is positive and $-A'$ if it is negative.

When the input signal-to-noise ratio is high enough for us to consider that the input consists essentially of the sinusoidal signal $A \sin(\omega t + \theta_i)$, the hard

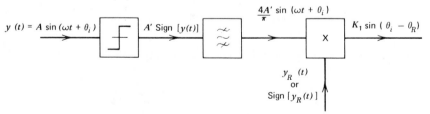

FIGURE 9.4. Model of sinusoidal phase detector following a band-pass limiter.

limiter output is a square signal, of constant amplitude A', independent of the variations of A. In particular, if the input signal is amplitude-modulated, the hard limiter suppresses this modulation. The purpose of the bandpass filter is to extract the fundamental of this square signal to obtain a sinusoidal signal $(4A'/\pi)\sin(\omega t + \theta_i)$. If this signal is applied to a multiplier with a reference signal $y_R = \cos(\omega t + \theta_R)$, or a multiplier with reference signal $\text{Sign}[\cos(\omega t + \theta_R)]$, the result in both cases is a sinusoidal phase detector. But the sensitivity K_1 is independent of the input signal amplitude A $(K_1 = 2A'/\pi$ or $K_1 = 8A'/\pi^2)$.

When the signal $A\sin(\omega t + \theta_i)$ is accompanied by a very strong additive noise $n(t)$, the hard limiter is seriously disturbed, to the extent that it is essentially the noise which decides the sign of the hard limiter output signal, so that it is as if the input signal had disappeared.

The phenomenon is, of course, progressive and, for intermediary values of the signal-to-noise ratio, as in the case of the AGC, a certain limiter suppression factor α can be considered. This factor depends on the input signal-to-noise ratio and, when the bandwidth preceding the limiter W can be kept as parameter, we can to a certain extent adjust the value of the phase-locked loop parameters (through K_1) versus the signal-to-noise ratio.

Let us consider the expression of the signal applied to the limiting device when accompanied by an additive noise $n(t)$ with a narrow bandwidth $(W \ll 2f)$ and featuring otherwise the same properties as described in Section 7.1.1:

$$y(t) = A\sin(\omega t + \theta_i) + n(t)$$

where θ_i is a constant or slowly varying quantity independent of $n(t)$. As in Section 7.1.4b, the noise can be split up into

$$n(t) = n_1'(t)\sin(\omega t + \theta_i) + n_2'(t)\cos(\omega t + \theta_i)$$

leading to

$$y(t) = \sqrt{\left[A + n_1'(t)\right]^2 + n_2'^2(t)} \; \sin\left[\omega t + \theta_i + \varphi_i(t)\right]$$

with

$$\cos\varphi_i(t) = \frac{A + n_1'(t)}{\sqrt{\left[A + n_1'(t)\right]^2 + n_2'^2(t)}}$$

and

$$\sin\varphi_i(t) = \frac{n_2'(t)}{\sqrt{\left[A + n_1'(t)\right]^2 + n_2'^2(t)}}$$

One can show (20) that the limiter output signal can be expressed as

$$s(t) = \frac{4A'}{\pi} \sum_{p=0}^{\infty} \frac{1}{2p+1} \sin\left[(2p+1)(\omega t + \theta_i + \varphi_i(t))\right]$$

and the output signal of the bandpass filter centered on frequency f is simply given by

$$u(t) = \frac{4A'}{\pi} \sin\left[\omega t + \theta_i + \varphi_i(t)\right]$$

which is the expression of a sinusoidal signal, phase modulated by the process $\varphi_i(t)$.

The power spectrum of this signal consists of a discrete line representing the "useful" signal and a continuous spectrum representing the modulation by the noise $\varphi_i(t)$.

The amplitude of the useful signal is given by (20)

$$A'' = \frac{2A'}{\sqrt{\pi}} \sqrt{\rho}\ {}_1F_1(\tfrac{1}{2}, 2; -\rho)$$

where the function ${}_1F_1$ is the confluent hypergeometric series function and ρ the input signal-to-noise ratio:

$$\rho = \frac{S_i}{N_i} = \frac{A^2}{2N_0 W}$$

When ρ approaches infinity, the function

$${}_1F_1(\tfrac{1}{2}, 2; -\rho) \cong \frac{\Gamma(2)}{\Gamma(\tfrac{3}{2})\sqrt{\rho}} = \frac{2}{\sqrt{\pi\rho}}$$

and consequently

$$A'' \cong \frac{2A'}{\sqrt{\pi}} \sqrt{\rho}\ \frac{2}{\sqrt{\pi\rho}} = \frac{4A'}{\pi}$$

which does, in fact, represent the amplitude of the fundamental of a square signal of amplitude A'.

When ρ approaches zero, the function ${}_1F_1(\tfrac{1}{2}, 2; -\rho) \cong 1$ and consequently

$$A'' \cong \frac{2A'}{\sqrt{\pi}} \sqrt{\rho}$$

The limiter suppression factor α is equal to the ratio of the useful signal amplitude for a given input signal-to-noise ratio, to the amplitude of the same signal, for a very high input signal-to-noise ratio:

$$\alpha = \frac{\left(2A'/\sqrt{\pi}\,\right)\sqrt{\rho}\,\,_1F_1\!\left(\tfrac{1}{2},2;\,-\rho\right)}{4A'/\pi}$$

that is,

$$\alpha = \tfrac{1}{2}\sqrt{\pi\rho}\,\,_1F_1\!\left(\tfrac{1}{2},2;\,-\rho\right) \tag{9.2}$$

The limiter suppression factor can be determined by another method, which consits in calculating the phase detector output dc component. We shall assume, for the purposes of this calculation, that the phase detector behaves like a true multiplier. If we multiply signal

$$u(t) = \frac{4A'}{\pi}\,\sin\left[\,\omega t + \theta_i + \varphi_i(t)\,\right]$$

by the reference signal $y_R(t) = \cos(\omega t + \theta_R)$, we get, after eliminating the 2ω angular frequency term,

$$u_1(t) = \frac{2A'}{\pi}\,\sin\left[\,\theta_i - \theta_R + \varphi_i(t)\,\right]$$

that is,

$$u_1(t) = \frac{2A'}{\pi}\,\sin\left(\theta_i - \theta_R\right)\cos\varphi_i(t)$$

$$+ \frac{2A'}{\pi}\,\cos\left(\theta_i - \theta_R\right)\sin\varphi_i(t)$$

or

$$u_1(t) = \frac{2A'}{\pi}\left[\,\sin\left(\theta_i - \theta_R\right)\frac{A + n_1'(t)}{\sqrt{\left[A + n_1'(t)\right]^2 + n_2'^2(t)}}\right.$$

$$\left. + \cos\left(\theta_i - \theta_R\right)\frac{n_2'(t)}{\sqrt{\left[A + n_1'(t)\right]^2 + n_2'^2(t)}}\,\right]$$

The phase detector characteristic is found by calculating the mean value $\overline{u_1(t)}$, for a given phase difference value $(\theta_i - \theta_R)$:

$$\overline{u_1(t)} = \frac{2A'}{\pi}\left[\,\sin\left(\theta_i - \theta_R\right)\overline{\cos\varphi_i(t)} + \cos\left(\theta_i - \theta_R\right)\overline{\sin\varphi_i(t)}\,\right]$$

It has been shown (34) that when $n_1'(t)$ and $n_2'(t)$ are two independent

Gaussian processes, the above expression leads to

$$u_1(t) = \frac{2A'}{\pi} \sin(\theta_i - \theta_R) \frac{\sqrt{\pi\rho}}{2} e^{-\rho/2} \left[I_0\left(\frac{\rho}{2}\right) + I_1\left(\frac{\rho}{2}\right) \right]$$

with $\rho = S_i/N_i = A^2/2N_0W$, where I_0 and I_1 are modified Bessel functions of the first kind and of order 0 and 1. The sensitivity K_1 of the phase detector is then expressed as

$$K_1 = A'\sqrt{\frac{\rho}{\pi}} \, e^{-\rho/2}\left[I_0\left(\frac{\rho}{2}\right) + I_1\left(\frac{\rho}{2}\right) \right]$$

When $\rho \to \infty$,

$$I_0\left(\frac{\rho}{2}\right) \cong I_1\left(\frac{\rho}{2}\right) \cong \frac{e^{\rho/2}}{\sqrt{\pi\rho}}$$

and consequently

$$K_1 \cong A'\sqrt{\frac{\rho}{\pi}} \, \frac{2}{\sqrt{\pi\rho}} = \frac{2A'}{\pi}$$

which corresponds to the sensitivity of a sinusoidal phase detector when the reference signal is a sinusoidal signal of amplitude 1 and the input signal a sinusoidal signal of amplitude $4A'/\pi$.

When $\rho \to 0$, $I_0(\rho/2) \cong 1$, $I_1(\rho/2) \cong 0$, and

$$K_1 \cong A'\sqrt{\frac{\rho}{\pi}}$$

it is then as if the input signal amplitude had become $A'' = 2A'\sqrt{\rho/\pi}$.

Then, the limiter suppression factor α is written as

$$\alpha = \frac{\sqrt{\pi\rho}}{2} e^{-\rho/2}\left[I_0\left(\frac{\rho}{2}\right) + I_1\left(\frac{\rho}{2}\right) \right] \qquad (9.3)$$

We can check that the two expressions obtained for the suppression factor by using either Davenport's power-spectrum-analysis results (20) (Eq. 9.2) or the Springett and Simon method of direct calculation (34) (Eq. 9.3) are identical, since

$$_1F_1(\tfrac{1}{2}, 2; -\rho) = e^{-\rho/2}\left[I_0\left(\frac{\rho}{2}\right) + I_1\left(\frac{\rho}{2}\right) \right]$$

The variations of α versus the input signal-to-noise ratio $\rho = S_i/N_i$, are shown in Fig. 9.2. For certain problems, we may require an analytical expression of α easier to handle than those given above. We can in this case use the following approximate formula (35):

$$\alpha^2 \cong \frac{0.7854\rho + 0.4768\rho^2}{1 + 1.024\rho + 0.4768\rho^2}$$

It is evident from the curve in Fig. 9.2 that the limiter gain control is more efficient than the AGC for $\rho \geqslant 1$, but that when $\rho < 1$, the amplitude variations are greater for the limiter than for the AGC.

Another important result concerns the power of the continuous spectrum (representing the noise power) accompanying the discrete frequency line of the useful signal at the output of the bandpass filter centered on the fundamental. It can be shown (36) that the output signal-to-noise ratio S/N of the bandpass filter (assumed to be sufficiently narrow to filter out all the useful signal components except the fundamental but sufficiently wide to let through all the noise components around the fundamental*) is related to the input signal-to-noise ratio $\rho = S_i/N_i$ by a formula

$$\frac{S}{N} = \lambda \frac{S_i}{N_i} = \lambda \rho \tag{9.4}$$

The coefficient λ is itself a function of S_i/N_i and varies from $(\pi/4)(-1.05$ dB) to $2(+3$ dB), when the signal-to-noise ratio varies from 0 to infinity.

The coefficient λ can be related to the suppression factor α. As shown above, the bandpass filter output signal is a sinusoidal signal, of amplitude $4A'/\pi$, phase modulated by the process $\varphi_i(t)$. The signal power is constant and consequently

$$S + N = \mathbb{S}_0 = \frac{1}{2}\left(\frac{4A'}{\pi}\right)^2$$

But

$$\frac{S}{\mathbb{S}_0} = \alpha^2 = \frac{S}{S + N} = \frac{S/N}{1 + S/N} = \frac{\lambda S_i/N_i}{1 + \lambda S_i/N_i} \tag{9.5}$$

that is,

$$\alpha^2 = \frac{\lambda \rho}{1 + \lambda \rho}$$

*In any case the bandwidth of this filter (sometimes called zonal filter) is assumed wider than the bandwidth W of the bandpass filter preceding the hard limiter.

Inversely,

$$\lambda = \frac{\alpha^2}{\rho(1-\alpha^2)}$$

Remark. It is possible, using one or other of the analytical formulas given above for α, to retrace the asymptotic values of λ. Thus when $\rho \to 0$, $\alpha^2 \cong \pi\rho/4$ and $\lambda \to \pi/4$.

When $\rho \to \infty$, we can use, for instance, the asymptotical expansions of $I_0(\rho/2)$ and $I_1(\rho/2)$:

$$I_0\left(\frac{\rho}{2}\right) \cong \frac{e^{\rho/2}}{\sqrt{\pi\rho}}\left(1+\frac{1}{4\rho}\right)$$

$$I_1\left(\frac{\rho}{2}\right) \cong \frac{e^{\rho/2}}{\sqrt{\pi\rho}}\left(1-\frac{3}{4\rho}\right)$$

Consequently,

$$\alpha \cong \frac{\sqrt{\pi\rho}}{2}e^{-\rho/2}\frac{e^{\rho/2}}{\sqrt{\pi\rho}}2\left(1-\frac{1}{4\rho}\right)=1-\frac{1}{4\rho}$$

implying that

$$\lambda \cong \frac{1-1/2\rho}{\rho-\rho(1-1/2\rho)} = \frac{1-1/2\rho}{\frac{1}{2}}$$

which quantity approaches 2 when ρ approaches infinity.

In fact, if we consider the signal-to-noise density ratio rather than the signal-to-total noise power ratio, it can be shown (37) that a low input signal-to-noise ratio corresponds to a slight widening of the continuous spectrum band at the output. The result of this is that the signal-to-noise density ratio deterioration is not -1.05 dB but only -0.65 dB. When the input signal-to-noise ratio S_i/N_i is high, the output noise spectrum does not widen (but stays the same bandwidth W as in front of the limiter) with the result that the signal-to-noise density ratio is improved by $+3$ dB.

This should by no means be interpreted as indicating that the VCO phase variance of a loop, following a bandpass limiter, under the same signal-to-noise ratio conditions, would be twice as low as without a limiter. The continuous spectrum accompanying the signal component at the output is not, in fact, an additive noise spectrum but represents the phase modulation of the signal by the process $\varphi_i(t)$. But we have seen in Section 7.1.4 that when the input signal-to-noise ratio is high, $\varphi_i(t) \cong n_2'(t)/A$ and, consequently, the two-sided power spectral density of $\varphi_i(t)$ is $S_{\varphi_i}(f) = N_0/A^2$. The VCO phase variance is given by Eq. 8.3:

$$\sigma_{\varphi_0}^2 = \int_{-\infty}^{+\infty} S_{\varphi_i}(f)|H(j2\pi f)|^2\,df$$

$$= \frac{N_0}{A^2}\int_{-\infty}^{+\infty}|H(j2\pi f)|^2\,df = \frac{N_0}{A^2}(2B_n)$$

that is, it is exactly identical with Eq. 7.15 obtained without a limiter.

A satisfactory physical explanation resides in the fact that the additive noise can be split up into an in-phase component (amplitude noise) and a quadrature component (phase noise). Since these two components have the same power, the +3 dB gain on the signal-to-noise ratio corresponds to complete suppression by the limiter of the amplitude noise when the signal-to-noise ratio is high. We have not then to take this gain into account when the limiter is followed by a phase-sensitive device, as is the case for a phase-locked loop.

When the signal-to-noise ratio is low, as mentioned in Section 7.1.4, this simple line of argument no longer applies, since the limiter operation is far more complex [in particular, the noise $\varphi_i(t)$ is no longer a Gaussian process]. Recent calculations (34) have, in fact, shown that if we consider the noise $n'(t)$ at the output of a sinusoidal phase detector preceded by a limiter, the spectral density of this noise depends on the shape of the filter preceding the limiter. For a "rectangular" filter, maximum deterioration (when $\rho = S_i/N_i \rightarrow 0$) as compared with what would be achieved without a limiter, for the same input signal-to-noise ratio, is -0.65 dB, when the phase detector static phase error is null. Under the same conditions, for a one-pole bandpass filter, the deterioration would only be -0.25 dB. Practically speaking, we can thus conclude that, disregarding the signal suppression phenomenon, the phase-locked loop performance in the presence of additive noise is unaffected by the limiter and the formulas of Section 7.2 are applicable without risk of serious error.

NUMERICAL EXAMPLE.   A second-order phase-locked loop has to operate in the two following extreme cases:

• When the input signal is low $(S/N_0 = 30$ dB-Hz) and perfectly stable.
• When the input signal is high $(S/N_0 = 60$ dB-Hz) and affected by a linear frequency drift $R = 40$ Hz/s.

In the first case, the standard deviation of the noise error should not exceed 0.1 rad, and in the second case the error induced by the linear frequency drift should not exceed 0.1 rad.

In the first case, we get (Eq. 7.15)

$$\sigma_{\varphi_{01}}^2 = \frac{1}{2}\left(\frac{N_0}{S}\right)_1 (2B_n)_1 = \frac{1}{2} \times 10^{-3}(2B_n)_1 = 10^{-2}\ \text{rad}^2$$

Thus

$$(2B_n)_1 = 20\ \text{Hz}$$

As in this case, only the noise effect is to be taken into account, we shall take $\zeta_1 = 0.5$ (Fig. 7.8) and $\omega_{n1} = 20\ \text{rad/s}$.

In the second case, we get (Eq. 5.13 for $t \to \infty$)

$$\frac{\mathcal{R}}{\omega_{n2}^2} = \frac{2\pi R}{\omega_{n2}^2} = 0.1\ \text{rad},$$

implying

$$\omega_{n2}^2 = \frac{2\pi \times 40}{0.1} = 2.5 \times 10^3\ (\text{rad/s})^2$$

that is,

$$\omega_{n2} = 50\ \text{rad/s}$$

By means of a limiter preceding the loop, we can reconcile these extreme conditions. If we call α_1 and α_2 the limiter suppression factor in each of the two cases, the bandwidth W preceding the limiter must be so chosen that $\sqrt{\alpha_2/\alpha_1} = 50/20 = 2.5$, that is, $\alpha_2/\alpha_1 = 6.25$.

It is apparent from Fig. 9.2 that if we have $W = 10^4$ Hz, then

$$\left(\frac{S}{N}\right)_1 = \left(\frac{S}{N_0 W}\right)_1 = -10\ \text{dB}, \qquad \text{implying } \alpha_1 = 0.26$$

$$\left(\frac{S}{N}\right)_2 = \left(\frac{S}{N_0 W}\right)_2 = +20\ \text{dB}, \qquad \text{implying } \alpha_2 = 1$$

The ratio $\alpha_2/\alpha_1 = 3.85$ is insufficient.

If we take $W = 10^5$ Hz, we get

$$\left(\frac{S}{N}\right)_1 = -20\ \text{dB}, \qquad \text{implying } \alpha_1 = 0.1$$

$$\left(\frac{S}{N}\right)_2 = +10\ \text{dB}, \qquad \text{implying } \alpha_2 = 0.98$$

after which the ratio $\alpha_2/\alpha_1 = 9.8$ is too high.

By a successive approximation approach, we find $W = 2.5 \times 10^4$ Hz, leading to

$$\left(\frac{S}{N}\right)_1 = -14 \text{ dB}, \qquad \text{implying } \alpha_1 = 0.165$$

$$\left(\frac{S}{N}\right)_2 = +16 \text{ dB}, \qquad \text{implying } \alpha_2 = 0.99$$

The ratio $\alpha_2/\alpha_1 = 6$ is very close to the result required. Considering the value required for α_1, we can also use the approximation obtained from Eq. 9.3 for $\rho = S_i/N_i$ very small

$$\alpha_1 \cong (\sqrt{\pi}/2)\sqrt{(S/N)_1}$$

which, taking $\alpha_2 = 0.99$, leads to $\alpha_1 = 0.159$, $(S/N)_1 = -15$ dB, and $W = 3.2 \times 10^4$ Hz.

Consequently, to solve the problem, the loop will be preceded by a bandpass filter, having a bandwidth of around 25 kHz, and a limiter (or AGC). The loop parameters K, τ_1, and τ_2 will be chosen so that $\omega_n = 20$ rad/s and $\zeta = \frac{1}{2}$ for the low input signal ($S/N_0 = 30$ dB-Hz). For the high input signal ($S/N_0 = 60$ dB-Hz) we shall have $\omega_n = 20\sqrt{6} = 49$ rad/s and $\zeta = 1.23$.

PART THREE
NONLINEAR
OPERATION

INTRODUCTION

As previously mentioned, nearly all the cases of phase-locked loop practical application are covered by the linear analysis exposed in Chapters 4–9.

However, for one important field of application, the acquisition phase, we have to refer to the nonlinear differential equations of Chapter 3. As will be seen in Chapter 10, the loop natural performances as regards acquisition are not outstanding and most of the time an acquisition subsidiary device will be required, of the type described in Chapter 11.

Once the acquisition phase is over, the loops generally revert to their linear operation zone. But the engineer using a method that he knows to be only approximative tends to seek a means of checking the validity of his calculation. It is for this reason that we describe in Chapter 12 certain extreme operating conditions, rarely actually encountered, but useful as references indicative of the safety margins available under normal operating conditions.

CHAPTER 10
NATURAL ACQUISITION
OF THE INPUT SIGNAL

In Part II, we managed, to a certain extent, to disregard the phase detector nonlinearity, by assuming the phase error to be constantly fairly small. The phenomena involved were then governed by constant-coefficient linear differential equations, which are particularly easy to handle as long as the phase-locked loop order is not too high. The low phase error hypothesis implies VCO synchronism with the input signal prior to application of the various disturbances described in the previous chapters.

But, when the signal actually arrives at the loop input, an unknown phase difference exists between the input signal and the VCO signal, and the linear approximation is no longer valid. We have to take into account the real characteristic of the phase detector, which leads to a nonlinear differential equation, for which, generally speaking, no analytical solution is known (see, for instance, Eqs. 3.12, 3.34, or 3.40).

To simplify matters, we shall only deal with the case of an unmodulated input signal, without additive noise. We shall limit our analysis to the case of a sinusoidal phase detector. We have mentioned the possibility of other types of characteristics, but in the majority of the cases they are periodic, with a a 2π period. Apart from a few minor points, which will be indicated as they arise, the operating principle is not basically different for these types of phase detector.

10.1 SIGNAL ACQUISITION BY A FIRST-ORDER LOOP

Let us consider the case of a first-order loop, as shown in Fig. 10.1. The input and VCO signals are expressed, respectively, as

$$y_i(t) = A \cos(\omega_i t + \theta_i)$$

$$y_0(t) = B \cos(\omega_0 t + \phi_0)$$

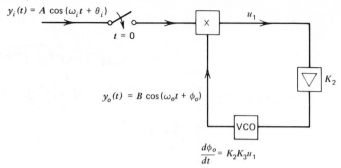

FIGURE 10.1.   First-order phase-locked loop.

with

$$\frac{d\phi_0}{dt} = K_2 K_3 u_1(t)$$

or again

$$\phi_0 = K_2 K_3 \int u_1(t)\, dt + C$$

C being an arbitrary constant.

The signal $y_i(t)$ is applied to the phase detector of sensitivity K_1 at instant $t=0$. For $t<0$, $u_1=0$ and $\phi_0 = C$. We shall assume that this constant is null by including it in the value of θ_i. At a given positive instant t, the phase detector output signal is written

$$u_1(t) = K_1 \cos\left[(\omega_i - \omega_0)t + \theta_i - \phi_0(t) \right]$$

and consequently

$$\frac{d\phi_0}{dt} = K_1 K_2 K_3 \cos\left[(\omega_i - \omega_0)t + \theta_i - \phi_0(t) \right]$$

The law governing the VCO output phase variation $\phi_0(t)$ from instant $t=0$, when the input signal is applied, is described by differential equation 10.1:

$$\frac{d\phi_0}{dt} = K \cos\left[(\omega_i - \omega_0)t + \theta_i - \phi_0(t) \right] \tag{10.1}$$

This equation is identical with Eq. 3.12, making allowances for the notations used for y_i and y_0. If we state that $\Omega_0 = \omega_i - \omega_0$ and $\phi(t) = \Omega_0 t + \theta_i - \phi_0(t)$, then

$$\frac{d\phi}{dt} = \Omega_0 - \frac{d\phi_0}{dt}$$

Equation 10.1 becomes

$$\frac{d\phi}{dt} = \Omega_0 - K\cos\phi(t)$$

which can be written, separating the variables,

$$dt = \frac{d\phi}{\Omega_0 - K\cos\phi} \qquad (10.2)$$

10.1.1   PHASE ACQUISITION

Let us first suppose, although the case is highly improbable, that $\Omega_0 = 0$. This means that, at instant $t = 0$, the input signal is at the same frequency as the VCO central frequency. These two signals are then only differentiated by a phase difference within the range $(0, 2\pi)$. Equation 10.2 becomes

$$\frac{d\phi}{\cos\phi} = -K\,dt$$

which is easily integrated to give

$$\tfrac{1}{2}\ln\frac{1+\sin\phi}{1-\sin\phi} = -K\,(t-t_0)$$

or

$$\frac{1+\sin\phi}{1-\sin\phi} = e^{-2Kt}e^{+2Kt_0}$$

The integration constant t_0 is eliminated if we state that at instant $t = 0$, the value of the phase difference ϕ is θ_i; this yields

$$\sin\phi = \frac{(1+\sin\theta_i)e^{-2Kt}-(1-\sin\theta_i)}{(1+\sin\theta_i)e^{-2Kt}+(1-\sin\theta_i)} \qquad (10.3)$$

Let us write $\sin\phi = x$; when $t \to \infty$, two cases may be encountered:

$$K > 0, \qquad x = \sin\phi \to -1, \qquad \phi \to -\frac{\pi}{2}$$

$$K < 0, \qquad x = \sin\phi \to +1, \qquad \phi \to +\frac{\pi}{2}$$

To find the direction of the variation of x between $t = 0$ and $t \to \infty$, we can calculate dx/dt. If we call D the denominator of expression 10.3, we obtain

$$\frac{dx}{dt} = -\frac{4K(1+\sin\theta_i)(1-\sin\theta_i)e^{-2Kt}}{D^2}$$

which means that the sign of dx/dt is constant and opposite to the sign of K. The variation of ϕ can be represented graphically for both cases (see Fig. 10.2).

CASE a. $K>0$. Using the trigonometric circle of Fig. 10.2a, x describes the vertical axis from $\sin\theta_i$ (for $t=0$) to -1 (for $t\to\infty$). According to the initial value of θ_i, $\phi(t)$ describes the corresponding arc, to the right or to the left.

It can be noted that if $\theta_i=\pi/2$, $\sin\theta_i=1$ leads to $\sin\phi=1$, so $\phi=\pi/2$. However, this equilibrium point is in fact unstable, for as soon as θ_i diverges

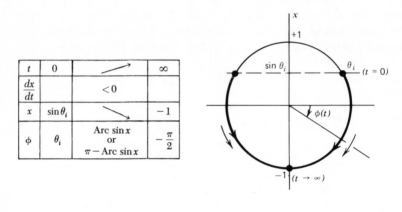

FIGURE 10.2. Trigonometrical representation of the phase error of a first-order loop during phase acquisition: (a) $K>0$; (b) $K<0$.

from value $\pi/2$ by a very small quantity ε, the sign of dx/dt is such that ϕ deviates definitively from the point $\pi/2$ and moves to the left or to the right (according to the sign of ε) towards the stable point of equilibrium $-\pi/2$.

CASE b. $K<0$. The results are here reversed, as is evident from the diagram of Fig. 10.2b. In this case the point $\phi=-\pi/2$ is an unstable equilibrium point.

Remark. Owing to the sinusoidal characteristic of the phase detector, the direction of the phase detector diodes or the sign of the VCO frequency variation versus the command signal can be disregarded. In standard control devices, such as AFC or AGC, the sign of the error signal must be taken into consideration, since one sign corresponds to a possible operating zone and the other to an impossible operating zone (the AFC decreases or increases the frequency offset). But here, this is not the case: the loop selects the zone of the characteristic having a slope with a suitable sign. According to circumstances, this may correspond to a VCO phase-lead or -lag with respect to the input signal phase.

This remark is also valid for detectors having a linear characteristic over an interval $(-\pi/2, +\pi/2)$ but does not apply to detectors linear over an interval $(-\pi, +\pi)$.

10.1.2 FREQUENCY ACQUISITION

We shall now suppose that, in addition to the initial phase difference θ_i, there is an angular frequency difference Ω_0 between the two signals. In this case, we refer to Eq. 10.2 and calculate $\int d\phi/(\Omega_0 - K\cos\phi)$.

Two cases can arise in connection with this calculation:

CASE a. $|\Omega_0| > |K|$. Then

$$\int \frac{d\phi}{\Omega_0 - K\cos\phi} = \frac{2}{\sqrt{\Omega_0^2 - K^2}} \arctan \frac{(\Omega_0 + K)\tan(\phi/2)}{\sqrt{\Omega_0^2 - K^2}}$$

leading to

$$t - t_0 = \frac{2}{\sqrt{\Omega_0^2 - K^2}} \arctan \frac{(\Omega_0 + K)\tan(\phi/2)}{\sqrt{\Omega_0^2 - K^2}} \tag{10.4}$$

The integration constant t_0 can be eliminated, noting that $\phi = \theta_i$ at instant

$t = 0$. But we also note that Eq. 10.4 above can be written

$$\frac{\Omega_0 + K}{\sqrt{\Omega_0^2 - K^2}} \tan \frac{\phi}{2} = \tan \left[\frac{\sqrt{\Omega_0^2 - K^2}}{2} (t - t_0) \right]$$

or

$$\phi(t) = 2 \arctan \left\{ \frac{\sqrt{\Omega_0^2 - K^2}}{\Omega_0 + K} \tan \left[\frac{\sqrt{\Omega_0^2 - K^2}}{2} (t - t_0) \right] \right\} \qquad (10.5)$$

When t varies, each time the quantity between the square brackets increases by π, $\phi(t)$ reverts to the same value. This means that $\phi(t)$ is a periodic time function of period

$$T = \frac{2\pi}{\sqrt{\Omega_0^2 - K^2}}$$

Consequently, when $|\Omega_0| > |K|$ there is no steady-state solution. The phase difference $\phi(t)$ travels within the phase detector characteristic according to a periodic law versus time. The resulting "beat" frequency decreases as Ω_0 approaches K.

CASE b. $|\Omega_0| < |K|$. Two cases can arise in the solutions of Eq. 10.2: If

$$\left| (\Omega_0 + K) \tan \frac{\phi}{2} \right| > \sqrt{K^2 - \Omega_0^2}$$

we have to use

$$t - t_0 = \frac{1}{\sqrt{K^2 - \Omega_0^2}} \ln \frac{(\Omega_0 + K) \tan (\phi/2) - \sqrt{K^2 - \Omega_0^2}}{(\Omega_0 + K) \tan (\phi/2) + \sqrt{K^2 - \Omega_0^2}} \qquad (10.6)$$

If

$$\left| (\Omega_0 + K) \tan \frac{\phi}{2} \right| < \sqrt{K^2 - \Omega_0^2}$$

the solution is

$$t - t_0 = \frac{1}{\sqrt{K^2 - \Omega_0^2}} \ln \frac{\sqrt{K^2 - \Omega_0^2} - (\Omega_0 + K) \tan(\phi/2)}{\sqrt{K^2 - \Omega_0^2} + (\Omega_0 + K) \tan(\phi/2)} \tag{10.7}$$

with, in both cases, $\phi = \theta_i$ for $t = 0$.

For the purposes of simplification, we shall restrict our discussion to the case where Ω_0 and K are both positive. The results corresponding to other possibilities are easily derived from this example. The limit condition becomes

$$\left| \tan \frac{\phi}{2} \right| > \frac{\sqrt{K^2 - \Omega_0^2}}{\Omega_0 + K} \quad \text{or} \quad \left| \tan \frac{\phi}{2} \right| < \frac{\sqrt{K^2 - \Omega_0^2}}{\Omega_0 + K}$$

It is then possible to determine on the trigonometric circle of Fig. 10.3 the arcs where either Eq. 10.6 or Eq. 10.7 apply, according to the value of $\phi/2$.

Equation 10.6, applicable for values of $\phi/2$ around $\pi/2$ and $3\pi/2$, can be written

$$e^{\sqrt{K^2 - \Omega_0^2}\,(t - t_0)} = \frac{(\Omega_0 + K) \tan(\phi/2) - \sqrt{K^2 - \Omega_0^2}}{(\Omega_0 + K) \tan(\phi/2) + \sqrt{K^2 - \Omega_0^2}}$$

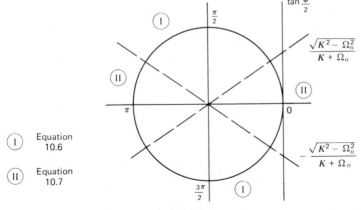

FIGURE 10.3.   Trigonometrical circle for the frequency acquisition study of a first-order loop.

from which we derive

$$\tan\frac{\phi}{2} = \frac{\sqrt{K^2-\Omega_0^2}}{\Omega_0+K}\,\frac{1+e^{\sqrt{K^2-\Omega_0^2}\,(t-t_0)}}{1-e^{\sqrt{K^2-\Omega_0^2}\,(t-t_0)}}$$

If we assume that at time $t=0$,

$$\left|\tan\frac{\theta_i}{2}\right| > \frac{\sqrt{K^2-\Omega_0^2}}{\Omega_0+K}$$

so that it is, in fact, Eq. 10.6 which applies, then

$$e^{-\sqrt{K^2-\Omega_0^2}\,t_0} = \frac{(K+\Omega_0)\tan(\theta_i/2)-\sqrt{K^2-\Omega_0^2}}{(K+\Omega_0)\tan(\theta_i/2)+\sqrt{K^2-\Omega_0^2}}$$

Substituting this expression in the formula giving $\tan(\phi/2)$, we can eliminate the integration constant t_0:

$$\tan\frac{\phi}{2} = \frac{\sqrt{K^2-\Omega_0^2}}{K+\Omega_0}$$

$$\times\;\frac{\tan\dfrac{\theta_i}{2}+\dfrac{\sqrt{K^2-\Omega_0^2}}{K+\Omega_0}+\left[\tan\dfrac{\theta_i}{2}-\dfrac{\sqrt{K^2-\Omega_0^2}}{K+\Omega_0}\right]e^{\sqrt{K^2-\Omega_0^2}\,t}}{\tan\dfrac{\theta_i}{2}+\dfrac{\sqrt{K^2-\Omega_0^2}}{K+\Omega_0}-\left[\tan\dfrac{\theta_i}{2}-\dfrac{\sqrt{K^2-\Omega_0^2}}{K+\Omega_0}\right]e^{\sqrt{K^2-\Omega_0^2}\,t}}\qquad(10.8)$$

In order to obtain the variations of ϕ versus time, we can state that

$$x = \frac{K+\Omega_0}{\sqrt{K^2-\Omega_0^2}}\tan\frac{\phi}{2}$$

and determine the sign of dx/dt; dx/dt has the same sign as its numerator,

that is,

$$2\sqrt{K^2-\Omega_0^2}\left[\tan\frac{\theta_i}{2}+\frac{\sqrt{K^2-\Omega_0^2}}{K+\Omega_0}\right]\left[\tan\frac{\theta_i}{2}-\frac{\sqrt{K^2-\Omega_0^2}}{K+\Omega_0}\right]e^{\sqrt{K^2-\Omega_0^2}\,t}$$

So dx/dt has the same sign as the quantity

$$\tan^2\frac{\theta_i}{2}-\left[\frac{\sqrt{K^2-\Omega_0^2}}{K+\Omega_0}\right]^2$$

Consequently, if

$$\left|\tan\frac{\theta_i}{2}\right|>\frac{\sqrt{K^2-\Omega_0^2}}{K+\Omega_0},$$

$(dx)/dt$ is positive, so x is an increasing function of time. When $t\rightarrow\infty$,

$$x\rightarrow-1\quad\text{and}\quad\tan\frac{\phi}{2}\rightarrow-\frac{\sqrt{K^2-\Omega_0^2}}{K+\Omega_0}$$

The variations of $\tan(\phi/2)$ and $\phi/2$ are shown in the diagram of Fig. 10.4a. The instant t_1, corresponding to the passage of $\phi/2$ through the values $\pi/2$ or $3\pi/2$, only exists if $\tan(\theta_i/2)>0$.

Equation 10.7, which applies for the values of $\phi/2$ around points 0 and π, leads to

$$\tan\frac{\phi}{2}=\frac{\sqrt{K^2-\Omega_0^2}}{K+\Omega_0}\frac{1-e^{\sqrt{K^2-\Omega_0^2}\,(t-t_0)}}{1+e^{\sqrt{K^2-\Omega_0^2}\,(t-t_0)}}$$

If we assume that at time $t=0$, the phase difference θ_i is such that

$$\left|\tan\frac{\theta_i}{2}\right|<\frac{\sqrt{K^2-\Omega_0^2}}{K+\Omega_0}$$

250

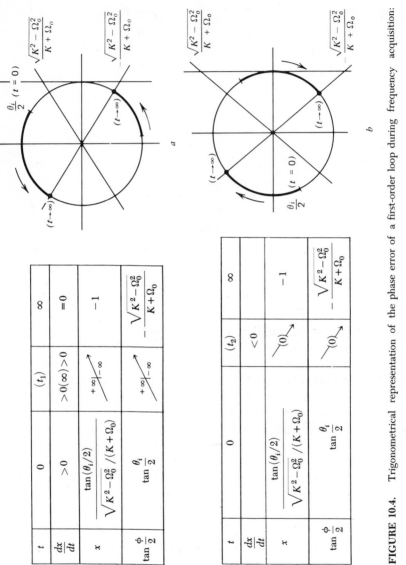

FIGURE 10.4. Trigonometrical representation of the phase error of a first-order loop during frequency acquisition:

$$\text{(a)} \quad \left|\tan\frac{\theta_i}{2}\right| > \frac{\sqrt{K^2-\Omega_0^2}}{K+\Omega_0} \qquad \text{(b)} \quad \left|\tan\frac{\theta_i}{2}\right| < \frac{\sqrt{K^2-\Omega_0^2}}{K+\Omega_0}$$

Table a:

t	0	(t_1)	∞
$\dfrac{dx}{dt}$	>0	$>0 \;(\infty)>0$	$=0$
x	$\dfrac{\tan(\theta_i/2)}{\sqrt{K^2-\Omega_0^2}/(K+\Omega_0)}$	$+\infty\big|{-}\infty$	-1
$\tan\dfrac{\phi}{2}$	$\tan\dfrac{\theta_i}{2}$	$+\infty\big|{-}\infty$	$-\dfrac{\sqrt{K^2-\Omega_0^2}}{K+\Omega_0}$

Table b:

t	0	(t_2)	∞
$\dfrac{dx}{dt}$		<0	
x	$\dfrac{\tan(\theta_i/2)}{\sqrt{K^2-\Omega_0^2}/(K+\Omega_0)}$	$(0)\nearrow$	-1
$\tan\dfrac{\phi}{2}$	$\tan\dfrac{\theta_i}{2}$	$(0)\nearrow$	$-\dfrac{\sqrt{K^2-\Omega_0^2}}{K+\Omega_0}$

we get

$$\tan\frac{\phi}{2} = \frac{\sqrt{K^2-\Omega_0^2}}{K+\Omega_0}$$

$$\times \frac{\left[\dfrac{\sqrt{K^2-\Omega_0^2}}{K+\Omega_0} + \tan\dfrac{\theta_i}{2}\right] - \left[\dfrac{\sqrt{K^2-\Omega_0^2}}{K+\Omega_0} - \tan\dfrac{\theta_i}{2}\right] e^{\sqrt{K^2-\Omega_0^2}\,t}}{\left[\dfrac{\sqrt{K^2-\Omega_0^2}}{K+\Omega_0} + \tan\dfrac{\theta_i}{2}\right] + \left[\dfrac{\sqrt{K^2-\Omega_0^2}}{K+\Omega_0} - \tan\dfrac{\theta_i}{2}\right] e^{\sqrt{K^2-\Omega_0^2}\,t}} \qquad (10.9)$$

The quantity dx/dt is then negative, that is, $\tan(\phi/2)$ is a decreasing function of time. When $t\to\infty$,

$$x\to -1 \quad \text{and} \quad \tan\frac{\phi}{2} \to -\frac{\sqrt{K^2-\Omega_0^2}}{K+\Omega_0}$$

The variations of $\tan(\phi/2)$ and $\phi/2$ are shown in the diagram of Fig. 10.4b. Instant t_2, corresponding to the possible passage of $\tan(\phi/2)$ through the value 0, only exists if $\tan(\theta_i/2)$ verifies the condition $\tan(\theta_i/2) > 0$.

Having considered all the possibilities for $\theta_i/2$, we can derive the variations of $\theta/2$, since $\tan(\phi/2)$ is a monotonic function, increasing or decreasing according to the initial value of $\theta_i/2$.

First of all, whatever the initial phase difference θ_i, the final value of ϕ, expressed ϕ_∞, is given by (see Eqs. 10.8 and 10.9)

$$\tan\frac{\phi_\infty}{2} = -\frac{\sqrt{K^2-\Omega_0^2}}{K+\Omega_0}$$

leading to $\cos\phi_\infty = \Omega_0/K$. Consequently, taking into account the sign of $\tan(\phi/2)$, the stable-state phase difference is

$$\phi_\infty = -\text{Arc}\cos\frac{\Omega_0}{K}$$

(In particular, for $\Omega_0 = 0$, we find $\phi_\infty = -\pi/2$, corresponding to the phase acquisition for $K > 0$ discussed in Section 10.1.1.)

The stable-state VCO phase is then

$$\phi_{0\infty} = \Omega_0 t + \theta_i - \phi_\infty = \Omega_0 t + \theta_i + \text{Arc}\cos\frac{\Omega_0}{K}$$

that is, the VCO output signal is written

$$y_0(t) = B\cos\left(\omega_0 t + \Omega_0 t + \theta_i + \text{Arc}\cos\frac{\Omega_0}{K}\right)$$

$$= B\cos\left(\omega_i t + \theta_i + \text{Arc}\cos\frac{\Omega_0}{K}\right)$$

Thus the VCO phase leads the input signal $y_i(t)$ by an angle equal to $\text{Arc}\cos(\Omega_0/K)$.

Secondly, we observe that the condition

$$\tan\frac{\theta_i}{2} = \frac{\sqrt{K^2 - \Omega_0^2}}{K + \Omega_0}$$

gives in both cases (Eq. 10.8 or 10.9) and whatever the value of t

$$\tan\frac{\phi}{2} = \tan\frac{\theta_i}{2}$$

The point $\phi = \text{Arc}\cos(\Omega_0/K)$ is therefore an equilibrium point. But it is also observed to be an unstable point. The slightest deviation causes $\phi/2$ to move either into the zone covered by Eq. 10.6 or that covered by Eq. 10.7, and in both cases the $\phi/2$ trajectory on the trigonometric circle ends at the point where $\phi = -\text{Arc}\cos(\Omega_0/K)$.

The diagram in Fig. 10.5 shows the path followed along the phase detector characteristic by the phase difference between the input signal and the VCO signal, according to the initial value of this phase difference.

From this we can derive the four types of curves, for the phase detector output signal $u_1 = f(t)$, based on the initial phase error value with respect to the final value. These curves are given in Fig. 10.6.

The results for $\Omega_0 < 0$ and $K > 0$ are easily derived from those obtained above and the stable equilibrium points correspond to $\text{Arc}\cos(\Omega_0/K) + \pi +$

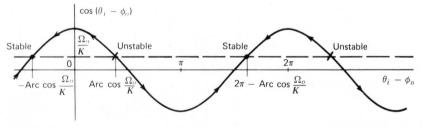

FIGURE 10.5. Path along the phase detector characteristic during the frequency acquisition of a first-order loop.

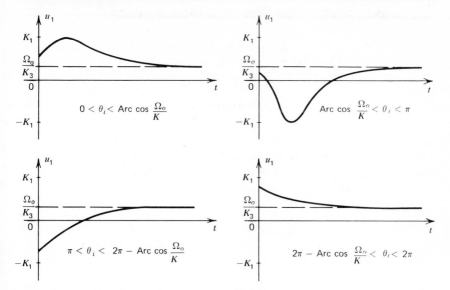

FIGURE 10.6. Phase detector output signal during the frequency acquisition of a first-order loop. (If $K_2 \neq 1$, replace K_3 by $K_2 K_3$).

$k2\pi$. When K is negative, the results are reversed, that is, the unstable equilibrium points become stable and vice versa.

In conclusion, for a first-order phase-locked loop, if the difference between the input signal angular frequency and the VCO central angular frequency is lower than the loop gain K, after the transients shown in Fig. 10.6, the VCO reaches synchronism with the input signal. The loop is then said to be locked or the VCO acquisition of the input signal is said to be complete. The acquisition range then corresponds to the angular frequency interval $(\omega_0 - K, \omega_0 + K)$, where ω_0 is the VCO central angular frequency. For every angular frequency ω_i of the input signal within this interval, the steady-state phase error, corresponding to the phase detector dc output signal necessary to shift the VCO angular frequency from ω_0 to ω_i, is given by

$$- \operatorname{Arc} \cos \frac{\omega_i - \omega_0}{K} \qquad \left(\text{if } \omega_i > \omega_0 \text{ and } K > 0\right)$$

For a first-order loop, the acquisition range is identical with the synchronization range defined in Chapter 1.

If the initial angular frequency difference exceeds the loop gain K in absolute value, the VCO does not become synchronous with the input signal.

This is indicated by the appearance of an angular frequency $\sqrt{\Omega_0^2 - K^2}$ beat signal at the phase detector output.

10.2 SIGNAL ACQUISITION BY A SECOND-ORDER LOOP

The circuit considered is shown in Fig. 10.7. The transfer function $F(j\omega)$ of the loop filter is given by Eqs. 2.7, 2.8, or 2.10. The signals applied to the phase detector are expressed as

$$y_i(t) = A\cos(\omega_i t + \theta_i)$$

$$y_0(t) = B\cos(\omega_0 t + \phi_0)$$

where $d\phi_0/dt = K_2 K_3 u_2(t)$, $u_2(t)$ being the phase detector output signal $u_1(t)$ filtered by the loop filter and, if necessary amplified in K_2.

Calculations are facilated by taking the expression used in Chapter 3 for the input signal $y_i(t)$:

$$y_i(t) = A\sin\left[\omega_0 t + \phi_i(t)\right]$$

with

$$\phi_i(t) = (\omega_i - \omega_0)t + \theta_i + \frac{\pi}{2} = \Omega_0 t + \theta_i + \frac{\pi}{2}$$

where $\theta_i + \pi/2$ is the instantaneous phase difference between the two signals at instant $t = 0$ and Ω_0 the initial angular frequency difference. The phase error $\phi(t) = \phi_i(t) - \phi_0(t)$ is then given by Eq. 3.34 for a filter $F(s) = (1 + \tau_2 s)/\tau_1 s$, by Eq. 3.37 for a filter $F(s) = 1/(1 + \tau_1 s)$, and by Eq. 3.40 for a filter $F(s) = (1 + \tau_2 s)/(1 + \tau_1 s)$.

These equations are second-order nonlinear differential equations, analytically insoluble. We shall first briefly describe the graphic method, known as the phase-plane method, and then an approximate analytical method based on the physical behavior of the device. We shall concentrate on the cases involving loop filters $F(s) = (1 + \tau_2 s)/\tau_1 s$ and $F(s) = (1 + \tau_2 s)/(1 + \tau_1 s)$, which are the most frequently used.

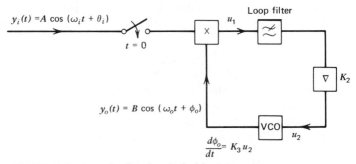

FIGURE 10.7. Second-order phase-locked loop.

10.2.1   PHASE-PLANE METHOD

Taking into account the expression for $\phi_i(t)$ given above, Eqs. 3.34 and 3.40 become, respectively,

$$\tau_1 \frac{d^2\phi}{dt^2} + K\tau_2 \cos\phi \cdot \frac{d\phi}{dt} + K\sin\phi = 0 \qquad (10.10)$$

$$\tau_1 \frac{d^2\phi}{dt^2} + (1 + K\tau_2\cos\phi)\frac{d\phi}{dt} + K\sin\phi = \Omega_0 \qquad (10.11)$$

If we divide each member by $\tau_1\dot{\phi}$, with $\dot{\phi} = d\phi/dt$ and $\ddot{\phi} = d^2\phi/dt^2$, we get for these equations

$$\frac{\ddot{\phi}}{\dot{\phi}} + \frac{K\tau_2\cos\phi}{\tau_1} + \frac{K\sin\phi}{\tau_1\dot{\phi}} = 0$$

$$\frac{\ddot{\phi}}{\dot{\phi}} + \frac{1 + K\tau_2\cos\phi}{\tau_1} + \frac{K\sin\phi}{\tau_1\dot{\phi}} = \frac{\Omega_0}{\tau_1\dot{\phi}}$$

Note that

$$\ddot{\phi} = \frac{d}{dt}(\dot{\phi}) = \frac{d}{d\phi}(\dot{\phi})\cdot\dot{\phi}$$

that is,

$$\ddot{\phi} = \frac{d\dot{\phi}}{d\phi}\dot{\phi}$$

For each of the two cases in question, we obtain

$$\frac{d\dot{\phi}}{d\phi} = -\frac{K\sin\phi}{\tau_1\dot{\phi}} - \frac{K\tau_2\cos\phi}{\tau_1} \qquad (10.12)$$

for a loop with integrator and correction network, and

$$\frac{d\dot{\phi}}{d\phi} = \frac{\Omega_0 - K\sin\phi}{\tau_1\dot{\phi}} - \frac{1 + K\tau_2\cos\phi}{\tau_1} \qquad (10.13)$$

for a loop with low-pass filter and correction network.

We can then make a plot representing loop operation. For each value of ϕ (x axis) and $\dot{\phi}$ (y axis), using Eqs. 10.12 and 10.13, we can calculate the slope of the trajectory and thus obtain an overall trajectory graph. Since the constants K, τ_1, and τ_2 are included, this calculation will necessitate the use of a numerical or analog computer. We finally obtain a trajectory network similar to those shown in Figs. 10.8 to 10.11. The main results reported here are taken from an article by Viterbi (34).

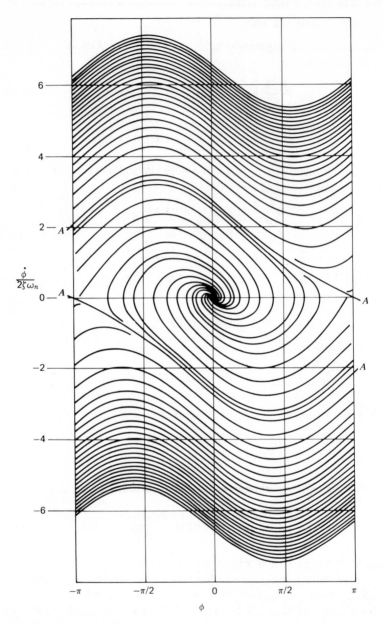

FIGURE 10.8. Phase-plane trajectories for a loop with $F(s) = (1 + \tau_2 s)/(\tau_1 s)$ when $\zeta = 0.5$ (from A. J. Viterbi; furnished through the courtesy of the Jet Propulsion Laboratory, California Institute of Technology).

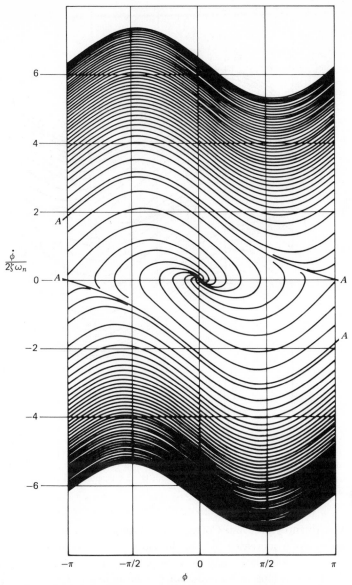

FIGURE 10.9. Phase-plane trajectories for a loop with $F(s) = (1 + \tau_2 s)/(\tau_1 s)$ when $\zeta = \sqrt{2}/2$ (from A. J. Viterbi; furnished through the courtesy of the Jet Propulsion Laboratory, California Institute of Technology).

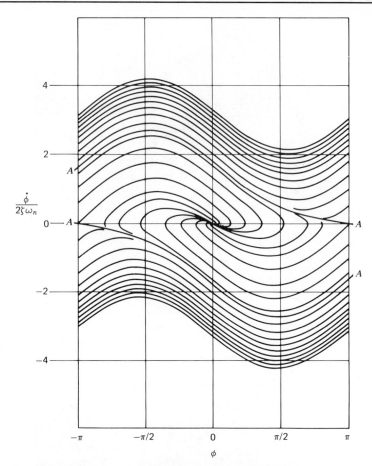

FIGURE 10.10. Phase-plane trajectories for a loop with $F(s) = (1 + \tau_2 s)/(\tau_1 s)$ when $\zeta = 1$ (from A. J. Viterbi; furnished through the courtesy of the Jet Propulsion Laboratory, California Institute of Technology).

a. Loop with Integrator and Correction Network. It is apparent from Eq. 10.12 that when ϕ increases by 2π, the slope $d\dot{\phi}/d\phi$ returns to the same value; the trajectories need then only be plotted in an interval of value 2π for ϕ. Moreover, as long as the value of $\dot{\phi}$ is fairly high,

$$\frac{d\dot{\phi}}{d\phi} \cong -K\frac{\tau_2}{\tau_1}\cos\phi = -2\zeta\omega_n\cos\phi$$

and the trajectories are practically sinusoidal. But when ϕ passes from $-\pi$ to $+\pi$, $\dot{\phi}$ is observed to decrease slightly. Since the diagram follows the ϕ

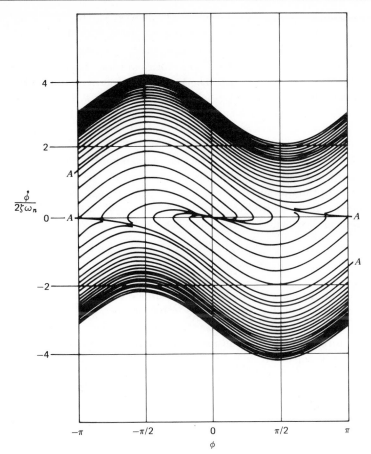

FIGURE 10.11. Phase-plane trajectories for a second-order loop with $F(s) = (1 + \tau_2 s)/(\tau_1 s)$ when $\zeta = \sqrt{2}$ (from A. J. Viterbi; furnished through the courtesy of the Jet Propulsion Laboratory, California Institute of Technology).

period, when the point $\phi = \pi$ is reached on any of the trajectories, since this point corresponds to a certain value of $\dot\phi$, this value must be kept when we start again from $\phi = -\pi$ (as regards the trajectories in the upper half-plane, traversed from left to right), up to the moment when the trajectory is represented by the separatrix A.A. (see Figs. 10.8 to 10.11). Below this separatrix the trajectory ends at point $(\phi_\infty = 0,\ \dot\phi_\infty = 0)$ without ϕ varying by more than 2π; the frequency acquisition is then said to be complete.

The separatrix A.A. of the upper half-plane ends at point $(\phi = \pi, \dot\phi = 0)$ corresponding to an unstable operating point.

Depending on the value of the damping factor, the stable operating point $(\phi_\infty = 0, \dot\phi_\infty = 0)$ is reached without being encircled $(\zeta > 1)$ or after being encircled several times $(\zeta < 1)$ by the trajectory.

Equation 10.12 indicates that the trajectories in Figs. 10.8 to 10.11 depend only on the parameters ω_n and ζ. The loop gain K plays no role in this case since the loop filter is assumed to use a perfect integrator. It is for the same reason that the acquisition range is infinite: whatever the initial phase and frequency difference between the input and VCO signals, the trajectories end at a stable point $(\phi_\infty = 0 + k2\pi, \dot\phi_\infty = 0)$.

By calculating the decrease in $\dot\phi$ for each variation of ϕ between $-\pi$ and $+\pi$, Viterbi succeeded in calculating approximately the acquisition time, defined as the time required to reach the separatrix A.A., starting from any given initial condition $[\phi(0) = \theta_i + \pi/2, \dot\phi(0) = \Omega_0 - 2\zeta\omega_n \sin\phi(0)]$. This acquisition time is approximated by the formula

$$T_{acq} = \frac{\Omega_0^2}{2\zeta\omega_n^3} \tag{10.14}$$

It will be noted that the initial phase difference $\phi(0)$ is disregarded in this expression, and that the greater the initial frequency difference Ω_0 the greater the validity of the formula.

The acquisition is generally considered as complete once the trajectory has gone beyond the separatrix A.A., since there can be no further cycle-slipping because ϕ remains inside the interval $(-\pi, +\pi)$ considered. Strictly speaking, acquisition can only be considered as complete when ϕ and $\dot\phi$ are both very close to zero or, at least, when their value is compatible with the intended application. However, for high values of Ω_0, the time required to complete acquisition after crossing the separatrix A.A. can be considered as small as compared with the time required to reach the separatrix.

When the initial difference Ω_0 is slight, it is preferable to use the phase-plane trajectory diagram directly, with the addition of the isochronous curves as in Fig. 10.12 (39). This graph corresponds to a damping factor $\zeta = \sqrt{2}/2$. The isochronous curves are dashed lines and the distance between two consecutive curves is $1/4\omega_n$. Starting from any given point on the diagram, we can follow the corresponding trajectory and count the number of isochronous curves crossed before reaching the zone where acquisition can be considered as complete for the application concerned (depending on circumstances, this could be a phase error condition or an angular frequency error condition or even both at once).

For intermediate values of Ω_0, we can use the curves in Fig. 10.13, taken from the same reference (39). It should be noted that these curves are only valid for a null initial phase difference $\phi(0)$ (and still for $\zeta = \sqrt{2}/2$) and correspond to a final phase error of 0.2 rad.

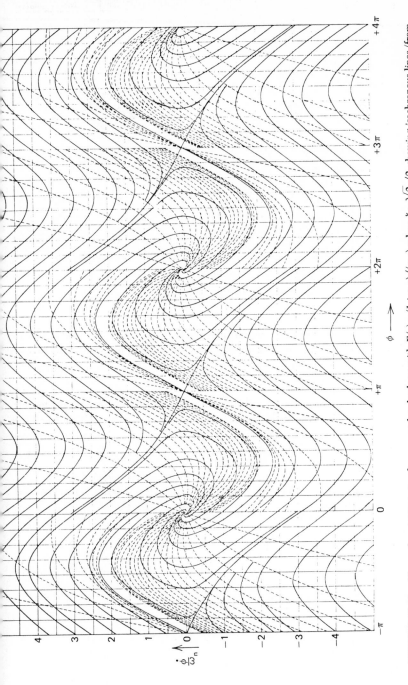

FIGURE 10.12. Phase-plane trajectories for a second-order loop with $F(s) = (1 + \tau_2 s)/(\tau_1 s)$ when $\zeta = \sqrt{2}/2$, showing isochronous lines (from R. W. Sanneman and J. R. Rowbotham, *IEEE Trans. on Aerospace and Navigational Electronics*, March 1964 ANE-11, p. 19).

$\phi \longrightarrow$

$\dfrac{\dot{\phi}}{\omega_n} \longleftarrow$

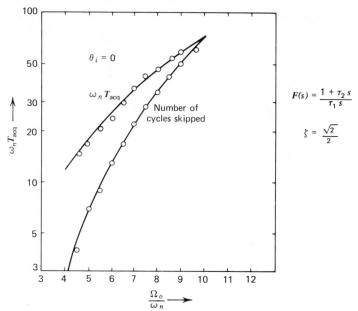

FIGURE 10.13. Acquisition time versus initial frequency error (from R. W. Sanneman and J. R. Rowbotham, *IEEE Trans. on Aerospace and Navigational Electronics*, March 1964 ANE-11, p. 20).

We can check that the acquisition-time-curve results are very close to those obtained with Eq. 10.14, even for values of Ω_0 as low as 4 or 5 ω_n. But it is important to bear in mind that when Ω_0 is small, the acquisition time value is sensitive to the initial phase difference value $\phi(0)$, as is evident in Fig. 10.12.

NUMERICAL EXAMPLE. Let us take a second-order loop $[F(s) = (1 + \tau_2 s)/\tau_1 s]$ such that

$$\zeta = \frac{\sqrt{2}}{2}, \qquad \omega_n = 100 \text{ rad/s}$$

(a) The initial frequency difference between the input signal and the VCO is 150 Hz:

$$\Omega_0 = 2\pi \times 150 = 942 \text{ rad/s}$$

The acquisition time obtained from Eq. 10.14 is

$$T_{acq} = \frac{942^2}{\sqrt{2} \times 10^6} = 628 \text{ ms}$$

The curve in Fig. 10.13 leads to $\omega_n T_{acq} \cong 65$, that is $T_{acq} \cong 650$ ms.

(b)   If the initial angular frequency difference is $\Omega_0 = 400$ rad/s, the application of Eq. 10.14 leads to

$$T_{acq} = \frac{16 \times 10^4}{\sqrt{2} \times 10^6} = 113 \text{ ms}$$

Using the curve in Fig. 10.13, we get $\omega_n T_{acq} \cong 12$, that is $T_{acq} \cong 120$ ms.

In fact, if we take into account the initial phase difference $\phi(0)$, we have to use the curves of Figs. 10.9 and 10.12. Starting with $\phi(0)$ and $\dot{\phi}(0) = \Omega_0 - 2\zeta\omega_n \sin\phi(0)$, the phase error stays below 0.2 rad after a time:

$$T \cong \frac{43}{4\omega_n} = 108 \text{ ms}, \quad \text{for } \phi(0) = \pi,$$

with 3 slipping cycles

$$T \cong \frac{40}{4\omega_n} = 100 \text{ ms}, \quad \text{for } \phi(0) = \frac{\pi}{2},$$

with 2 or 3 slipping cycles (according to

the accuracy of the plot)

$$T \cong \frac{33}{4\omega_n} = 83 \text{ ms}, \quad \text{for } \phi(0) = 0,$$

with 2 slipping cycles

$$T \cong \frac{42}{4\omega_n} = 105 \text{ ms}, \quad \text{for } \phi(0) = -\frac{\pi}{2},$$

with 3 slipping cycles.

In the case of a loop with integrator and correction network, we can also make a phase-plane study of loop operation when the input signal is affected by a linear frequency drift. The input signal phase $\phi_i(t)$ can then be expressed as

$$\phi_i(t) = \tfrac{1}{2}\Re\, t^2 + \Omega_0 t + \theta_i + \frac{\pi}{2}$$

and Eq. 3.34 becomes

$$\tau_1 \ddot{\phi} + K\tau_2 \cos\phi \cdot \dot{\phi} + K \sin\phi = \tau_1 \mathcal{R} \qquad (10.15)$$

The main conclusions obtained (38) are as follows. First of all, the diagrams all have a ϕ period basis but, within the interval $(-\pi < \phi < +\pi)$, the curves cease to be symmetrical with respect to the origin. The phenomenon differs according to whether the slope $\mathcal{R}$ tends to increase or decrease the difference between the input signal frequency and the VCO central frequency.

The stable operating point coordinates when this point exists, are: $\phi_\infty = \text{Arc}\sin(\mathcal{R}/\omega_n^2)$ and $\dot{\phi}_\infty = 0$. The various cases encountered can be analyzed with reference to the parameter $\mathcal{R}/\omega_n^2$:

- When $\mathcal{R}/\omega_n^2$ is small, if the sign of $\mathcal{R}$ is such that the difference between the signal frequencies tends to decrease, acquisition always takes place. If the sign of $\mathcal{R}$ is such that the difference widens, acquisition may take place if the initial angular frequency difference Ω_0 is not too large, or may not take place if Ω_0 is too large.

- When $\mathcal{R}/\omega_n^2 = \frac{1}{2}$, if the sign of $\mathcal{R}$ is such that the frequency difference widens, acquisition cannot take place. If the sign of $\mathcal{R}$ is such that the difference tends to decrease, acquisition still takes place.

- When $\frac{1}{2} < \mathcal{R}/\omega_n^2 < 1$, even in the case of frequency difference decrease, acquisition does not necessarily take place: the phase plane diagram comprises a "corridor" where acquisition is precluded. This corridor is relatively narrow if $\mathcal{R}/\omega_n^2$ is around 0.5 but widens gradually as $\mathcal{R}/\omega_n^2$ approaches 1.

- For $\mathcal{R}/\omega_n^2 \geqslant 1$, none of the trajectories can lead to acquisition, since, in this case, the steady-state phase error (corresponding to the stable operating point) $\text{Arc}\sin(\mathcal{R}/\omega_n^2)$ has no significance. Not only the acquisition cannot take place, but if the loop was initially locked, it falls definitively out of lock.

b. Loop with Low-Pass Filter and Corrrection Network $[F(s) = (1 + \tau_2 s)/(1 + \tau_1 s)]$. The phase plane trajectories are derived from Eq. 10.13. Viterbi (38) has shown that the coordinates of the stable operating point, when this point exists, are

$$\phi_\infty = \text{Arc}\sin\frac{\Omega_0}{K}, \qquad \dot{\phi}_\infty = 0$$

Consequently, a loop initially in lock falls definitively out of lock if the angular frequency difference Ω_0 exceeds the loop gain K, even when this frequency difference is obtained by very delicate, almost imperceptible variations of the input signal frequency. We thus return to the notion of synchronization range introduced in Chapter 1. When $\Omega_0 > K$, without a perfect integrator, the loop is incapable of producing the dc signal required by the VCO to follow the input signal frequency.

Moreover, when $\Omega_0 < K$, acquisition does not necessarily take place. It only occurs if the initial angular frequency difference is below, in absolute value, a certain value Ω_{acq}. An approximate formula for Ω_{acq} is as follows (38):

$$\Omega_{acq} = 2\sqrt{K\left(\zeta\omega_n + \frac{1}{2\tau_1}\right)} \qquad (10.16)$$

For most second-order loops, the loop gain K far exceeds the natural angular frequency ω_n and consequently the quantity $1/2\tau_1$ is negligible as compared with $\zeta\omega_n$. We can then use the approximate formula

$$\Omega_{acq} = 2\sqrt{K\zeta\omega_n} \qquad (10.17)$$

When the phase detector characteristic is not sinusoidal but linear within a certain interval, a certain number of results can be obtained by analytical approximative methods. Thus for a linear detector over an interval $(-\pi/2, +\pi/2)$ (sometimes called a "triangular" phase detector), the limit value of the angular frequency compatible with acquisition is given by (40)

$$\Omega_{acq} = \pi\sqrt{K\zeta\omega_n}$$

and the synchronization range is defined by

$$\Omega_{syn} = K\frac{\pi}{2}$$

As compared with the results obtained for the sinusoidal detector, we would here seem to benefit directly from the $\pi/2$ gain obtained on the maximum dc-component value furnished by the phase detector.

NUMERICAL EXAMPLE. A second-order loop $[F(s) = (1 + \tau_2 s)/(1 + \tau_1 s)]$ is such that

$$\zeta = \frac{\sqrt{2}}{2}, \qquad \omega_n = 100 \text{ rad/s}, \qquad K = 2 \times 10^5 \text{ rad/s}$$

The time constant $\tau_1 = K/\omega_n^2 = 20$ s is such that the quantity $1/2\tau_1$ is negligible, compared to $\zeta\omega_n$. The acquisition range can then be derived from

Eq. 10.17, if the phase detector is sinusoidal:

$$\Omega_{acq} = 2\sqrt{2 \times 10^5 \frac{\sqrt{2}}{2} 100}$$

that is,

$$\Omega_{acq} = 7.52 \times 10^3 \text{ rad/s}$$

If the input signal frequency differs from the VCO central frequency by less than 1.2 kHz, acquisition will occur naturally. The acquisition range will be 1.88 kHz with a triangular phase detector.

c. Second-Order Loop with Low-Pass Filter $[F(s) = 1/(1 + \tau_1 s)]$. In the case where the loop filter consists of a single-pole, low-pass filter, of time constant $\tau_1 = RC$, a certain number of analytical results are available.

For a linear phase detector over an interval $(-\pi/2, +\pi/2)$ the ratio γ_P between the acquisition Ω_{acq} and synchronization Ω_{syn} ranges $[\Omega_{syn} = K(\pi/2)$ is this case] is given by (41)

$$\frac{1 - \gamma_P}{1 + \gamma_P} = \frac{\sqrt{\Delta + 1} - 1}{\sqrt{\Delta + 1} + 1} \exp\left[\frac{-\pi + \text{Arc} \tan \sqrt{\Delta^2 - 1}}{\sqrt{\Delta - 1}} \right]$$

$$\text{if } \Delta = 4K\tau_1 > 1$$

and

$$\gamma_P = 1 \qquad \text{if } 4K\tau_1 \leqslant 1$$

For a linear phase detector over an interval $(-\pi, +\pi)$ (sometimes called a "sawtooth" detector) the ratio γ_P is given by (42)

$$\gamma_P = \tanh \frac{\pi}{4} \left(K\tau_1 - \tfrac{1}{4} \right)^{-1/2}, \quad \text{if } 4K\tau_1 > 1$$

and

$$\gamma_P = 1 \qquad\qquad\qquad \text{if } 4K\tau_1 \leqslant 1$$

It must of course be remembered that for this type of detector the synchronization range Ω_{syn} is then given by $\Omega_{syn} = K\pi$.

It should also be borne in mind that, for this type of loop filter, when $\tau_1 \leqslant 1/4K$, the loop works very much like a first-order loop, and the time constant τ_1 can be considered as a low-value parasitic time constant (see curves in Fig. 6.19); it is consequently not unexpected to find that the acquisition and synchronization ranges are equal, as for a first-order loop (Section 10.1.2).

10.2.2 ANALYTICAL APPROXIMATIVE METHOD

We can duplicate analytically the main results established by Viterbi using the phase-plane procedure for a second-order phase-locked loop with a sinusoidal detector. We can use an approximate method, which is interesting in that it is based on the physical behavior and produces results that are a very good approximation.

At instant $t = 0$ when the input signal is applied to the phase detector, the detector output signal is a sinusoidal signal of angular frequency Ω_0, where Ω_0 is the difference between the input signal angular frequency ω_i and the VCO central angular frequency ω_0 (the $\omega_i + \omega_0$ angular frequency term is assumed to have been suppressed by filtering). After going through the loop filter, this signal of angular frequency Ω_0 modulates the VCO. The VCO signal spectrum is then composed of discrete frequency lines and one of the lines is at the same frequency as the input signal. The result is that the phase detector output, in addition to the Ω_0 beat signal, comprises a dc component that tends to move the average VCO frequency towards the input signal frequency. If we study this phenomenon in relation to time we can obtain a clear conception of acquisition.

Let us consider a sinusoidal phase detector driven by the input signal $y_i(t)$ and the VCO output signal $y_0(t)$:

$$y_i(t) = A \cos \omega_i t$$

$$y_0(t) = B \cos (\omega_0 t + \phi_0)$$

Let us suppose that the VCO is frequency modulated by a signal expressed as

$$v = u + \Delta u \cos (\Omega t - \psi) \qquad (10.18)$$

If K_3 is the VCO modulation sensitivity,

$$y_0(t) = B \cos \left[(\omega_0 + K_3 u)t + \theta_0 + K_3 \frac{\Delta u}{\Omega} \sin (\Omega t - \psi) \right]$$

Let us also suppose that the angular frequency deviation $K_3 \Delta u$ is small as compared with the modulation angular frequency Ω, so that

$$\cos \left[\frac{K_3 \Delta u}{\Omega} \sin (\Omega t - \psi) \right] \cong J_0 \left(\frac{K_3 \Delta u}{\Omega} \right)$$

$$\sin \left[\frac{K_3 \Delta u}{\Omega} \sin (\Omega t - \psi) \right] \cong 2 J_1 \left(\frac{K_3 \Delta u}{\Omega} \right) \sin (\Omega t - \psi)$$

If we disregard the angular frequency terms around $\omega_i + \omega_0$, the phase detector output signal is expressed as

$$u_1(t) = K_1 J_0\left(\frac{K_3 \Delta u}{\Omega}\right) \cos\left[(\omega_i - \omega_0 - K_3 u)t - \theta_0\right]$$

$$+ K_1 J_1\left(\frac{K_3 \Delta u}{\Omega}\right)\{\cos\left[(\omega_i - \omega_0 - K_3 u - \Omega)t - \theta_0 + \psi\right]$$

$$- \cos\left[(\omega_i - \omega_0 - K_3 u + \Omega)t - \theta_0 - \psi\right]\} + \cdots$$

Let us now state that the angular frequency Ω of the modulation signal is in fact equal to the difference in average angular frequency between $y_i(t)$ and $y_0(t)$:

$$\Omega = \omega_i - \omega_0 - K_3 u \qquad (10.19)$$

Then

$$u_1(t) = K_1 J_0\left(\frac{K_3 \Delta u}{\Omega}\right) \cos(\Omega t - \theta_0)$$

$$+ K_1 J_1\left(\frac{K_3 \Delta u}{\Omega}\right)\left[\cos(\psi - \theta_0) - \cos(2\Omega t - \theta_0 - \psi)\right]$$

That is, signal $u_1(t)$ comprises three terms:

- a dc component, of value

$$u_{10} = K_1 J_1\left(\frac{K_3 \Delta u}{\Omega}\right) \cos(\psi - \theta_0)$$

- an ac component at angular frequency Ω:

$$u_{11} = K_1 J_0\left(\frac{K_3 \Delta u}{\Omega}\right) \cos(\Omega t - \theta_0)$$

- an ac component at angular frequency 2Ω:

$$u_{12} = -K_1 J_1\left(\frac{K_3 \Delta u}{\Omega}\right) \cos(2\Omega t - \theta_0 - \psi)$$

We assumed previously that the modulation index $K_3(\Delta u / \Omega)$ is fairly small. Consequently, the amplitude of the 2Ω angular frequency component is smaller than that of the Ω angular frequency component and will be disregarded in the subsequent equations. We thus obtain for the phase detector output signal, applied to the loop filter, the expression

$$u_1(t) \cong K_1 J_1\left(K_3 \frac{\Delta u}{\Omega}\right)\cos\left(\psi - \theta_0\right)$$

$$+ K_1 J_0\left(K_3 \frac{\Delta u}{\Omega}\right)\cos\left(\Omega t - \theta_0\right) \qquad (10.20)$$

The signal $u_2(t)$ obtained at the loop filter output is related to signal $u_1(t)$ by the differential equations 3.31 or 3.38:

$$\tau_1 \frac{du_2}{dt} = K_2\left[\tau_2 \frac{du_1}{dt} + u_1(t)\right], \quad \text{if } F(s) = \frac{1 + \tau_2 s}{\tau_1 s}$$

$$\tau_1 \frac{du_2}{dt} + u_2(t) = K_2\left[\tau_2 \frac{du_1}{dt} + u_1(t)\right], \quad \text{if } F(s) = \frac{1 + \tau_2 s}{1 + \tau_1 s}$$

Since these differential equations are linear, we can calculate separately the response u_{20} to the dc component u_{10} (we shall do this further on), and the response u_{21} to the sinusoidal component u_{11} of angular frequency Ω. In both cases, the latter is expressed, as long as Ω is larger than $1/\tau_1$ and $1/\tau_2$ (in other words larger than the quantity $\omega_n/2\zeta$):

$$u_{21} = K_1 K_2 J_0\left(K_3 \frac{\Delta u}{\Omega}\right)\frac{\tau_2}{\tau_1}\cos\left(\Omega t - \theta_0\right)$$

Since u_2 is the VCO command signal, the term u_{21} can be identified with the ac component of the second member of Eq. 10.18:

$$\Delta u \cos\left(\Omega t - \psi\right) \equiv K_1 K_2 J_0\left(K_3 \frac{\Delta u}{\Omega}\right)\frac{\tau_2}{\tau_1}\cos\left(\Omega t - \theta_0\right)$$

And consequently,

$$\psi = \theta_0$$

$$\Delta u = K_1 K_2 J_0\left(K_3 \frac{\Delta u}{\Omega}\right)\frac{\tau_2}{\tau_1}$$

The initial hypothesis by which sine and cosine expansions are limited to their first term is verified for $K_3(\Delta u/\Omega) < 1$. Since Δu is smaller than $K_1 K_2(\tau_2/\tau_1)$, the preceding condition is fulfilled if

$$\frac{K}{\Omega}\frac{\tau_2}{\tau_1} < 1$$

that is,

$$\Omega > K \frac{\tau_2}{\tau_1}$$

or

$$\Omega > 2\zeta\omega_n$$

As mentioned above, the most stringent condition for the calculation of u_{21} is $\Omega > \omega_n/2\zeta$. This condition is automatically fulfilled when $\Omega > 2\zeta\omega_n$ and when $\zeta > \frac{1}{2}$. For values of ζ below $\frac{1}{2}$ (infrequent), the condition $\Omega > \omega_n/2\zeta$ is the most constraining.

We can now turn our attention to the "dc component" $u_{20}{}^*$ applied at the VCO modulation input, which can be identified with the dc component of the second member of Eq. 10.18. Consequently Eq. 10.19 becomes

$$\Omega = \omega_i - \omega_0 - K_3 u_{20} \tag{10.21}$$

where u_{20} is the loop filter response to the "dc component" u_{10},

$$u_{10} = K_1 J_1 \left(K_3 \frac{\Delta u}{\Omega} \right) \cos(\psi - \theta_0)$$

$$= K_1 J_1 \left(K_3 \frac{\Delta u}{\Omega} \right) \cong K_1 J_1 \left(\frac{K\tau_2}{\Omega\tau_1} \right)$$

The modulation index, as long as $\Omega > 2\zeta\omega_n$, is such that

$$u_{10} = K_1 \frac{K\tau_2}{2\tau_1} \frac{1}{\Omega} \tag{10.22}$$

The loop filter response u_{20} to component u_{10} depends on the loop filter used

$$\tau_1 \frac{du_{20}}{dt} = K_2 \left[\tau_2 \frac{du_{10}}{dt} + u_{10} \right] \quad \text{if } F(s) = \frac{1 + \tau_2 s}{\tau_1 s} \tag{10.23}$$

$$\tau_1 \frac{du_{20}}{dt} + u_{20} = K_2 \left[\tau_2 \frac{du_{10}}{dt} + u_{10} \right] \quad \text{if } F(s) = \frac{1 + \tau_2 s}{1 + \tau_1 s} \tag{10.24}$$

a.   Loop with Integrator and Correction Network.   Equations 10.21, 10.22, and 10.23 lead to the differential equation below representing the variations of Ω versus time:

$$\left(\frac{\tau_2}{\Omega} - \frac{\tau_1 \Omega}{\alpha^2} \right) d\Omega = dt \tag{10.25}$$

*"dc component" for slowly varying as regards u_{21}.

with

$$\alpha^2 = \frac{K^2}{2}\frac{\tau_2}{\tau_1}$$

The general solution of Eq. 10.25 is

$$\tau_2 \ln \Omega - \frac{\tau_1}{\alpha^2}\frac{\Omega^2}{2} = t - t_0$$

The integration constant t_0 is calculated bearing in mind that at instant $t = 0$, $\Omega = \omega_i - \omega_0 = \Omega_0$, and consequently

$$t = \frac{\tau_1}{2\alpha^2}(\Omega_0^2 - \Omega^2) - \tau_2 \ln \frac{\Omega_0}{\Omega}$$

Substituting for α^2 the expression $\alpha^2 = (K^2/2)(\tau_2/\tau_1)$ and replacing $K(\tau_2/\tau_1)$ by $2\zeta\omega_n$, we obtain

$$\frac{t}{\tau_2} = \frac{\Omega_0^2 - \Omega^2}{(2\zeta\omega_n)^2} - \ln \frac{\Omega_0}{\Omega} \tag{10.26}$$

The variations of t versus $x = \Omega/2\zeta\omega_n$ (with $x_0 = \Omega_0/2\zeta\omega_n$) are given in Fig. 10.14a. The variations of x versus time t are derived by rotating the x and y axis (Fig. 10.14.b). These curves should, of course, be restricted to the zone where the hypotheses are valid, that is $x > 1$.

We observe that when x_0 exceeds a certain figure (say, $x_0 \geqslant 10$), the second term of the right-hand member of Eq. 10.26 is negligible as compared with the first term as long as x is greater than 1; the larger x_0, the more this is true. Consequently, Eq. 10.26, under these conditions, can be approximated by Eq. 10.27:

$$\frac{t}{\tau_2} = \frac{\Omega_0^2 - \Omega^2}{(2\zeta\omega_n)^2} \tag{10.27}$$

For x_0 below this figure and in any case for $x \leqslant 1$, we have to use the phase plane to obtain an acceptable representation of the acquisition phenomenon.

Furthermore, still within the zone where $x > 1$, Ω is a monotonic decreasing function of time between value Ω_0 and value $2\zeta\omega_n$. Since this is true what ever the values of Ω_0, we conclude that the loop acquisition range is infinite.

The acquisition time, defined as the time necessary for the frequency difference between input and VCO signals to pass from the initial value Ω_0 to $2\zeta\omega_n$, is given by

$$T_{\text{acq}} = \tau_2 \left[\left(\frac{\Omega_0}{2\zeta\omega_n} \right)^2 - 1 - \ln \frac{\Omega_0}{2\zeta\omega_n} \right] \tag{10.28}$$

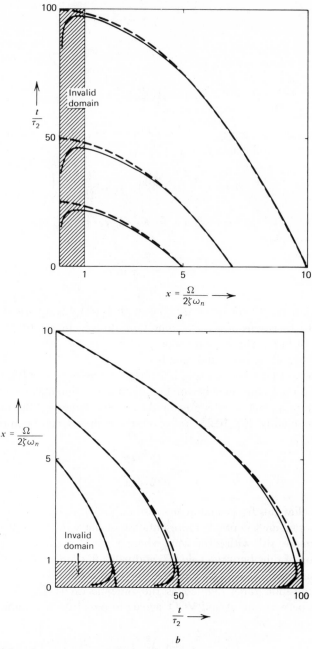

FIGURE 10.14. Representation of the frequency difference between input and VCO signals during acquisition for a loop with $F(s) = (1 + \tau_2 s)/(\tau_1 s)$: (a) time versus angular frequency difference; (b) angular frequency difference versus time.

272

According to the curves of Fig. 10.14, it is reasonable to disregard the last two terms as soon as Ω_0 is large enough; Eq. 10.28 is then very well approximated by

$$T_{\text{acq}} = \tau_2 \frac{\Omega_0^2}{(2\zeta\omega_n)^2} = \frac{\Omega_0^2}{2\zeta\omega_n^3} \qquad (10.29)$$

which is, in fact, identical with Eq. 10.14 given by Viterbi.

NUMERICAL EXAMPLE. Let us take a second-order loop $[F(s) = (1 + \tau_2 s)/\tau_1 s]$ such that $\zeta = \sqrt{2}/2$ and $\omega_n = 100$ rad/s. What is the acquisition time for an initial frequency detuning of 150 Hz?

The time constant τ_2 is given, using Eqs. 3.41 and 3.42, by $\tau_2 = 2\zeta/\omega_n$, that is, $\tau_2 = \sqrt{2}/100$ s. Equation 10.28 leads to

$$T_{\text{acq}} = \frac{\sqrt{2}}{100}\left[\left(\frac{2\pi\,150}{\sqrt{2}\,100}\right)^2 - 1 - \ln\frac{2\pi\,150}{\sqrt{2}\,100}\right]$$

that is,

$$T_{\text{acq}} = \frac{\sqrt{2}}{100}(44.4 - 1 - 1.9) = 587 \text{ ms}$$

whereas Eq. 10.14 leads to 628 ms. For a 600-Hz frequency difference, Eq. 10.28 gives an acquisition time of 9.98 s, while Eq. 10.14 gives $T_{\text{acq}} = 10.05$ s.

b. Loop with Low-Pass Filter and Correction Network. In this case, we use Eqs. 10.21, 10.22, and 10.24 to obtain the differential equation representing the angular frequency difference variations versus time:

$$\left(\alpha^2\tau_2 - \Omega^2\tau_1\right)\frac{d\Omega}{dt} = \Omega^3 - \Omega_0\Omega^2 + \alpha^2\Omega \qquad (10.30)$$

The general solution of this equation is given by

$$t - t_0 = \alpha^2\tau_2 \int \frac{d\Omega}{\Omega(\Omega^2 - \Omega_0\Omega + \alpha^2)} - \tau_1 \int \frac{\Omega\,d\Omega}{\Omega^2 - \Omega_0\Omega + \alpha^2}$$

where determination of the integration constant t_0 takes into account the fact that at instant $t = 0$, $\Omega \cong \Omega_0$.

There are two possible cases, according to whether equation $\Omega^2 - \Omega_0\Omega + \alpha^2 = 0$ has real roots or not.

First Case: $\Omega_0^2 > 4\alpha^2 = K^2(2\tau_2/\tau_1)$. Both roots are real and expressed

$$\Omega_1 = \frac{\Omega_0 + \sqrt{\Omega_0^2 - 4\alpha^2}}{2}$$

$$\Omega_2 = \frac{\Omega_0 - \sqrt{\Omega_0^2 - 4\alpha^2}}{2}$$

The solution of Eq. 10.30 can then be formulated as

$$t = \tau_2 \ln \frac{\Omega}{\Omega_0} + \frac{\tau_1 + \tau_2}{2} \ln \frac{\alpha^2}{(\Omega - \Omega_1)(\Omega - \Omega_2)}$$

$$+ \frac{\Omega_0(\tau_2 - \tau_1)}{2\sqrt{\Omega_0^2 - 4\alpha^2}} \left[\ln \frac{\Omega - \Omega_1}{\Omega - \Omega_2} - \ln \frac{\Omega_2}{\Omega_1} \right] \tag{10.31}$$

The curve representing the variations of t versus Ω is given in Fig. 10.15a, from which can be derived the curve representing the variations of Ω versus time, given in Fig. 10.15b. Since at time $t = 0$, $\Omega = \Omega_0$, the path of Ω follows the upper branch of curve, and when $t \to \infty$,

$$\Omega \to \Omega_1 = \frac{\Omega_0 + \sqrt{\Omega_0^2 - 4\alpha^2}}{2}$$

Consequently, when $|\Omega_0| > K\sqrt{2\tau_2/\tau_1}$, acquisition does not take place. The VCO frequency is attracted towards the input signal frequency but fails to reach it. When $|\Omega_0|$ approaches quantity $K\sqrt{2\tau_2/\tau_1}$, the final value of the angular frequency difference Ω_1 is very close to $\Omega_0/2$; that is to say, the VCO angular frequency covers half the distance separating it from the input signal angular frequency but goes no further (this phenomenon is very clearly experimentally verified).

Second Case: $\Omega_0^2 < 4\alpha^2 = K^2(2\tau_2/\tau_1)$. The solution of differential equation 10.30 is in this case expressed as

$$t = \tau_2 \ln \frac{\Omega}{\Omega_0} + \frac{\tau_1 + \tau_2}{2} \ln \frac{\alpha^2}{\Omega^2 - \Omega_0 \Omega + \alpha^2}$$

$$+ \frac{\Omega_0(\tau_2 - \tau_1)}{\sqrt{4\alpha^2 - \Omega_0^2}} \left[\text{Arc tan} \frac{2\Omega - \Omega_0}{\sqrt{4\alpha^2 - \Omega_0^2}} - \text{Arc tan} \frac{\Omega_0}{\sqrt{4\alpha^2 - \Omega_0^2}} \right] \tag{10.32}$$

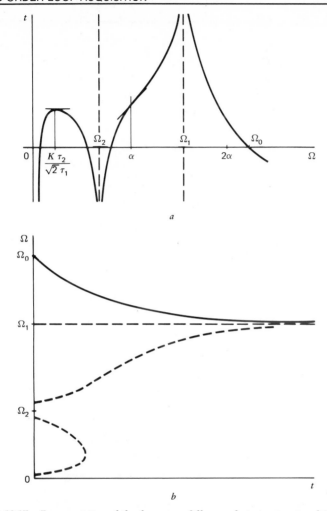

FIGURE 10.15. Representation of the frequency difference between input and VCO signals when acquisition cannot occur for a loop with $F(s) = (1 + \tau_2 s)/(1 + \tau_1 s)$: (a) time versus angular frequency difference; (b) angular frequency difference versus time.

The curve representing the variations of t versus Ω is given in Fig. 10.16a and the variations of Ω versus time are shown in Fig. 10.16b. These curves, of course, are only valid in the zone covered by the hypothesis $\Omega > K\,(\tau_2/\tau_1)$. If we admit that when the angular frequency difference is below $K\,(\tau_2/\tau_1)$, acquisition is practically complete, we have a curve and an analytical expression representing the VCO angular frequency variation during acquisition. We shall take as acquisition time value T_{acq} the time required for Ω to move from Ω_0 to $K\,(\tau_2/\tau_1)$ (generally, $K\,(\tau_2/\tau_1) \cong 2\zeta\omega_n$).

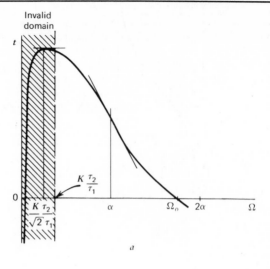

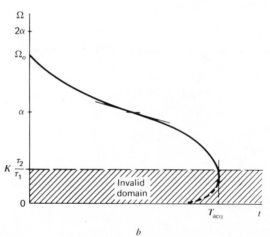

FIGURE 10.16. Representation of the frequency difference between input and VCO signals during acquisition for a loop with $F(s) = (1 + \tau_2 s)/(1 + \tau_1 s)$: (a) time versus angular frequency difference; (b) angular frequency difference versus time.

If Ω_0 is much larger than $K(\tau_2/\tau_1)$, we get

$$T_{\text{acq}} \cong \frac{2(\tau_1 - \tau_2)\Omega_0}{\sqrt{4\alpha^2 - \Omega_0^2}} \operatorname{Arc\,tan} \frac{\Omega_0}{\sqrt{4\alpha^2 - \Omega_0^2}} + \frac{\tau_1}{2} \ln \frac{K}{K - 2\Omega_0} \quad (10.33)$$

the second term of the right-hand member very often being negligible (particularly if $\Omega_0 \ll K$).

If we also require Ω_0 to be much smaller than 2α, which owing to the first condition, implies $\tau_2 \ll \tau_1$ (very frequently encountered in practice), we get

$$T_{\text{acq}} \cong \frac{2\tau_1 \Omega_0^2}{4\alpha^2} = \frac{\Omega_0^2}{K^2(\tau_2/\tau_1^2)} = \frac{\Omega_0^2}{2\zeta\omega_n^3} \qquad (10.34)$$

which is the approximate acquisition time expression for a second-order loop with integrator and correction network given by Eq. 10.14.

The limit condition between possible and impossible acquisition cases is given by

$$\Omega_0^2 < 4\alpha^2 = 2K^2\frac{\tau_2}{\tau_1} = (2K)\left(K\frac{\tau_2}{\tau_1}\right) = 2K\left(2\zeta\omega_n - \frac{1}{\tau_1}\right)$$

that is,

$$|\Omega_0| < 2\sqrt{K\left(\zeta\omega_n - \frac{1}{2\tau_1}\right)} \qquad (10.35)$$

The limit condition given by Viterbi is (Eq. 10.16)

$$|\Omega_0| < 2\sqrt{K\left(\zeta\omega_n + \frac{1}{2\tau_1}\right)}$$

As in practice, $1/\tau_1$ is generally negligible as compared with $2\zeta\omega_n = K(\tau_2/\tau_1) + 1/\tau_1$, the approximate analytical method appears to produce a result comparable with that obtained by the phase-plane method.

The method described in this paragraph is inadequate when the angular frequency difference Ω is close to $K(\tau_2/\tau_1) \cong 2\zeta\omega_n$. However, we have every reason to suppose that once in this zone, acquisition is practically complete since the phase-plane trajectory is close to the separatrix.

If the initial angular frequency difference is below $2\zeta\omega_n$, the only valid method consists in using phase-plane trajectories, taking into account the initial phase difference. Acquisition will be very fast and may occur without cycle slipping. For a more precise description of the phenomenon involved and an evaluation of the acquisition time it is still advisable to refer to the trajectories in Figs. 10.8 to 10.12.

NUMERICAL EXAMPLE. A second-order loop $[F(s) = (1 + \tau_2 s)/(1 + \tau_1 s)]$ is such that

$$\zeta = \frac{\sqrt{2}}{2}, \qquad \omega_n = 100 \text{ rad/s}, \qquad K = 2 \times 10^5 \text{ rad/s}$$

With an initial frequency difference of 600 Hz, use of Eq. 10.14 leads to an acquisition time of 10.05 s.

The value of time constants τ_1 and τ_2 is derived from Eqs. 3.41 and 3.52: $\tau_1 = 20$ s and $\tau_2 = \sqrt{2}\ /100$ s. Use of Eq. 10.33 leads to

$$T_{acq} = \frac{2\left(20 - \sqrt{2}\ \times 10^{-2}\right) 2\pi \times 600}{\sqrt{4\sqrt{2}\ \times 10^7 - \left(2\pi \times 600\right)^2}}$$

$$\times \text{Arc} \tan \frac{2\pi \times 600}{\sqrt{4\sqrt{2}\ \times 10^7 - \left(2\pi \times 600\right)^2}}$$

$$+ \frac{20}{2}\ \ln \frac{2 \times 10^5}{2 \times 10^5 - 2 \times 2\pi \times 600}$$

that is,

$$T_{acq} = 12.13 + 0.38 = 12.51 \text{ s}$$

CHAPTER 11
ACQUISITION
SUBSIDIARY DEVICES

We have seen that for a first-order loop, the acquisition range is equal to the synchronization range. As the latter is always above the possible variations of the input signal frequency with respect to the VCO central frequency, acquisition is no problem: it occurs automatically without any cycle slipping of the VCO with respect to the input signal. The time required to complete acquisition, corresponding for instance to a phase error smaller than 0.1 or 0.01 rad, can be derived from Eqs. 10.8 or 10.9.

Unfortunately, first-order loops are rarely used, at least for reception, because of their high sensitivity to noise and their inadequacy as demodulators. Preference is generally given to second-order loops with low-pass filter and phase-lead correction. If the loop gain is very high, the formulas corresponding to the loop with integrator and correction can be used, except for acquisition problems. Under these conditions, the synchronization range remains equal to the loop gain, but the acquisition range, especially for narrow loops, becomes much smaller, so that in some cases the initial input signal frequency may be beyond the acquisition range. Moreover, even if the input signal frequency is within the acquisition range, the acquisition time formula (Eqs. 10.14 or 10.33) shows that it may become very large if the initial angular frequency difference is large as compared with the natural angular frequency of the loop.

Several techniques have been devised to remedy these serious defects.

The first consists in increasing the natural angular frequency ω_n of the loop during acquisition, so as to increase the acquisition range and reduce the acquisition time. We can increase ω_n either by commutation of the loop filter components (preferably the resistors, so that the capacitor charge remains unchanged, thus avoiding a transient that could hinder the commutation) or, as mentioned in Chapter 9, by adjusting the loop gain K. For instance, K can be increased by applying a high-amplitude input signal during acquisition. This is accomplished automatically if a coherent AGC device is used: as long as acquisition has not occurred, the AGC does not operate and the gain of the amplifying chain preceding the phase detector is at its maximum value.

The drawback with this technique is that it is only possible if the signal-to-noise ratio conditions are such that the phase error due to thermal noise, for the "wide" loop equivalent noise bandwidth value, remains small.

A second technique consists in modifying the VCO central frequency or the input signal frequency by means of a frequency-tunable local oscillator, so as to obtain a low initial angular frequency difference.

If the VCO central frequency is very stable and the input signal frequency varies according to a known law (like the Doppler effect, for instance), we can attempt to tune the VCO according to this law (acquisition following computer information or duplication of acquisition in another loop) or have an operator tune the VCO to a frequency corresponding to that expected.

If the input signal frequency does not follow a known law (random oscillator frequency drift, wide variety of Doppler effects depending on transmitter-receiver positioning, and so on), an automatic device must be used to reduce the initial frequency difference.

A first solution consists in using an automatic frequency-control device (AFC) which divides the initial difference Ω_0 by the AFC loop gain. Here again, this type of device is only usable for a fairly high signal-to-noise ratio. The bandwidth of the bandpass filter preceding the AFC discriminator, has to equal the maximum frequency offset, which is by definition greater than the phase-locked loop bandwidth, for otherwise the device would be unnecessary. But if we situate the AFC discriminator operating threshold between 0 and a few decibels (instead of about 10–12 dB for a demodulation discriminator), the signal-to-noise ratio condition for the AFC will rarely be compatible with the phase-locked loop performances. For example, if the frequency offset is 10 kHz, a signal-to-noise density ratio of at least 40 dB will be required for AFC. But if the phase-locked loop has a 50-Hz equivalent noise bandwidth, we can expect it to operate up to a signal-to-noise density ratio of about 23–25 dB, which is beyond the scope of the AFC discriminator.

A second solution, when the frequency offset is very high, consists in searching by linear sweeping of the VCO frequency (or the input signal frequency through a frequency-tunable local oscillator). Various methods can be used, producing a fairly wide range of performances, differing as to search time, successful acquisition probability, or operating threshold. These methods can be grouped, fairly arbitrarily, in two categories, as in the following sections.

Remark. A linear sweep of the VCO frequency can be obtained by applying a linear variation command signal at the VCO modulation input or by applying a voltage step at the loop filter input. The closer the loop filter transfer function approximation to the formula $F(s) = (1 + \tau_2 s)/\tau_1 s$, the greater will be the linearity of the frequency variation obtained.

11.1   OPEN LOOP FREQUENCY SWEEPING

Since the problem consists in reducing the initial frequency error, it can be assimilated to the search for a signal of unknown frequency, within a certain range, under given signal-to-noise ratio conditions. Once this frequency has been estimated, we have simply to tune the VCO accordingly before applying the signal to the loop input (or before closing the loop if the VCO was itself the search device).

The loop properties have direct influence on the quality of the frequency estimation: when the loop is closed, acquisition has to proceed naturally as quickly as possible. Thus the error on the estimated angular frequency must be low, if possible below $2\zeta\omega_n$, and in any case below $\Omega_{acq} \cong 2\sqrt{K\zeta\omega_n}$.

A first solution consists in surveying the possible range of the input frequency using a bank of bandpass filters. A detector placed at each filter output will indicate which of the filters is excited by the input signal. Making allowances for the noise generally accompanying the input signal in this type of problem, the filter bandwidth will depend on the probability of making a correct decision and the quality of the frequency estimation. If the possible range is large, the number of filters required is considerable and this solution, although interesting from the accuracy point of view, is little used because of the cost involved (particularly as the stability required of the transfer functions of the different filters, despite all the possible causes of drift, poses major implementation problems).

A less expensive solution would be to use a single filter and perform a linear frequency sweep. A possible configuration is shown in Fig. 11.1.

The local oscillator central frequency must be such that the signal resulting from the frequency translation with the input signal frequency at nominal value, is at the decision filter central frequency f_0 which is also the VCO central frequency. Since the local oscillator frequency variation range corresponds with the input frequency possible range, during a sweep, the frequency of the mixer output signal has to pass through the value f_0. A dc component is obtained at the detector output; if it is above a certain predetermined threshold δ, the linear frequency sweep is stopped and the signal is applied to the loop input.

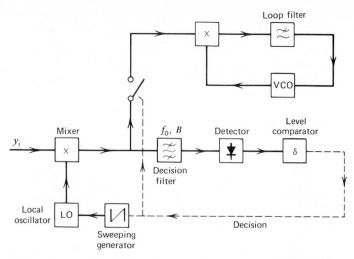

FIGURE 11.1. Block diagram of an open-loop frequency searching circuit.

A variation consists in using the loop VCO for the linear frequency search operation, as shown in Fig. 11.2. The VCO signal expression is then:

$$y_0(t) = B \cos\left(\omega_0 t + \tfrac{1}{2}\Re t^2 + \theta_0\right)$$

where $\Re/2\pi = R$ is the slope of the linear frequency variation (R in Hz/s, $\Re$ in rad/s^2).

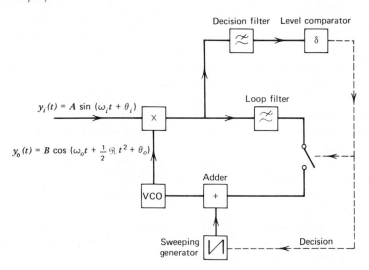

FIGURE 11.2. Other type of open-loop frequency searching circuit.

The decision filter is a low-pass filter because the mixer output signal

$$u_1 = K_1 \sin\left[(\omega_i - \omega_0)t - \tfrac{1}{2}\Re\, t^2 + \theta_i - \theta_0 \right]$$

is a very low-frequency signal when the input and VCO signal angular frequencies are close together, that is, in the vicinity of time $t_d = (\omega_i - \omega_0)/\Re$ (this signal is, in fact, a dc component once the acquisition is over). The loop phase detector can be used as a mixer since the loop will only start operating after the decision.

As long as the input signal–VCO signal beat frequency is high enough, the amplitude of the decision filter output signal is very low, below decision level δ, and the sweep signal is applied to the VCO. When the beat frequency falls within the decision filter bandwidth, if the sweep speed is not excessive, the filter output signal exceeds the decision threshold δ: sweeping is stopped as soon as this threshold is crossed and the loop is closed. The VCO central frequency is then very close to the input signal frequency and acquisition is completed automatically. For given signal-to-noise ratio conditions at the mixer input, the main problem is generally the choice of decision filter (type and bandwidth), decision level δ, and frequency slope R to ensure that acquisition will take place in the minimum time, with a good chance of success.

The signal obtained at the decision filter output during a VCO frequency sweep can, in fact, have any of the several shapes shown in Fig. 11.3 (single-pole, low-pass filter and absence of noise) when the sweep speed— the frequency ramp—is fairly high. If the decision level is set too high ($\delta = \delta_H$), the sweeping only stops in case a and, consequently, the probability of making the decision to stop sweeping is low (unless the sweep speed is very low). But if the decision level is set too low ($\delta = \delta_L$), although the probability that the decision will be made is improved, there is a risk of the decision being premature, corresponding to a beat signal frequency that is still too high. Furthermore, in the presence of noise, the probability of false decision is considerable. Since the sweeping stops as soon as the decision threshold is crossed, the threshold has to be set fairly high to provide against noise effects.

For a relatively high decision threshold (level δ_H in Fig. 11.3), it will be noted that the probability of success is enhanced if the polarity of the signal corresponding to case d is reversed prior to comparison with the threshold level. It is also worth noting that if we use a second mixer for which the VCO signal has undergone a $\pi/2$ phase shift (phase detector operating as "coherent amplitude detector" when acquisition is complete; see Section 11.2.1), the case b and c signals resemble those of cases a and d. The appropriate circuit diagram is that of Fig. 11.4. The results obtained with this device (curves of Fig. 11.5) show clearly that in the operating zone

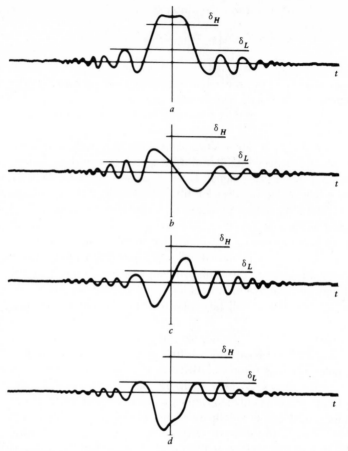

FIGURE 11.3. Waveforms at the decision filter output of Fig. 11.2.

corresponding to a low probability of decision for a single decision channel, the probability is doubled if the rectified signal is used and quadrupled by the addition of a second mixer. It should be noted that these curves were obtained in the absence of noise; the random appearance of the decision is due entirely to the fact that at the instant when sweeping begins, the phase difference between the input and VCO signals is a random variable, distributed uniformly over the interval $(0, 2\pi)$ (before the decision filter, the amplitude of the signal at instant t_d when the angular frequency cancels out is $K_1 \sin[\theta_i - \theta_0 + (\omega_i - \omega_0)^2 / \Re]$). Even in the absence of noise and assuming the decision filters to be simple, one-pole, low-pass filters, it does not seem

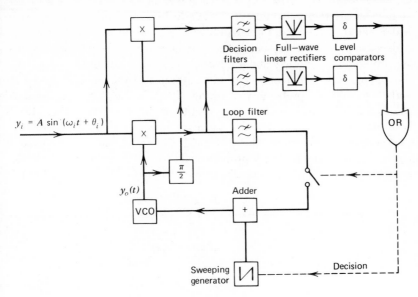

FIGURE 11.4. Implementation example of an open-loop searching circuit.

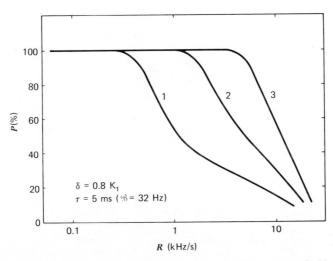

FIGURE 11.5. Successful acquisition probability versus the sweeping speed: (1) one mixer, one half-wave rectifier; (2) one mixer, one full-wave rectifier; (3) two mixers, two full-wave rectifiers.

easy to calculate the probability of correct decision. It depends on the sweep speed R (in Hz/s), on the filter bandwidth $\mathcal{B}$ and on the level $\delta = \alpha K_1$ of the decision threshold (K_1 being the maximum signal amplitude obtained when sweeping is very slow and α a parameter comprised between 0 and 1).

The main experimental survey results (44) are that the value of the sweep speed R, for a given probability of success P and a fixed decision threshold value δ, is proportional to $\mathcal{B}^2$. These results are summarized in Fig. 11.6. It is also interesting to note that the mean frequency difference between the VCO at the instant when the sweeping stops and the input signal is proportional to R and inversely proportional to $\mathcal{B}$. The frequency dispersion around this mean value is practically $\pm \mathcal{B}$ (even in the absence of noise).

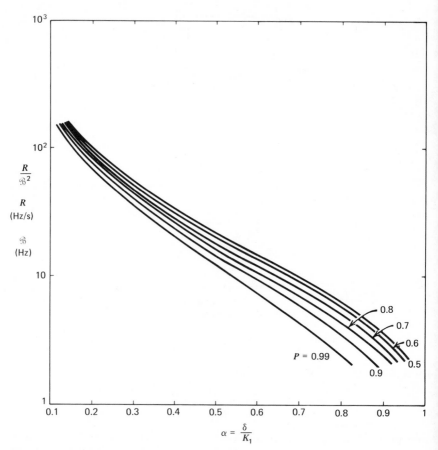

FIGURE 11.6. Sweeping speed versus decision level (with the successful acquisition probability as a parameter).

With favorable signal-to-noise ratio conditions, the threshold δ can be fixed at a fairly low value, which will lead to a very high probability that sweeping will stop, even for high sweeping speeds (although the loop acquisition has still to be successfully accomplished). But, when the signal-to-noise ratio conditions are bad, the decision threshold δ has to be high if we wish to prevent the false decision probability (false alarm) from being too large; in this case we have to sweep very slowly to keep a high probability of making the right decision.

NUMERICAL EXAMPLE. We want to find with a probability $P = 0.99$ the frequency of an unknown signal in a band $W = 10$ kHz in one second or less. The precision of the frequency estimation need only be 20 or 30 Hz.

The frequency slope to be applied to the local search oscillator is $R = 10^4$ Hz/s.

If the decision level is set at $\alpha = \delta / K_1 = 0.5$, the bandwidth of the decision filter will be (Fig. 11.6) such that $R / \mathcal{B}^2 \cong 12$, so $\mathcal{B}^2 = 833$ Hz^2, that is $\mathcal{B} = 29$ Hz. The decision filter time constant $\tau = 1/2\pi\mathcal{B} = 5.5$ ms. The frequency estimation will thus comprise a mean error of about $R\tau = 10^4 \times 5.5 \times 10^{-3} = 55$ Hz and a random error within ± 29 Hz around this mean error.

11.2 CLOSED-LOOP FREQUENCY SWEEPING

We have seen in Section 10.2.1 that if a second-order phase-locked loop input signal is affected by a frequency ramp (linear sweeping), acquisition of the input signal frequency by the VCO may take place under certain conditions. The results are identical when the VCO is subjected to linear frequency sweeping: from the moment when acquisition occurs, the loop produces an error signal, of opposite sign to the sweep signal, so that the VCO frequency remains equal to the constant frequency of the input signal.

If we start sweeping at one extremity of the possible frequency error range, the configuration will necessarily be that where the frequency ramp draws the frequencies together. Under these conditions, Viterbi (38), using the phase-plane method, was able to establish that, in absence of noise, acquisition is certain if the angular frequency ramp $\mathcal{R}$ is below $\frac{1}{2}\omega_n^2$. The probability lessens for a higher value (acquisition takes place or not, depending on initial phase and frequency difference conditions). The probability is null for a ramp $\mathcal{R}$ equal to ω_n^2. These results were obtained for a loop with perfect integrator and correction network, but seem valid for a loop with low-pass filter and correction network, at least if the loop gain K is high.

In the presence of additive noise, the acquisition probability is likely to fall below 1 for ramp values lower than $\omega_n^2/2$. This would seem to be confirmed by the experimental results of Frazier and Page (45). The influence of noise enters into a formula given by these authors to calculate a ramp guaranteeing an acquisition probability of 0.9. This formula (modified in accordance with Ref. 43) is as follows:

$$\mathcal{R} = \frac{\left(1 - \sqrt{2}\,\sigma_{\varphi_0}\right)\omega_n^2}{1 + d} \tag{11.1}$$

where σ_{φ_0} is the standard deviation of the VCO phase fluctuations induced by additive noise, ω_n the natural angular frequency and d a factor related as follows to the damping factor ζ:

$$d = e^{-\pi\left(\zeta/\sqrt{1-\zeta^2}\right)} \quad \text{for } \zeta < 1$$
$$d = 0 \qquad\qquad\quad \text{for } \zeta \geqslant 1$$

(However, we note that this empirical formula gives, for $\zeta \geqslant 1$, in the absence of noise, a probability of 0.9 for $\mathcal{R} = \omega_n^2$, which is in disagreement with the theoretical results of Viterbi.)

The theoretical results of Viterbi (in the absence of noise) and the experimental results of Frazier and Page concern acquisition probability, that is the probability that the VCO will synchronize with the input signal, despite the sweeping. Once the acquisition is complete, for a loop with perfect integrator and correction network, there remains a steady-state phase error, the value of which is $\text{Arc}\sin(\mathcal{R}/\omega_n^2)$. For a loop with low-pass filter and correction network, the steady-state phase error is expressed as

$$\text{Arc}\sin\left(\frac{\mathcal{R}}{\omega_n^2} + \frac{\mathcal{R}t}{K}\right)$$

(where the time origin is the instant corresponding to acquisition completion). The term $\mathcal{R}t$ represents the difference between the input signal angular frequency and the angular frequency that the VCO would have because of the sweeping if acquisition had not occurred.

It may seem preferable to eliminate this error in the steady state, by stopping the sweeping after acquisition. We are thus again confronted with a problem of decision to stop sweeping, as in the previous subsection. But this time, the decision only concerns stopping the sweeping and not closing the loop, since the loop is already closed during the search operation. The configuration here applicable is that shown in Fig. 11.7, where the signal applied to the decision filter possesses a "dc component," the value of which, in the absence of noise, is

$$K_1\sqrt{1 - \left(\frac{\mathcal{R}}{\omega_n^2} + \frac{\mathcal{R}t}{K}\right)^2}$$

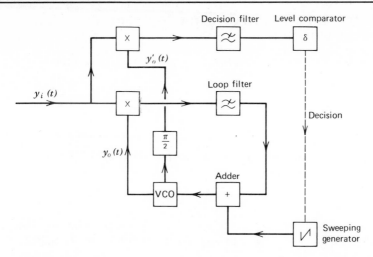

FIGURE 11.7.   Block diagram of a closed-loop frequency searching circuit.

if K_1 is the sensitivity of the auxiliary phase detector serving as coherent amplitude detector. In the configuration of Fig. 11.4, on the other hand, the signal applied to the decision filter only has a "dc component" during the brief instant where the VCO frequency is very close to the input signal frequency. The dc component obtained in Fig. 11.7 configuration arises, of course, because acquisition of the input frequency by the VCO has already taken place; that is, the VCO is in synchronism with the input signal, to within the phase error $\text{Arc}\sin(\mathcal{R}/\omega_n^2 + \mathcal{R}t/K)$.

In this case, we can use as decision filter a very-narrow-bandwidth filter, thus ensuring a better performance if the additive noise is high. The narrow bandwidth of this filter can involve a relatively large decision delay t_d, but this is not very important since the loop is already in lock (although the error term $\mathcal{R}t_d/K$ must not become too large). When sweeping is stopped, the dc-component decision filter output is approximately equal to K_1 (assuming $\mathcal{R}t_d \ll K$) and remains at this value as long as the loop phase error is low: it then becomes an indicator of the state of the loop. The level detector, which was used to make the decision to stop sweeping, can also be used to detect the state of the loop. For example, should the dc component K_1 disappear, it may be decided to start sweeping again for a further signal search.

In view of the importance of the role played by the auxiliary phase detector used as a coherent amplitude detector (that is, with a $\pi/2$ phaseshift included on one of the inputs of a device similar to that used as phase detector in the loop), it is of interest to analyze in greater detail its operating principle.

11.2.1   OPERATING PRINCIPLE OF THE COHERENT AMPLITUDE DETECTOR

The following expressions designate the input signal $y_i(t)$ and the VCO signal $y_0(t)$:

$$y_i(t) = A \sin(\omega_i t + \theta_i)$$

$$y_0(t) = B \cos(\omega_0 t + \theta_0)$$

The signal $y_0'(t)$ obtained by shifting the phase of signal $y_0(t)$ by $\pi/2$ (see Fig. 11.7) is expressed as

$$y_0'(t) = B \sin(\omega_0 t + \theta_0)$$

The output signal obtained from the auxiliary "phase detector" working as a multiplier, followed by a filter to suppress the high frequency components, is expressed as

$$u_1'(t) = \frac{AB}{2} \cos\left[(\omega_i - \omega_0)t + \theta_i - \theta_0 \right]$$

that is,

$$u_1'(t) = K_1 \cos\left[(\omega_i - \omega_0)t + \theta_i - \theta_0 \right]$$

When the loop is out of lock, the auxiliary phase detector output signal is a sinusoidal signal similar to that obtained at the main detector output, to within the $\pi/2$ phase shift.

If the loop is in lock, with a null phase error, that is, if $\omega_i = \omega_0$, $u_1' = K_1$. Since K_1 is directly proportional to the input signal amplitude A, whether the detector works like a real multiplier or like a multiplier by the sign of the reference signal $y_0'(t)$, the output signal u_1' is a duplicate of the input signal amplitude. It is for this reason that the auxiliary phase detector in the Fig. 11.7 configuration is said to work like a coherent amplitude detector.

When the loop is in lock, with a steady-state phase error, for example,

$$\theta_i - \theta_0 = \text{Arc} \sin \frac{\omega_i - \omega_0}{K}$$

in the case of a loop with low-pass filter and correction and an initial frequency difference between the input and VCO signals, or

$$\theta_i - \theta_0 = \text{Arc} \sin \frac{\mathcal{R}}{\omega_n^2}$$

in the case of a loop with integrator and correction and with a linear frequency drift $R = \mathcal{R}/2\pi$ in the input signal; the auxiliary detector output signal is then expressed as

$$u_1' = K_1 \cos\left(\text{Arc}\sin\frac{\omega_i - \omega_0}{K}\right) = K_1 \sqrt{1 - \left(\frac{\omega_i - \omega_0}{K}\right)^2}$$

or

$$u_1' = K_1 \cos\left(\text{Arc}\sin\frac{\mathcal{R}}{\omega_n^2}\right) = K_1 \sqrt{1 - \left(\frac{\mathcal{R}}{\omega_n^2}\right)^2}$$

Consequently, as long as the phase error $\theta_i - \theta_0$ is small enough, we can consider that $u_1' \cong K_1$.

Using the analytical approximate method described in Section 10.2.2, it can be shown that the dc component of signal u_1' remains null during acquisition. If we refer to the notation used in this subsection,

$$y_0(t) = B\cos\left[(\omega_0 + K_3 u)t + \theta_0\right.$$
$$\left. + K_3 \frac{\Delta u}{\Omega}\sin(\Omega t - \psi)\right]$$

then

$$y_0'(t) = B\cos\left[(\omega_0 + K_3 u)t + \theta_0 - \frac{\pi}{2}\right.$$
$$\left. + K_3 \frac{\Delta u}{\Omega}\sin(\Omega t - \psi)\right]$$

After multiplication by signal $y_i(t)$ we obtain

$$u_1'(t) = K_1 \left\{ J_0\left(K_3 \frac{\Delta u}{\Omega}\right)\sin(\Omega t - \theta_0) + J_1\left(K_3 \frac{\Delta u}{\Omega}\right)\right.$$
$$\left. \times \left[\sin(\theta_0 - \psi) + \sin(2\Omega t - \theta_0 - \psi)\right] + \cdots \right\}$$

But we observed in the course of the acquisition survey that $\theta_0 = \psi$. Consequently, the dc component of $u_1'(t)$ is null.

When the input signal is accompanied by a Gaussian additive noise $n(t)$, having a null mean value and a one-sided spectral density N_0 in a band W centered on f_0, the results given in Section 7.1 are still valid. The coherent amplitude detector produces a noise term $K_1 n'(t)$, where $n'(t)$ is a Gaussian process, having a null mean value and a uniform two-sided spectral density N_0/A^2 from $-W/2$ to $+W/2$.

Finally, if in addition to the static phase error $\theta_i - \theta_0$, there is a random phase error that we shall call $\varphi(t)$, when the signals are synchronized, $u_1'(t)$ is a random signal, expressed as

$$u_1'(t) = K_1 \cos \left[\theta_i - \theta_0 + \varphi(t) \right]$$

that is,

$$u_1'(t) = K_1 \cos (\theta_i - \theta_0) \cos \varphi(t) - K_1 \sin (\theta_i - \theta_0) \sin \varphi(t)$$

The value of the dc component $\overline{u_1'(t)}$ is

$$\overline{u_1'(t)} = K_1 \, \overline{\cos \varphi(t)} \, \cos (\theta_i - \theta_0) - K_1 \, \overline{\sin \varphi(t)} \, \sin (\theta_i - \theta_0)$$

If $\varphi(t)$ is a Gaussian process having a null mean value,

$$\overline{\sin \varphi(t)} = 0$$

$$\overline{\cos \varphi(t)} = e^{-\frac{1}{2} \sigma_\varphi^2}$$

then

$$\overline{u_1'(t)} = K_1 e^{-\frac{1}{2} \sigma_\varphi^2} \cos (\theta_i - \theta_0)$$

Thus, when the input signal is accompanied by an additive noise, the VCO phase becomes a random process $\varphi_n(t)$ that can be approximated (Section 12.1.3) by a Gaussian process with a null mean value and a variance of $\sigma_{\varphi_n}^2$ (Eq. 12.25). Then

$$\overline{u_1'(t)} = K_1 e^{-\frac{1}{2} \sigma_{\varphi_n}^2} \cos (\theta_i - \theta_0)$$

We shall also see in Section 12.1.3 that the static phase error itself is affected by the random modulation $\varphi_n(t)$. The phase error in the case of a frequency ramp $R = \Re / 2\pi$ becomes

$$\theta_i - \theta_0 = \text{Arc} \sin \left(\frac{\Re}{\omega_n^2} e^{\frac{1}{2} \sigma_{\varphi_n}^2} \right).$$

The coherent amplitude detector dc component becomes

$$\overline{u_1'(t)} = K_1 e^{-\frac{1}{2} \sigma_{\varphi_n}^2} \sqrt{1 - \left(\frac{\Re}{\omega_n^2} \right)^2 e^{\sigma_{\varphi_n}^2}}$$

that is,

$$\overline{u_1'(t)} = K_1 \sqrt{e^{-\sigma_{\varphi_n}^2} - \left(\frac{\Re}{\omega_n^2} \right)^2} \qquad (11.2)$$

11.2.2 EXPERIMENTAL RESULTS

An extremely thorough operational analysis of the device shown in Fig. 11.7 was made by Vialle (46), using the approximate analytical method described in Section 10.2.2 and the Booton quasi-linearization method (see Chapter 12) to evaluate the influence of additive noise. Unfortunately, it is impossible to derive a formula from this analysis enabling us to calculate the probability P of successful acquisition versus the different parameters. The stop-sweeping decision has a random character, due essentially to the fact that the time during which signal $u_1'(t)$ has a non-null dc component is a highly variable magnitude when the input additive noise is high. As this dc signal is integrated by the low-pass filter of time constant $\tau = RC$ (Fig. 11.7 decision filter) before being compared with the reference level δ, a correct decision to stop sweeping is only made if the duration of the dc component $u_1'(t)$ is sufficient.

The experimental results shown in Figs. 11.8–11.12 were obtained with fairly common parameter values:

$$K = 2 \times 10^5 \text{ rad/s}, \qquad \omega_n = 10^3 \text{ rad/s}, \qquad \zeta = 1$$

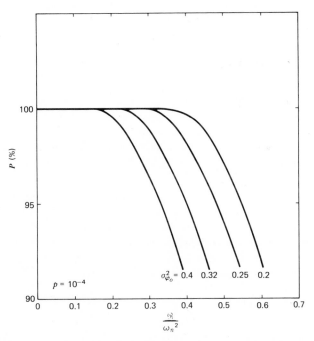

FIGURE 11.8. Successful acquisition probability for a second-order loop when the false-alarm probability is $p = 10^{-4}$.

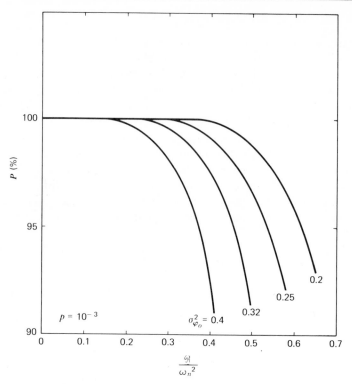

FIGURE 11.9. Successful acquisition probability for a second-order loop when $p = 10^{-3}$.

and these curves would serve as a basis for accurate performance approximation for a wide range of loop parameter values.

The curves in Figs. 11.8–11.11 give the probability P of successful acquisi-
tion versus the variable $\mathcal{R}/\omega_n^2$ ($\mathcal{R}$ being in rad/s² when ω_n is expressed in
rad/s), the parameter being $\sigma_{\varphi_0}^2 = N_0/A^2(2B_n)$ expressed in rad², for different
values of p, which is the probability that the auxiliary phase detector output
noise, filtered by the decision filter, will exceed the decision level δ before
the acquisition occurs. Since sweeping stops during the instants when the
filtered noise exceeds δ, the effective sweeping speed is multiplied by $(1 - p)$,
which is of no great consequence as long as $p < 10^{-2}$ or 10^{-1}. In all cases,
the decision level was set at half the value of the auxiliary detector output dc
component at the moment of acquisition (Eq. 11.2):

$$\delta = \tfrac{1}{2} K_1 \sqrt{e^{-\sigma_{\varphi n}^2} - \left(\frac{\mathcal{R}}{\omega_n^2}\right)^2}$$

this value leading to a minimum dispersion on the stop-sweeping instant.

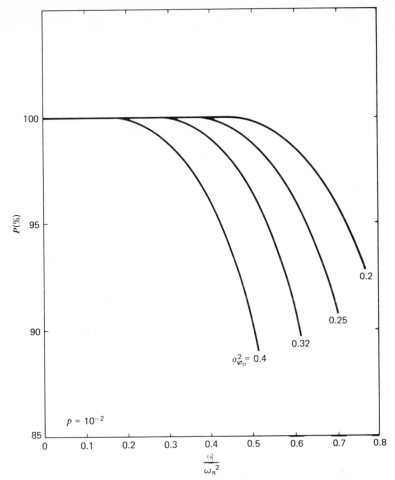

FIGURE 11.10. Successful acquisition probability for a second-order loop when $p = 10^{-2}$.

Finally, the decision filter time constant $\tau = RC$ is related to probability p by

$$p = \frac{1}{\sqrt{2\pi}} \int_x^{\infty} e^{-u^2/2}\, du$$

with

$$x = \frac{\delta}{K_1\sqrt{(N_0/A^2)(1/2\tau)}}$$

This expression shows that the time constant τ should be as low as possible,

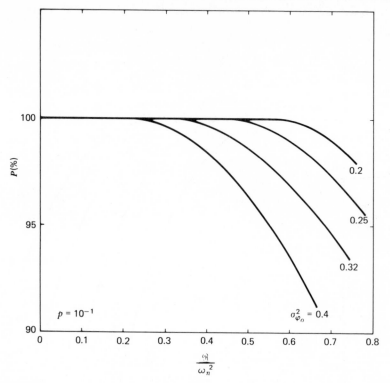

FIGURE 11.11. Successful acquisition probability for a second-order loop when $p = 10^{-1}$.

so that p will be as high as possible (for example, $p = 10^{-1}$), which will lead to the maximum value of P for a given $\mathcal{R}/\omega_n^2$ or the maximum value of $\mathcal{R}/\omega_n^2$ for a given probability of successful acquisition P. But the choice of a low time constant value τ (that is, a wide bandwidth for the decision filter) also leads to a high probability p' of an accidental sweeping restart once acquisition is complete, which is obviously an extremely serious event for the device. The accidental sweeping restart probability is the probability that the decision filter output signal will drop below the decision level after the decision has been made, that is,

$$p' = \frac{1}{\sqrt{2\pi}} \int_y^\infty e^{-u^2/2} \, du$$

with

$$y = \frac{K_1 e^{-\frac{1}{2}\sigma_{\varphi n}^2} - \delta}{K_1 \sqrt{(N_0/A^2)(1/2\tau)}}$$

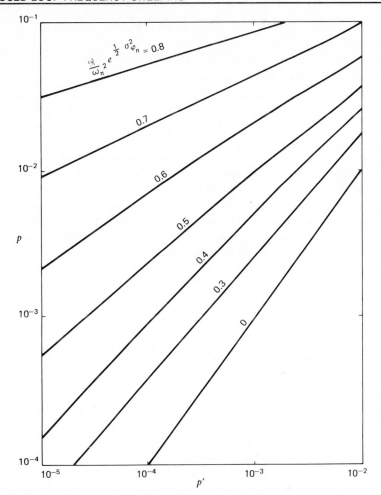

FIGURE 11.12. False-alarm probability p versus "accidental sweeping restart" probability p' (with the quantity $(\mathcal{R}/\omega_n^2)e^{\frac{1}{2}\sigma_{\varphi n}^2}$ as a parameter).

Although the risk of this occurrence can be limited by switching over the decision filter immediately after the decision and replacing it by a narrower filter by means of which the state of the loop can be checked, it is advisable to make sure that the value of p' before commutation is acceptable. This can be done by means of the curves in Fig. 11.12, relating p to p' for different values of the parameter $[(\mathcal{R}/\omega_n^2)e^{\frac{1}{2}\sigma_{\varphi n}^2}]$.

Remark. The parameter of the Figs. 11.8–11.11 curves is the VCO phase variance calculated with the help of the linear analysis formula (Section 7.2). We have used the expression $\sigma_{\varphi_0}^2$ to designate it, for in the subsequent formulas $\sigma^2_{\varphi_n}$ represents the real VCO phase variance, which can be adequately approximated using the quasi-linearization method (Section 12.1.3). In this Chapter, the value of the loop natural angular frequency ω_n is that corresponding to the linear analysis, that is, unaffected by the $e^{-\frac{1}{2}\sigma^2_{\varphi_n}}$ apparent decrease in loop gain due to VCO phase modulation resulting from additive noise (Section 12.1.3). A loop gain decrease due to the action of an AGC or a limiter (Chapter 9) would, of course, have to be taken into account in the calculation of ω_n.

The previous analysis highlights a necessary condition for acquisition: the auxiliary detector output dc component must exist if a decision is to be made. As we have seen, this is not a sufficient condition: for the decision to be made, the duration of this component must be such that the decision level δ is reached at the decision low-pass filter output (which is not always the case, given the fluctuating character of this duration related to the VCO phase modulation due to additive noise). As this dc component is expressed by (Eq. 11.2)

$$\overline{u'_1(t)} = K_1 \sqrt{e^{-\sigma^2_{\varphi_n}} - \left(\frac{\mathcal{R}}{\omega_n^2}\right)^2}$$

it is necessary that

$$\frac{\mathcal{R}}{\omega_n^2} < e^{-\frac{1}{2}\sigma^2_{\varphi_n}} \tag{11.3}$$

For example, for a loop signal-to-noise ratio value ρ_L of 4 dB, leading to a phase variance of 0.4 rad^2, assuming linear operation, and of $\sigma^2_{\varphi_n} = 0.58$ rad^2 by the quasi-linearization method (see Fig. 12.2, curve for $\zeta = 1$), the probability P of successful acquisition should be null for any value of the ramp $\mathcal{R}$ above $0.75\omega_n^2$. If this seems fairly plausible for the curves of Figs. 11.8–11.10, this would not appear to be so for those of Fig. 11.11. Moreover, still in the case of Fig. 11.11, when the noise level is low, the probability of successful acquisition remains equal to 1 beyond the value $\mathcal{R} = 0.5\omega_n^2$, which is in contradiction with the theoretical result given by Viterbi. In fact, the results obtained when $p = 10^{-1}$ would appear to be a little optimistic. This is doubtless because sometimes, without acquisition actually occurring, a signal is obtained at the auxiliary phase detector output when the VCO frequency is very close to that of the input signal. If this signal is such that the decision level δ is reached at the decision filter output, sweeping is stopped. As the frequency difference between input and VCO signals is very slight at this moment, acquisition finishes naturally, which increases artificially the probability of successful acquisition.

Remark: Hybrid sweeping. A third method of sweeping, in a way intermediate between the two methods described, has also been suggested. The configuration is that of Fig. 11.7. This method consists in sweeping linearly the VCO frequency and detecting when the input signal–VCO beat frequency touches zero (using decision filter and level detector), as in the first method, the difference being that this time the loop is closed. Even if the sweeping speed is too high for acquisition to take place, the influence of the loop is felt. The VCO tends towards synchronism with the input signal when the two angular frequencies are very close, so that the auxiliary detector output signal resembles that shown in Fig. 11.3a with a far higher probability than for the three other cases in this figure.

When the additive noise level is not too high, this device can be used to obtain a faster frequency search result than in the case of closed-loop sweeping. The curves in Fig. 11.11 indicate fairly clearly that for very low values of $\sigma_{\varphi_0}^2$ we can obtain a probability of successful acquisition $P = 1$ for values of the angular frequency ramp well over $\frac{1}{2}\omega_n^2$.

But when the noise level is high, the probability P of success becomes only moderate. This is because in this method the auxiliary detector output signal is no longer a dc signal, since theVCO does not really synchronize with the input signal. To keep an acceptable probability of success value, the sweeping speed $\mathcal{R}$ has to be reduced so that the VCO can really synchronize with the input signal, that is, use the closed-loop sweeping method.

NUMERICAL EXAMPLE. The signal applied to the input of a phase-locked loop has a frequency of unknown value within a band $W = 10$ kHz. The device is a second-order loop such that $\zeta = \sqrt{2}/2$ and $\omega_n = 100$ rad/s. The conditions of signal-to-noise ratio are such that $S/N_0 = 24$ dB-Hz.

In view of the signal-to-noise ratio conditions, we decide to use the closed-loop linear frequency sweeping method.
The equivalent noise bandwidth (Eq. 7.18) is

$$(2B_n) = 100 \frac{1 + 4\frac{1}{2}}{4\sqrt{2}/2} = 106 \text{ Hz}$$

The VCO variance, derived from the linear analysis (Eq. 7.15) is

$$\sigma_{\varphi_0}^2 = \frac{1}{2}\frac{1}{250}106 = 0.212 \text{ rad}^2$$

A more realistic value for the phase error variance will be (Fig. 12.2) $\sigma_{\varphi_n}^2 = 0.26$ rad^2.
The maximum value of the angular frequency slope $\mathcal{R}$ is related to the natural angular frequency ω_n by Eq. 11.3. We should have

$$\frac{\mathcal{R}}{\omega_n^2} < e^{-0.13}, \qquad \text{that is } \frac{\mathcal{R}}{\omega_n^2} < 0.878$$

But if we fix $\mathcal{R}$ at a value too close to 0.878, the probability of successful acquisition will be very low.

If we set ourselves a probability of success $P = 0.99$, we can fix $\mathcal{R}$ so that (curves in Figs. 11.8–11.11)

$$\mathcal{R} = 0.41\ \omega_n^2 \quad \text{if } p = 10^{-3}$$
$$\mathcal{R} = 0.49\ \omega_n^2 \quad \text{if } p = 10^{-2}$$
$$\mathcal{R} = 0.60\ \omega_n^2 \quad \text{if } p = 10^{-1}$$

For each of the pairs of values $(\mathcal{R}, p)$ we can evaluate the accidental sweeping restart probability p' using the curves of Fig. 11.12. We obtain

$$p' = 4\text{–}5 \times 10^{-4} \quad \left(\frac{\mathcal{R}}{\omega_n^2} e^{\frac{1}{2}\sigma_{\varphi n}^2} = 0.47 \right) \quad \text{if } p = 10^{-3}$$

$$p' = 4\text{–}5 \times 10^{-3} \quad \left(\frac{\mathcal{R}}{\omega_n^2} e^{\frac{1}{2}\sigma_{\varphi n}^2} = 0.56 \right) \quad \text{if } p = 10^{-2}$$

$$p' < 10^{-2} \quad \left(\frac{\mathcal{R}}{\omega_n^2} e^{\frac{1}{2}\sigma_{\varphi n}^2} = 0.68 \right) \quad \text{if } p = 10^{-1}$$

We select the value $\mathcal{R} = 0.41\ \omega_n^2$, for it corresponds both to a low false alarm probability ($p = 10^{-3}$) and a low accidental sweeping restart probability ($p' = 4\text{–}5 \times 10^{-4}$). We then have

$$\mathcal{R} = 0.41 \times 10^4 = 4.1 \times 10^3 \text{ rad/s}^2$$

that is,

$$R = \frac{\mathcal{R}}{2\pi} = 650 \text{ Hz/s}$$

The signal search time will then be $T_{\text{search}} = W/R = 15.4$ s. The decision threshold δ will have to be

$$\frac{\delta}{K_1} = \frac{1}{2} \sqrt{ e^{-\sigma_{\varphi n}^2} - \left(\frac{\mathcal{R}}{\omega_n^2} \right)^2 } = \tfrac{1}{2} \sqrt{0.771 - 0.168} = 0.39$$

The decision filter time constant τ will then be chosen so that $p = 10^{-3}$, which implies

$$x = \frac{\delta}{K_1 \sqrt{(N/A^2)(1/2\tau)}} = 3.1$$

from which we derive $\tau = 64$ ms.

With these data we can verify that

$$y = \frac{K_1 e^{-\frac{1}{2}\sigma_{\varphi n}^2} - \delta}{K_1 \sqrt{(N_0/A^2)(1/2\tau)}} = 3.92$$

which leads to $p' = 4.6 \times 10^{-4}$.

CHAPTER 12
NONLINEAR OPERATION
IN THE PRESENCE
OF NOISE

All the results given in Part II of this book, and particularly in Chapter 7 concerning performances in the presence of additive noise, were based on the assumption that the phase detector was operating in the linear zone. For this hypothesis to be confirmed, the first condition is that the VCO has to have synchronized with the input signal, that is, acquisition has to have taken place, either naturally (Chapter 10) or with the help of a subsidiary device (Chapter 11).

It is then interesting to see what happens when, with the loop assumed to be in lock, the input signal phase variations or the level of additive noise accompanying the input signal are such as to invalidate the low-phase-error hypothesis. We shall now see that this in fact diminishes the possibilities of the loop: the VCO sensitivity to input signal phase variations or additive noise is heightened to the point where the VCO is no longer in synchronism with the input signal. This corresponds to what is known as the operating threshold of the device.

In this chapter, we shall use $\varphi_n(t)$ to designate the VCO random modulation due to additive noise, keeping the expression $\phi_0(t)$ for the VCO response to an input signal phase modulation $\phi_i(t)$ and the expression $\varphi_0(t)$ for the loop response to additive noise, within the linear hypothesis (see Chapter 7).

12.1 INFLUENCE OF ADDITIVE NOISE

Phase-locked loop operation in the presence of noise beyond the linear zone is a difficult problem, for which no precise mathematical solution has so far been found—at least as regards loops of the second order or higher. Several authors have arrived at results based on simplifying hypotheses; these results are sufficiently close to the experimental results to be valid for the purposes of the design engineer.

12.1.1 ANALYSIS OF THE PROBLEM INVOLVED

In order to well pose the problem, we shall refer to the general block diagram of a phase-locked loop given in Fig. 12.1. We shall use a sinusoidal phase detector. As we saw in Section 7.1, this instrument may be a real analog multiplier or a device multiplying the input signal and the noise by the sign of the reference signal. We shall limit our discussion to this type of detector, since, with a high additive noise, the operation of other types of detectors becomes extremely complex. We have already mentioned in Section 7.1 that the triangular detectors (linear over $\pm \pi/2$) and sawtooth detectors (linear over $\pm \pi$) tend to become sinusoidal when the signal-to-noise ratio preceding the phase detector is very low.

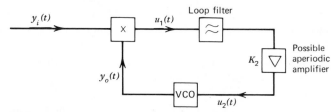

FIGURE 12.1.   Block diagram of a phase-locked loop.

In the absence of noise, the input signal expression is

$$y_i(t) = A \sin\left[\omega_0 t + \phi_i(t)\right]$$

and that of the VCO signal is

$$y_0(t) = B \cos\left[\omega_0 t + \phi_0(t)\right]$$

The phase detector output signal is expressed as

$$u_1(t) = K_1 \sin\left[\phi_i(t) - \phi_0(t)\right]$$

After filtering by the loop filter [impulse response $f(t)$] and after gain amplification K_2 when necessary, we get the signal

$$u_2(t) = K_2 u_1(t) * f(t)$$

This signal $u_2(t)$ is applied to the VCO modulation input, so that

$$\frac{d\phi_0}{dt} = K_3 u_2(t)$$

We derive from these three formulas a general equation representing the operation of a loop with sinusoidal detector (Eq. 3.4):

$$\frac{d\phi_0}{dt} = K \sin\left[\phi_i(t) - \phi_0(t)\right] * f(t)$$

When the input signal is accompanied by an additive noise $n(t)$, a noise term added to the useful signal $u_1(t)$, after filtering in the loop filter, induces a supplementary random VCO phase modulation $\varphi_n(t)$. If we split up the noise $n(t)$ as described in Section 7.1, we obtain

$$y_i(t) = A \sin\left[\omega_0 t + \phi_i(t)\right]$$
$$+ n_1(t) \sin\omega_0 t + n_2(t) \cos\omega_0 t$$

After multiplication by $y_0(t) = B\cos[\omega_0 t + \phi_0(t) + \varphi_n(t)]$ and low-pass filtering to eliminate the angular frequency terms in the vicinity of $2\omega_0$, the phase detector output signal becomes

$$u_1(t) = K_1\left\{ \sin\left[\phi_i(t) - \phi_0(t) - \varphi_n(t)\right]\right.$$
$$- \frac{n_1(t)}{A} \sin\left[\phi_0(t) + \varphi_n(t)\right]$$
$$\left. + \frac{n_2(t)}{A} \cos\left[\phi_0(t) + \varphi_n(t)\right]\right\} \qquad (12.1)$$

After filtering in the loop filter, the signal is applied to the VCO modulation input so as to obtain

$$\frac{d(\phi_0 + \varphi_n)}{dt} = K\left\{ \sin\left[\phi_i(t) - \phi_0(t) - \varphi_n(t)\right]\right.$$
$$- \frac{n_1(t)}{A} \sin\left[\phi_0(t) + \varphi_n(t)\right]$$
$$\left. + \frac{n_2(t)}{A} \cos\left[\phi_0(t) + \varphi_n(t)\right]\right\} * f(t) \qquad (12.2)$$

Equation 12.2 is the general differential equation representing the phase-locked loop operation when the input signal is modulated and accompanied by an additive noise.

If we assume that the total phase error $\phi_i(t) - \phi_0(t) - \varphi_n(t)$ remains sufficiently low, we can write

$$\sin\left[\phi_i(t) - \phi_0(t) - \varphi_n(t)\right] \cong \phi_i(t) - \phi_0(t) - \varphi_n(t)$$

For this condition to be verified, both the phase error $\phi_i(t) - \phi_0(t)$ from the modulation $\phi_i(t)$ of the input signal $y_i(t)$ and the VCO random phase modulation $\varphi_n(t)$ from additive noise must be small. On the other hand, if $\phi_0(t)$ and $\varphi_n(t)$ vary slowly enough as compared with $n_1(t)$ and $n_2(t)$, it is possible to replace the quantity

$$-\frac{n_1(t)}{A} \sin \left[\phi_0(t) + \varphi_n(t) \right] + \frac{n_2(t)}{A} \cos \left[\phi_0(t) + \varphi_n(t) \right]$$

by a noise term $n'(t)$ having the properties described in Section 7.1. The phase detector output signal can then be written, replacing $\varphi_n(t)$ by $\varphi_0(t)$ (since we are in the linear domain) as

$$u_1(t) = K_1 \left[\phi_i(t) - \phi_0(t) - \varphi_0(t) + n'(t) \right]$$

The additive noise $n(t)$ effect can then be interpreted as a supplementary phase modulation $n'(t)$ of the input signal phase. Under these conditions, the VCO phase becomes the sum of response $\phi_0(t)$ to input signal modulation $\phi_i(t)$ and response $\varphi_0(t)$ to input signal equivalent modulation $n'(t)$. Given the properties of $n'(t)$, we conclude that $\varphi_0(t)$ is a Gaussian random process having a null mean value and a spectral density $S_{\varphi_0}(f)$ (Eq. 7.8):

$$S_{\varphi_0}(f) = \frac{N_0}{A^2} |H(j2\pi f)|^2$$

Since the process is Gaussian with a null mean value, the fact that its variance $\sigma_{\varphi_0}^2$ is known (Eqs. 7.9 and 7.15),

$$\sigma_{\varphi_0}^2 = \int_{-\infty}^{+\infty} \frac{N_0}{A^2} |H(j2\pi f)|^2 \, df = \frac{N_0}{A^2} (2B_n)$$

enables us to define the probability density function:

$$p(\varphi_0) = \frac{1}{\sqrt{2\pi} \, \sigma_{\varphi_0}} e^{-\varphi_0^2/2\sigma_{\varphi_0}^2}$$

Further, the power spectral density data enable us to calculate the autocorrelation function:

$$R_{\varphi_0}(\tau) = \int_{-\infty}^{+\infty} S_{\varphi_0}(f) e^{j2\pi f\tau} \, df$$

This formula, in the case, for example, of a second-order loop with perfect integrator and correction when $\zeta < 1$, leads to Eq. 7.22:

$$R_{\varphi_0}(\tau) = \frac{N_0}{A^2}\frac{\omega_n}{4}e^{-\zeta\omega_n\tau}\left[\frac{1+4\zeta^2}{\zeta}\cos\omega_n\sqrt{1-\zeta^2}\,\tau\right.$$

$$\left. + \frac{1-4\zeta^2}{\sqrt{1-\zeta^2}}\sin\omega_n\sqrt{1-\zeta^2}\,\tau\right]$$

Consequently, adopting the linear hypothesis, the random process $\varphi_n(t)$ $= \varphi_0(t)$ is Gaussian, of null mean value and known autocorrelation function. The process is thus completely defined since we can calculate the joint probability density function of any set of random variables $\varphi_{01}, \varphi_{02}, \ldots, \varphi_{0k}$, obtained by sampling process $\varphi_0(t)$ at instant $t_1, t_2, \ldots, t_k$.

However, if we abandon the linear hypothesis, matters are very different. The output process $\varphi_n(t)$ must be derived from Eq. 12.2 and we can easily imagine the problems involved. We shall first give some results obtained by Viterbi (47) using the Fokker–Plank method, and then we shall apply the Booton's quasi-linearization method.

12.1.2 FOKKER—PLANCK METHOD

Let us consider Eq. 12.2 and suppose that the input signal is not modulated: $\phi_i(t) = \theta_i$. Given the notation used, this hypothesis implies that the loop is in lock with a null static phase error, that is, that the angular frequency difference between input and VCO signals before acquisition was null. Then $\phi_0(t) = \theta_i$, and Eq. 12.2 becomes

$$\frac{d\varphi_n}{dt} = K\left\{-\sin\varphi_n(t) - \frac{n_1(t)}{A}\sin\left[\theta_i + \varphi_n(t)\right]\right.$$

$$\left. + \frac{n_2(t)}{A}\cos\left[\theta_i + \varphi_n(t)\right]\right\} * f(t) \qquad (12.3)$$

It can be shown (48) that if the bandwidth W of the additive noise $n(t)$ is large, compared with the loop bandwidth, the term $n'(t)$, given by

$$n'(t) = -\frac{n_1(t)}{A}\sin\left[\theta_i + \varphi_n(t)\right] + \frac{n_2(t)}{A}\cos\left[\theta_i + \varphi_n(t)\right]$$

represents a Gaussian noise term, having a null mean value and a uniform spectral density equal to N_0/A^2 from $-W/2$ to $+W/2$. This means that the noise term $n'(t)$ obtained at the phase detector output still has the same properties as in Section 7.1, independently of the properties of the process $\varphi_n(t)$ modulating the VCO signal, providing that the variations of $\varphi_n(t)$ are slow as compared with those of processes $n_1(t)$ and $n_2(t)$. With respect to the loop, process $n'(t)$ can be considered as a Gaussian white noise having a two-sided power spectral density N_0/A^2.

The differential equation governing the loop operation is then

$$\frac{d\varphi_n}{dt} = K \left\{ -\sin\left[\varphi_n(t) \right] + n'(t) \right\} * f(t) \tag{12.4}$$

First-Order Loop. The loop filter in Fig. 12.1 is then an all-pass filter and calculating the product of convolution by $f(t)$ is the same as multiplying by 1. Equation 12.4 becomes

$$\frac{d\varphi_n}{dt} = -K\sin\varphi_n(t) + Kn'(t) \tag{12.5}$$

The process $\varphi_n(t)$, the derivative of which, at each instant, does not depend on the past history of the process [assuming $n'(t)$ to be strictly white] but only on the value of the process at the instant considered, is consequently a Markov process. The probability density function at each instant $p(\varphi_n, t)$ is given by the Fokker–Plank equation (47):

$$\frac{\partial p(\varphi_n, t)}{\partial t} = -\frac{\partial}{\partial \varphi_n} \left[A_1(\varphi_n)\, p(\varphi_n, t) \right]$$

$$+ \frac{1}{2} \frac{\partial^2}{\partial \varphi_n^2} \left[A_2(\varphi_n)\, p(\varphi_n, t) \right] \tag{12.6}$$

with

$$A_1(\varphi_n) = \lim_{\Delta t \to 0} \frac{E\left[\Delta\varphi_n/\varphi_n\right]}{\Delta t}$$

and

$$A_2(\varphi_n) = \lim_{\Delta t \to 0} \frac{E\left[\Delta\varphi_n^2/\varphi_n\right]}{\Delta t}$$

where $E[x/y]$ represents the mathematical conditional expectation of x, given the value of y. But, taking into account Eq. 12.5,

$$\Delta\varphi_n = -K\sin\varphi_n\,\Delta t + K\int_t^{t+\Delta t} n'(u)\, du$$

Consequently, since process $n'(t)$ has a null mean value,

$$E[\Delta\varphi_n/\varphi_n] = -K\sin\varphi_n\Delta t \quad \text{and} \quad A_1(\varphi_n) = -K\sin\varphi_n$$

Furthermore,

$$\Delta\varphi_n^2 = K^2\sin^2\varphi_n\Delta t^2 - 2K^2\sin\varphi_n\Delta t\int_t^{t+\Delta t} n'(u)\,du$$

$$+ K^2\int_t^{t+\Delta t}\int_t^{t+\Delta t} n'(u)n'(v)\,du\,dv$$

which implies

$$E[\Delta\varphi_n^2/\varphi_n] = K^2\sin^2\varphi_n\Delta t^2 - 2K^2\sin\varphi_n\Delta t\int_t^{t+\Delta t} E[n'(u)]\,du$$

$$+ K^2\int_t^{t+\Delta t}\int_t^{t+\Delta t} E[n'(u)n'(v)]\,du\,dv$$

$E[n'(u)]$ is null, and consequently

$$A_2(\varphi_n) = \lim_{\Delta t\to 0}\frac{K^2}{\Delta t}\int_t^{t+\Delta t}\int_t^{t+\Delta t} E[n'(u)n'(v)]\,du\,dv$$

Since $n'(t)$ is a Gaussian white process with a spectral density N_0/A^2,

$$E[n'(u)n'(v)] = \frac{N_0}{A^2}\delta(u-v)$$

$\delta(\cdot)$ being the Dirac delta function. Then

$$A_2(\varphi_n) = K^2\frac{N_0}{A^2}$$

The Fokker–Planck equation 12.6 becomes

$$\frac{\partial p(\varphi_n,t)}{\partial t} = -\frac{\partial}{\partial\varphi_n}\left[-K\sin\varphi_n p(\varphi_n,t)\right]$$

$$+ \frac{K^2}{2}\frac{N_0}{A^2}\frac{\partial^2 p(\varphi_n,t)}{\partial\varphi_n^2} \tag{12.7}$$

If we admit that a stationary solution exists for process $\varphi_n(t)$ and if we concentrate on this solution alone, the probability density function is also stationary (independent of time):

$$\frac{\partial p(\varphi_n, t)}{\partial t} = 0$$

and the steady-state Fokker–Planck equation is reduced to

$$0 = -\frac{d}{d\varphi_n}\left[-K\sin\varphi_n p(\varphi_n)\right] + \frac{K^2}{2}\frac{N_0}{A^2}\frac{d^2 p(\varphi_n)}{d\varphi_n^2}$$

The solution of this differential equation is (47)

$$p(\varphi_n) = \frac{e^{\rho_L \cos\varphi_n}}{2\pi I_0(\rho_L)}, \qquad -\pi < \varphi_n < +\pi \qquad (12.8)$$

where ρ_L represents the signal-to-noise ratio in the loop (see Section 7.2) or the reciprocal of the phase variance value resulting from linear analysis (Eq. 7.16):

$$\rho_L = \frac{A^2}{N_0(2B_n)} = \frac{A^2}{N_0(K/2)} = \frac{1}{\sigma_{\varphi_0}^2}$$

and $I_0(\rho_L)$ is the modified Bessel function, of the first kind, of order 0 and argument ρ_L.

Consequently, the process $\varphi_n(t)$, outside the linear hypothesis, is no longer a Gaussian process. Nevertheless, its mean value is null and its variance as derived from Eq. 12.8, is expressed as

$$\sigma_{\varphi_n}^2 = \frac{\pi^2}{3} + \sum_{k=1}^{\infty}(-1)^k \frac{I_k(\rho_L)}{I_0(\rho_L)}$$

When the signal-to-noise ratio in the loop is large, $p(\varphi_n)$ decreases sharply. For low values of φ_n, a series expansion of $\cos\varphi_n$ can be used, restricted to the first two terms, $\cos\varphi_n \cong 1 - \frac{1}{2}\varphi_n^2$; furthermore, using the asymptotic expansion of $I_0(\rho_L)$, Eq. 12.8 becomes

$$p(\varphi_n) = \frac{e^{\rho_L}e^{-\frac{1}{2}\rho_L\varphi_n^2}}{2\pi\left(e^{\rho_L}/\sqrt{2\pi\rho_L}\right)}$$

that is,

$$p(\varphi_n) = \frac{1}{\sqrt{2\pi}\sqrt{1/\rho_L}}\exp\left[-\frac{\varphi_n^2}{2(1/\rho_L)}\right]$$

Thus, when the signal-to-noise ratio in the loop ρ_L is large, the distribution of φ_n tends to be gaussian with a variance $1/\rho_L$, which concurs with the linear analysis result.

On the other hand, when the signal-to-noise ratio is very bad, that is, when $\rho_L \rightarrow 0$, the distribution of the random variable φ_n tends to be uniform:

$$p(\varphi_n) = \frac{1}{2\pi}, \qquad -\pi < \varphi_n < +\pi$$

and the variance tends towards $\pi^2/3$.

Second-Order Loop. We shall assume, for the time being, that the loop filter is $F(s) = (1 + \tau_2 s)/(1 + \tau_1 s)$, since this is the most frequently encountered transfer function. We then have to use Eq. 12.4, where $f(t)$ represents the loop filter impulse response. The convolution product introduces an integration between 0 and t in this equation. Process $\varphi_n(t)$ is not in this case a Markov process, since the value of its derivative at time t depends on the history of the process from $t = 0$ until instant t.

The basic differential equation underlying the operating principle is expressed as

$$\tau_1 \frac{d^2\varphi_n}{dt^2} + \frac{d\varphi_n}{dt} = K\left[-\sin\varphi_n(t) + n'(t) \right]$$

$$+ K\tau_2 \frac{d}{dt}\left[-\sin\varphi_n(t) + n'(t) \right] \qquad (12.9)$$

After changing the variable,

$$\varphi_n(t) = \varepsilon + \tau_2 \frac{d\varepsilon}{dt} \qquad (12.10)$$

Equation 12.9 becomes

$$\tau_1 \frac{d^2\varepsilon}{dt^2} + \frac{d\varepsilon}{dt} + \tau_1\tau_2 \frac{d^3\varepsilon}{dt^3} + \tau_2 \frac{d^2\varepsilon}{dt^2} = K\left[-\sin\varphi_n(t) + n'(t) \right]$$

$$+ K\tau_2 \frac{d}{dt}\left[-\sin\varphi_n(t) + n'(t) \right]$$

Each member of this differential equation is the sum of two terms, one of which is the derivative of the other, to within the multiplying constant τ_2. Then, to solve the above differential equation (49) it suffices to solve the differential equation

$$\tau_1 \frac{d^2\varepsilon}{dt^2} + \frac{d\varepsilon}{dt} = -K\sin\varphi_n(t) + Kn'(t)$$

Taking $\varepsilon = y_1$ and $d\varepsilon/dt = y_2$, we obtain the following equations:

$$\tau_1 \frac{dy_2}{dt} + y_2 + K\sin(y_1 + \tau_2 y_2) = Kn'(t) \tag{12.11}$$

$$\frac{dy_1}{dt} = y_2 \tag{12.12}$$

The two-component [$y_1(t)$ and $y_2(t)$] random process is a Markov process since each component is a Markov process. The joint probability density function of the two variables $y_1(t)$ and $y_2(t)$ is then given at each instant by the second-order Fokker–Planck equation:

$$\frac{\partial p(y_1, y_2, t)}{\partial t} = -\sum_{i=1}^{2} \frac{\partial}{\partial y_i} \left[A_i(y_1, y_2) p(y_1, y_2, t) \right]$$

$$+ \frac{1}{2} \sum_{i=1}^{2} \sum_{j=1}^{2} \frac{\partial^2}{\partial y_i \partial y_j} \left[A_{ij}(y_1, y_2) p(y_1, y_2, t) \right] \tag{12.13}$$

with

$$A_i(y_1, y_2) = \lim_{\Delta t \to 0} \frac{E\left[\Delta y_i / y_1, y_2 \right]}{\Delta t}$$

and

$$A_{ij}(y_1, y_2) = \lim_{\Delta t \to 0} \frac{E\left[\Delta y_i \Delta y_j / y_1, y_2 \right]}{\Delta t}$$

where $E[x/y, z]$ is the mathematical conditional expectation of x, given the values of y and z.

Considering that process $n'(t)$ is a Gaussian white noise, having a null mean value and spectral density N_0/A^2, if the calculation is performed as in the previous case discussed, it is easy to show that

$$A_1(y_1, y_2) = y_2, \qquad A_2(y_1, y_2) = -\frac{y_2 + K\sin(y_1 + \tau_2 y_2)}{\tau_1}$$

and

$$A_{11} = A_{12} = A_{21} = 0, \qquad A_{22}(y_1, y_2) = \frac{K^2}{\tau_1^2} \frac{N_0}{A^2}$$

The steady-state Fokker–Planck equation (that is, $\partial p(y_1, y_2, t)/\partial t = 0$) then becomes

$$y_2 \frac{\partial p}{\partial y_1} = \frac{1 + K\tau_2 \cos(y_1 + \tau_2 y_2)}{\tau_1} p + \frac{y_2 + K\sin(y_1 + \tau_2 y_2)}{\tau_1} \frac{\partial p}{\partial y_2}$$

$$+ \frac{1}{2} \frac{K^2}{\tau_1^2} \frac{N_0}{A^2} \frac{\partial^2 p}{\partial y_2^2} \tag{12.14}$$

Unfortunately, this partial derivative equation has no analytical solution in most cases. Apart from the case $\tau_1 = \tau_2$, corresponding to the first-order loop, we can only integrate this equation in the case $\tau_2 = 0$, corresponding to a second-order loop with loop filter

$$F(s) = \frac{1}{1 + \tau_1 s}$$

The steady-state Fokker–Planck equation is then written

$$y_2 \frac{\partial p}{\partial y_1} = \frac{p}{\tau_1} + \frac{y_2 + K\sin y_1}{\tau_1} \frac{\partial p}{\partial y_2} + \frac{1}{2} \frac{K^2}{\tau_1^2} \frac{N_0}{A^2} \frac{\partial^2 p}{\partial y_2^2} \tag{12.15}$$

and it can be verified that a solution such as

$$p(y_1, y_2) = C_1 e^{m \cos y_1} \cdot C_2 e^{l y_2^2} \tag{12.16}$$

with

$$l = -\frac{\tau_1 A^2}{K^2 N_0} \quad \text{and} \quad m = \frac{2A^2}{K N_0}$$

is adequate.

The two random variables $y_1 = \varphi_n$ and $y_2 = d\varphi_n/dt$ are statistically independent since their joint-probability density function can be formulated

$$p(y_1, y_2) = p(y_1) \cdot p(y_2)$$

with

$$p(y_1) = C_1 e^{m \cos y_1} \quad \text{and} \quad p(y_2) = C_2 e^{l y_2^2}$$

The constants C_1 and C_2 are easily determined, when we observe that we must have

$$\int_{-\pi}^{+\pi} p(y_1)\,dy_1 = 1$$

since the random variable y_1 is a phase and is consequently only defined to within 2π, and

$$\int_{-\infty}^{+\infty} p(y_2)\,dy_2 = 1$$

These two conditions lead to

$$p(y_1) = \frac{1}{2\pi I_0(m)}\, e^{m\cos y_1} \tag{12.17}$$

and

$$p(y_2) = \frac{1}{\sqrt{2\pi}\,\sigma_{y_2}}\, \exp\left[-\frac{y_2^2}{2\sigma_{y_2}^2}\right] \tag{12.18}$$

with

$$\sigma_{y_2}^2 = -\frac{1}{2l} = \frac{1}{2}\frac{K^2}{\tau_1}\frac{N_0}{A^2} \tag{12.19}$$

Consequently, for a second-order loop using a one-pole low-pass filter as a loop filter, the VCO phase distribution law is identical with that for a first-order loop (Eq. 12.8), the parameter m being here again the loop signal-to-noise ratio:

$$m = \frac{A^2}{N_0}\frac{2}{K} = \frac{A^2}{N_0(2B_n)} = \frac{1}{\sigma_{\varphi_0}^2} = \rho_L$$

The new element in this case is that the VCO instantaneous angular frequency distribution is Gaussian (Eq. 12.18), having a null mean value and a variance (Eq. 12.19)

$$\sigma_{\varphi_n}^2 = \frac{K}{2}\frac{N_0}{A^2}\frac{K}{\tau_1} = \frac{N_0}{A^2}(2B_n)\omega_n^2$$

Consequently, for this type of loop, we can write

$$\sigma_{\varphi_n}^2 = \omega_n^2\sigma_{\varphi_0}^2 \tag{12.20}$$

calling $\sigma_{\varphi_0}^2$ the VCO phase variance resulting from linear analysis.

Another case where Eq. 12.14 can be solved corresponds to the linear operation case. If we assume that $\varphi_n = y_1 + \tau_2\, y_2$ is small enough for

$$\sin(y_1 + \tau_2\, y_2) \cong y_1 + \tau_2\, y_2$$

$$\cos(y_1 + \tau_2\, y_2) \cong 1$$

we can verify the form of the solution

$$p(y_1, y_2) = p(y_1) \cdot p(y_2)$$

with

$$p(y_1) = \frac{1}{\sqrt{2\pi}\ \sigma_{y_1}} \exp\left[-\frac{y_1^2}{2\sigma_{y_1}^2}\right]$$

$$p(y_2) = \frac{1}{\sqrt{2\pi}\ \sigma_{y_2}} \exp\left[-\frac{y_2^2}{2\sigma_{y_2}^2}\right]$$

and:

$$\sigma_{y_1}^2 = \frac{N_0}{A^2}\frac{\omega_n}{4\zeta}, \qquad \sigma_{y_2}^2 = \frac{N_0}{A^2}\frac{\omega_n^3}{4\zeta}$$

The two random variables y_1 and y_2 are Gaussian, of null mean value and statistically independent. The VCO phase modulation by the noise $\varphi_n = y_1 + \tau_2\, y_2$ is also a Gaussian random variable, having a null mean value and a variance

$$\sigma_{\varphi_n}^2 = \sigma_{y_1}^2 + \tau_2^2 \sigma_{y_2}^2$$

or

$$\sigma_{\varphi_n}^2 = \frac{N_0}{A^2}\frac{\omega_n}{4\zeta}\left(1 + \omega_n^2 \tau_2^2\right)$$

$$= \frac{N_0}{A^2}\frac{\omega_n}{4\zeta}\left[1 + \left(2\zeta - \frac{\omega_n}{K}\right)^2\right]$$

This expression is identical with Eq. 7.13 given in Section 7.2 for a loop with a loop filter transfer function $F(s) = (1 + \tau_2 s)/(1 + \tau_1 s)$.

Generally speaking, for loop filters $(1 + \tau_2 s)/\tau_1 s$ and $(1 + \tau_2 s)/(1 + \tau_1 s)$, we have no means of solving Eq. 12.14 and calculating $p(y_1, y_2)$. Certain authors (47, 49, 50) give approximate solutions for the phase probability density function. In particular the expression

$$p(\varphi_n) = \frac{e^{\rho_L \cos \varphi_n}}{2\pi I_0(\rho_L)}, \qquad -\pi < \varphi_n < +\pi$$

which is exact for a first-order loop and for a second-order loop with low-pass filter, also appears to be a good approximation for loop filters $F(s) = (1 + \tau_2 s)/\tau_1 s$ and $F(s) = (1 + \tau_2 s)/(1 + \tau_1 s)$. In the above expression ρ_L is the signal-to-noise ratio in the loop and must therefore be calculated using the corresponding equivalent noise bandwidth expressions given in Section 7.2. When ρ_L is large, as indicated for the first-order loop, the function $p(\varphi_n)$ tends towards a Gaussian probability density function, where the variance of the random variable corresponds to the linear analysis.

12.1.3 BOOTON QUASI-LINEARIZATION METHOD

This method, proposed by Develet for phase-locked loop study (51) shows clearly the increase of $\sigma_{\varphi_n}^2$ with respect to the linear analysis value $\sigma_{\varphi_0}^2$, when the VCO phase fluctuations $\varphi_n(t)$ are large.

The increase of $\sigma_{\varphi_n}^2$ is due to a decrease of the phase detector sensitivity K_1. This decrease in K_1 will be shown here following a slightly different way than Develet's method: we shall introduce a small static phase error in the loop as an indicator of the servo-device behavior modification in the presence of noise.

Let us consider the phase detector output signal expression (Eq. 12.1):

$$u_1(t) = K_1 \left\{ \sin\left[\phi_i(t) - \phi_0(t) - \varphi_n(t)\right] \right.$$

$$- \frac{n_1(t)}{A} \sin\left[\phi_0(t) + \varphi_n(t)\right]$$

$$\left. + \frac{n_2(t)}{A} \cos\left[\phi_0(t) + \varphi_n(t)\right] \right\}$$

As seen at the beginning of the previous subsection, the last two terms of the second member can be replaced by $n'(t)$, which is a Gaussian process having a null mean value and a spectral density N_0/A^2 from $-W/2$ to $+W/2$, provided that the bandwidth W of the additive noise $n(t)$ is large as compared with the loop bandwidth. Then

$$u_1(t) = K_1 \{ \sin\left[\phi_i(t) - \phi_0(t) - \varphi_n(t)\right] + n'(t) \} \qquad (12.21)$$

Let us suppose that the input signal is unmodulated but, for instance, that there is a small frequency offset with respect to the VCO central frequency, that is,

$$\phi_i(t) = (\omega_i - \omega_0)t + \theta_i$$

We have seen that, in the absence of noise, with the loop in lock, the VCO phase is reduced to the term

$$\phi_0(t) = (\omega_i - \omega_0)t + \theta_0$$

The small static phase difference $\theta_i - \theta_0$ is precisely that which initiates the small dc signal $K_1 \sin(\theta_i - \theta_0)$, by means of which the VCO moves from its central angular frequency ω_0 to the input signal angular frequency ω_i. In the presence of noise, the VCO phase becomes

$$\phi_0(t) + \varphi_n(t) = (\omega_i - \omega_0)t + \theta_0' + \varphi_n(t) \tag{12.22}$$

where $\theta_i - \theta_0'$ is the possible new mean phase difference between input and VCO signals. The phase detector output signal (Eq. 12.21) becomes

$$u_1(t) = K_1 \{ \sin[\theta_i - \theta_0' - \varphi_n(t)] + n'(t) \}$$

that is,

$$u_1(t) = K_1 [\sin(\theta_i - \theta_0') \cos\varphi_n(t)$$
$$- \cos(\theta_i - \theta_0') \sin\varphi_n(t) + n'(t)] \tag{12.23}$$

The component serving to keep the loop in lock is the phase detector output dc component:

$$\overline{u_1(t)} = K_1 [\sin(\theta_i - \theta_0') \overline{\cos\varphi_n(t)}$$
$$- \cos(\theta_i - \theta_0') \overline{\sin\varphi_n(t)} + \overline{n'(t)}]$$

If we suppose that $\varphi_n(t)$ remains a Gaussian random variable, having a null mean value and a variance $\sigma_{\varphi_n}^2$,

$$\overline{\cos\varphi_n(t)} = e^{-\frac{1}{2}\sigma_{\varphi_n}^2}$$

$$\overline{\sin\varphi_n(t)} = 0$$

Consequently, taking into account the fact that $\overline{n'(t)} = 0$,

$$\overline{u_1(t)} = K_1 \sin(\theta_i - \theta_0') e^{-\frac{1}{2}\sigma_{\varphi_n}^2} \tag{12.24}$$

We thus observe that when the VCO is phase modulated by the process

$\varphi_n(t)$, it is as if the phase detector sensitivity were multiplied by $e^{-\frac{1}{2}\sigma_{\varphi_n}^2}$ to become $K_1 e^{-\frac{1}{2}\sigma_{\varphi_n}^2}$. The loop gain K is also affected by this multiplication, with the consequences thus involved for the loop parameters.

But, as we observe that the noise term at the phase detector output $K_1 n'(t)$ has not varied, its relative value with respect to the signal is multiplied by $e^{\frac{1}{2}\sigma_{\varphi_n}^2}$ or, again, it is as if the power spectral density of $n'(t)$ were multiplied by $e^{\sigma_{\varphi_n}^2}$ to become $(N_0/A^2)e^{\sigma_{\varphi_n}^2}$.

Under these conditions, if we apply the linear analysis results, for example, for a second-order loop with perfect integrator and correction network (Eq. 7.18), we get

$$\sigma_{\varphi_n}^2 = \frac{N_0}{A^2} e^{\sigma_{\varphi_n}^2} (2B_n)$$

or

$$\sigma_{\varphi_n}^2 = \frac{N_0}{A^2} e^{\sigma_{\varphi_n}^2} \frac{\omega_n e^{-\frac{1}{4}\sigma_{\varphi_n}^2}\left(1 + 4\zeta^2 e^{-\frac{1}{2}\sigma_{\varphi_n}^2}\right)}{4\zeta e^{-\frac{1}{4}\sigma_{\varphi_n}^2}}$$

that is,

$$\sigma_{\varphi_n}^2 = \frac{N_0}{A^2} e^{\sigma_{\varphi_n}^2} \frac{\omega_n}{4\zeta}\left(1 + 4\zeta^2 e^{-\frac{1}{2}\sigma_{\varphi_n}^2}\right)$$

If we compare this expression with the expression that would have been obtained by application of the linear analysis, $\sigma_{\varphi_0}^2$, we get

$$\frac{\sigma_{\varphi_n}^2}{\sigma_{\varphi_0}^2} = e^{\sigma_{\varphi_n}^2} \frac{1 + 4\zeta^2 e^{-\frac{1}{2}\sigma_{\varphi_n}^2}}{1 + 4\zeta^2} \tag{12.25}$$

The results for different values of ζ are shown in Fig. 12.2 (the curves are easily obtained by calculating $\sigma_{\varphi_0}^2$ versus $\sigma_{\varphi_n}^2$). These curves show a σ_{φ_0} maximum reached for values of σ_{φ_n} corresponding to serious operational disturbances in the device and such as to lie beyond the scope of the calculation method. For practical application, we should then restrict the curves to the hard outlined lower branch. For values such that $\sigma_{\varphi_0}^2 < \frac{1}{2}$ rad^2, the curves are nevertheless a good reflection of the VCO phase fluctuation increase, which is to be feared as a consequence of phase detector nonlinearity.

A second conclusion can be drawn from Eq. 12.24. If the loop remains in lock (and on this point experience is categorical: a loop can stay in lock for values of $\sigma_{\varphi_n}^2$ such that the behavior is outside the linear hypothesis), it is

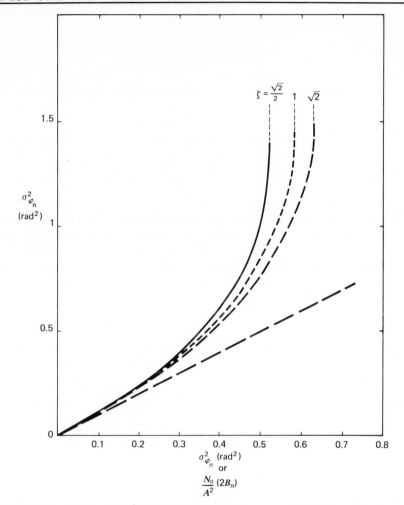

FIGURE 12.2.   Departure of $\sigma^2_{\varphi_n}$ from the linear theory for a second-order loop when $K \gg \omega_n$.

because the dc component by means of which the VCO remains in synchronism with the input signal has not varied with respect to the dc component in the absence of noise. Consequently, the steady-state phase difference $\theta_i - \theta'_0$ has increased to compensate the decrease of K_1.

In the absence of noise, let us call $K_1 \sin(\theta_i - \theta_0)$ the dc signal by means of which the VCO (central frequency $\omega_0/2\pi$) stabilizes at the input signal frequency $\omega_i/2\pi$:

$$\omega_i - \omega_0 = K_3 \,\overline{u_2} = K_2 K_3 \,\overline{u_1} = K \sin(\theta_i - \theta_0)$$

which means that, in the absence of noise, we have

$$\theta_i - \theta_0 = \text{Arc sin} \, \frac{\omega_i - \omega_0}{K}$$

In the presence of noise, the continued synchronism condition is

$$\omega_i - \omega_0 = K e^{-\frac{1}{2}\sigma^2_{\varphi_n}} \sin\left(\theta_i - \theta_0'\right)$$

Consequently,

$$\theta_i - \theta_0' = \text{Arc sin}\left(\frac{\omega_i - \omega_0}{K} e^{\frac{1}{2}\sigma^2_{\varphi_n}}\right) \qquad (12.26)$$

If the difference between input signal frequency and VCO central frequency is the same, in the presence of noise the steady-state phase error is larger, with the result that the servo-loop performances are diminished.

In the same way, for a second-order loop with $F(s) = (1 + \tau_2 s)/\tau_1 s$ in the presence of a frequency ramp $R = \Re/2\pi$ of the input signal, the steady-state phase error in the absence of noise is $(\theta_i - \theta_0) = \text{Arc sin}(\Re/\omega_n^2)$ leading to a dc signal at the phase detector output given by

$$u_1 = K_1 \sin\left(\theta_i - \theta_0\right) = K_1 \frac{\Re}{\omega_n^2}$$

This dc signal has been shown (46) to be independent of the phase detector sensitivity. This confirms the fact that when the phase detector sensitivity becomes $K_1 e^{-\frac{1}{2}\sigma^2_{\varphi_n}}$ in presence of noise, the natural angular frequency of the loop becomes $\omega_n e^{-\frac{1}{4}\sigma^2_{\varphi_n}}$ and the static phase error is

$$(\theta_i - \theta_0') = \text{Arc sin}\left(\frac{\Re}{\omega_n^2} e^{\frac{1}{2}\sigma^2_{\varphi_n}}\right)$$

12.2 OPERATING THRESHOLD IN THE PRESENCE OF ADDITIVE NOISE

The notion of operating threshold is fairly subjective and depends essentially on the function for which an instrument is intended. Generally speaking, we can say that the threshold is reached when the instrument ceases to fulfill this function or when it does so imperfectly, with degraded performances.

In our case, if the important factor is the VCO phase variance in the presence of additive thermal noise, the operating threshold is reached when the variance increases faster than expected according to the linear operating theory. We could then, by analogy with other devices, decide that when the variance exceeds by 1 dB the value given by the linear equations, the threshold of the device is reached. Using the results obtained by the Booton quasi-linearization method, this should occur around $\sigma^2_{\varphi_0} = 0.3$ rad^2 (for $\zeta = \sqrt{2}\,/2$) or for a signal-to-noise ratio in the loop ρ_L in the vicinity of $+5$ dB (Fig. 12.2).

There are other methods of checking this phenomenon, in addition to those indicated above (47, 49, 50), such as the linear spectrum approximation method (52), and the Volterra function expansion method (53). The results obtained by the different methods can be compared using the curves in Fig. 12.3 (54). Whatever the method chosen, we observe that the threshold at $+1$ dB from the linear equation value occurs for a signal-to-noise ratio in the loop ρ_L of $+4.5$ dB (± 1 dB) and for a "linear" VCO phase variance of around $\sigma^2_{\varphi_0} = 0.35$ rad^2, or a real phase variance of $\sigma^2_{\varphi_n} = 0.45$ rad^2.

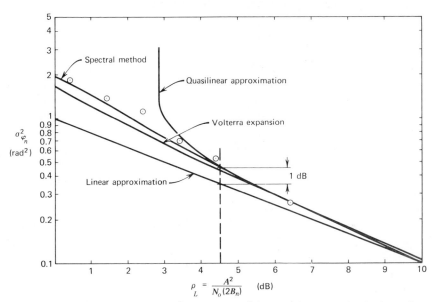

FIGURE 12.3. Different results for $\sigma^2_{\varphi_n}$ versus $\rho_L = A^2/N_0(2B_n)$ for a second-order loop when $K \gg \omega_n$ and $\zeta = \sqrt{2}\,/2$ (from W. C. Lindsey and R. C. Tausworthe; furnished through the courtesy of the Jet Propulsion Laboratory, California Institute of Technology).

But when the VCO phase variance is high, another phenomenon occurs, which is observed experimentally and for which only the Viterbi method makes allowances. Since the VCO phase is a random process, it can be sometimes increased to such an extent that the VCO slips one or several cycles with respect to the input signal. We have seen that the phase detector characteristic is a periodic function (period 2π) of the phase error. From time to time the loop jumps from one stable operating point to another similar point 2π further on (slip of one cycle) or $k2\pi$ further on (slip of k cycles). This is a random occurence, with a very low probability if the VCO variance is low, but with an increasing probability when the VCO variance increases.

If the only cause of phase error in the loop is the additive noise, that is, if the input signal is not modulated and if its frequency is equal to the VCO central frequency, cycle slipping either in the plus or minus direction has the same probability. If the input signal frequency is not equal to the VCO central frequency, for a given signal-to-noise ratio, the probability that the phenomenon will occur is heightened. Furthermore, cycle-slipping direction is no longer equiprobable: the cycle slips most frequently in the direction tending to bring the VCO frequency back to its central value. These results, observed experimentally, have been demonstrated in the case of a first-order loop (55).

When the cycle-slipping phenomenon occurs, the number of the cycles lost or gained depends on the signal-to-noise ratio and the difference between the input signal frequency and the VCO central frequency. For a second-order loop, we also observed a certain influence of the time constant τ_1 of the loop filter $F(s) = (1 + \tau_2 s)/(1 + \tau_1 s)$ (56). The equivalent noise bandwidth of this loop depends essentially on the natural angular frequency value ω_n. But $\omega_n^2 = K/\tau_1$ and a given value of ω_n can be obtained using a high loop gain K and a high time constant τ_1, or using a lower loop gain value and a lower time constant. But the time constant τ_1 represents, as it were, the VCO memory with respect to the input signal frequency. If the constant τ_1 is very high, when cycle slipping occurs, the VCO loses less cycles than with a low τ_1, because it returns much more slowly towards its central frequency. The experimental results for this type of loop would also appear to show that the smaller the damping factor ζ, the more frequent are the cycle slips, all other conditions being equal. This point can be connected with the fact that at a given $\sigma_{\varphi_0}^2$, the smaller the damping factor ζ, the greater is the real phase variance $\sigma_{\varphi_n}^2$ (see Fig. 12.2).

For a second-order-loop [with $F(s) = (1 + \tau_2 s)/(1 + \tau_1 s)$], experience also shows that if the difference between the input signal frequency and the VCO central frequency is below the acquisition range, periods, of variable duration, where the VCO is in synchronism with the input signal alternate with periods where it is not, to such an extent that we can speak of the percentage of time during which the loop is in lock or out of lock. This percentage of time can be used to define an operating threshold for certain applications (for instance, in the case of a loop used as a coherent phase demodulator). But, when the difference between the input signal frequency and the VCO frequency exceeds the acquisition range and if the signal-to-noise ratio is bad, the VCO, after an indeterminate period, systematically returns to its central frequency. It is no longer in synchronism with the input signal, even for a low percentage of time: the loop is definitively out of lock.

It is very difficult to estimate at what signal-to-noise ratio values the various phenomena (slipping of one cycle, slipping of several cycles, loop definitively out of lock) are likely to occur and with what probability. Theoretical considerations offer no assistance in this domain, where only practical experiments can provide data. Unfortunately, the experimental surveys that have been undertaken are too closely related to the specific character of the loop used and the technological problems arising from the difficult operating conditions are too serious for the results to be transposed. Take, for example, the fact that almost all the phase detectors tend to give a nonzero dc output component for $\phi = 0$ when the signal-to-noise ratio at the phase detector input is small. Consequently, even if the initial frequency difference between input signal and VCO is null, the dc component, depending on the signal-to-noise ratio, appears and modifies the apparent value of the VCO central frequency. This phenomenon considerably complicates experimental investigation and, in particular, can cause the loop to fall definitively out of lock even if, in the absence of noise, the initial frequency difference is zero. It is consequently advisable for the design engineer anticipating very bad signal-to-noise ratio conditions to make sure experimentally that the frequency and seriousness of the cycle slipping are compatible with the function the loop is required to fill.

For a first-order loop, a certain number of results can be derived from the Fokker–Planck equations (47). If the difference between the input signal frequency and the VCO central frequency is zero, the mean value of the time required for the first cycle slip to occur, starting from an initial zero phase error, is given by

$$T_m = \frac{\pi^2 \rho_L I_0^2(\rho_L)}{(2B_n)} \qquad (12.27)$$

In this expression, $I_0(\rho_L)$ is the modified Bessel function, of the first kind, of order 0 and argument ρ_L, ρ_L being the signal-to-noise ratio in the loop:

$$\rho_L = \frac{1}{\sigma_{\varphi_0}^2} = \frac{A^2}{N_0(2B_n)}$$

For the values of ρ_L generally encountered, $I_0(\rho_L)$ can be replaced by the first term of its asymptotic expansion,

$$I_0(\rho_L) \cong e^{\rho_L} (2\pi\rho_L)^{-1/2}.$$

Then

$$T_m \cong \frac{\pi}{2} \frac{e^{2\rho_L}}{(2B_n)} \tag{12.28}$$

After each cycle slip, for a first-order loop, the servo loop comes back to its departure condition (zero phase error) and consequently the mean frequency of cycle slipping is

$$\bar{F} = \frac{1}{T_m} = \frac{(2B_n)}{\pi^2 \rho_L I_0^2(\rho_L)} \cong \frac{2}{\pi} \frac{(2B_n)}{e^{2\rho_L}} \tag{12.29}$$

For second-order loops, certain experimental work would seem to indicate that Eqs. 12.28 and 12.29 can be used for T_m and $\bar{F}$, but using for $(2B_n)$ the expression corresponding to the second-order loop considered. The results thus obtained could provide an approximation of T_m and $\bar{F}$, at least if the initial frequency difference is null. If this is not the case, cycle slipping generally occurs by groups of several cycles and the time passed by the loop in losing or gaining cycles is no longer negligible against the correct operating time, so that $\bar{F} \neq 1/T_m$. For a second-order loop, with very high loop gain K and for a damping factor ζ between $\sqrt{2}/4$ and $\sqrt{2}$ the results for a simulation are given in Fig. 12.4 (57) representing variations of $\omega_n T_m$ versus σ_{φ_0}. However, it should be noted that for the authors of this simulation, T_m is not exactly the average time elapsed before slip of a cycle, but the time required, in the phase plane, for the trajectory, starting from a point $(\phi = 0, \dot{\phi} = 0)$ to cross the vertical axis at $\phi = \pm \pi$ and remain on the other side for a time exceeding or equal to $4/\omega_n$. The results in Fig. 12.4 seem pessimistic as compared with Eq. 12.28. This is perhaps because in Ref. 57, an approximation is made and the variable used is not really $\dot{\phi}$ but a simplified version of $\dot{\phi}$, partially cleared of noise terms.

Still using these simulation results, the authors suggest the following formula for the calculation of the probability $P(T)$ that the VCO will slip one cycle during time T, starting from a zero phase error:

$$P(T) = 1 - e^{-T/T_m} \tag{12.30}$$

This result seems to be confirmed by the result of another simulation (58), as well as by experiment (59).

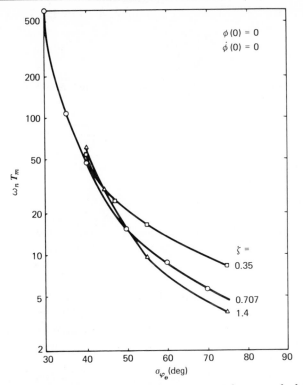

$\phi(0) = 0$
$\dot{\phi}(0) = 0$

$\zeta =$
0.35
0.707
1.4

σ_{φ_θ} (deg)

FIGURE 12.4. Mean time between cycle-slipping versus VCO linear standard deviation σ_{φ_0}, for a second-order loop with $K \gg \omega_n$ (with the damping ratio ζ as a parameter). (from J. R. Rowbotham and R. W. Sanneman, *IEEE Transactions on Aerospace and Electronic Systems*, AES-3, No. 4, 611, July 1967).

NUMERICAL EXAMPLE.   A loop with equivalent noise band $(2B_n) = 10^3$ Hz is in lock at instant $t = 0$ with a zero static phase error. What is the probability that this loop will stay in lock for 10 seconds if the signal-to-noise ratio conditions are such that

$$\frac{S}{N_0} = 37 \text{ dB-Hz}$$

The signal-to-noise ratio in the loop (Eq. 7.16) is

$$\rho_L = \frac{1}{\sigma_{\varphi_0}^2} = \frac{2S}{N_0} \frac{1}{(2B_n)} = \frac{2 \times 5 \times 10^3}{10^3} = 10$$

The mean time before the first cycle slip is given by Eq. 12.28, leading to

$$T_m = \frac{\pi}{2} \frac{e^{20}}{10^3} = 7.62 \times 10^5 \text{ s} \qquad (212 \text{ hours})$$

The probability that the VCO will slip 1 cycle during the first ten seconds is given by Eq. 12.30:

$$P(10) = 1 - \exp\left[-\left(\frac{10}{7.62} 10^{-5} \right) \right] = 1.3 \times 10^{-5}$$

The probability that the loop will stay in lock, without any cycle slipping, during 10 s is 0.999987.

If the signal-to-noise density ratio is only $S/N_0 = 34$ dB-Hz, we obtain respectively $T_m = 34.6$ s, $P(10) = 0.25$, and a probability of the loop remaining in lock equal to 0.75. These figures provide a good example of the suddenness of the threshold phenomenon in a phase-locked loop.

12.3 INFLUENCE OF THE INPUT SIGNAL MODULATION

Fairly generally, when the input signal is modulated, two categories of loop are used. Narrowband loops are used when we want to use the carrier alone, despite the modulation, or when we are attempting to build a coherent phase or amplitude demodulator of the input signal. This implies a low index for the input signal phase modulation. Wideband loops, on the contrary, are used when the input signal phase modulation index is high and when it is absolutely necessary for the VCO to follow the greater part of this modulation if the phase error is to remain low enough for normal operation to continue. This solution corresponds more particularly to the case where the input signal is frequency modulated and to the use of a phase-locked loop as a discriminator.

12.3.1 CASE OF PHASE MODULATION

Let us first suppose that the input signal frequency is equal to the VCO central frequency:

$$y_i(t) = A \sin \left[\omega_0 t + \theta_i + \phi_i(t) \right]$$

If the variations of $\phi_i(t)$ are fast enough not to pass into the loop, the VCO output signal is expressed as

$$y_0(t) = B \cos (\omega_0 t + \theta_i)$$

Consequently, the phase detector output signal is written

$$u_1(t) = AB \sin \left[\omega_0 t + \theta_i + \phi_i(t) \right] \cos (\omega_0 t + \theta_i)$$

that is, if we eliminate the angular frequency components in the vicinity of $2\omega_0$,

$$u_1(t) = K_1 \sin \left[\phi_i(t) \right]$$

If the modulation is sinusoidal, $\phi_i(t) = m \sin (\Omega t + \Theta)$, the demodulated signal obtained is expressed as

$$u_1(t) = K_1 \sin \left[m \sin (\Omega t + \Theta) \right]$$

that is,

$$u_1(t) = 2K_1 \left[J_1(m) \sin (\Omega t + \Theta) + J_3(m) \sin (3\Omega t + 3\Theta) + \cdots \right]$$

The phase detector nonlinearity is revealed in the appearance of distortion terms on all the odd harmonics of the modulation signal.

It may be interesting to note that if the bandpass filter preceding the phase detector only lets through the first components arising from the modulation, the input signal becomes

$$y_i(t) = A \left\{ J_0(m) \sin (\omega_0 t + \theta_i) + J_1(m) \sin \left[(\omega_0 + \Omega) t + \theta_i + \Theta \right] \right.$$
$$\left. - J_1(m) \sin \left[(\omega_0 - \Omega) t + \theta_i - \Theta \right] \right\}$$

When this signal is applied to the phase detector, the result of the multiplication by $B \cos (\omega_0 t + \theta_i)$ is, after elimination of the angular frequency terms around $2\omega_0$,

$$u_1(t) = 2K_1 J_1(m) \sin (\Omega t + \Theta)$$

In other words, in this particular case, there is no distortion of the demodulated signal. The nonlinearity of the phase detector compensates exactly for the distortion, which arises from the fact that the filter bandwidth preceding the detector is not wide enough, and which would have been obtained with a perfectly linear phase demodulator.

If the modulation consists of a multiplex of sinusoidal signals,

$$\phi_i(t) = \sum_{k=1}^{n} m_k \sin (\Omega_k t + \Theta_k)$$

the expression of the demodulated signal is derived from the expansion of the expression $\sin\left[\sum_k m_k \sin(\Omega_k t + \Theta_k)\right]$, that is,

$$\text{Im}\left[\prod_{k=1}^{n}\sum_{l_k=-\infty}^{+\infty} J_{l_k}(m_k) e^{jl_k(\Omega_k t + \Theta_k)}\right]$$

where the symbol $\text{Im}[x]$ represents the imaginary part of the complex quantity x. In this case, in addition to the angular frequency harmonic terms $l_k\Omega_k$ representing the distortion on each modulation sinusoidal signal, we observe the appearance of $(r\Omega_p + s\Omega_q)$ angular frequency terms, generally known as intermodulation products (r and s are any integers, positive or negative).

Using the same procedure as in Section 12.1.3, we now show that the modulation, still because of the nonlinearity of the phase detector, reduces possible loop performances.

Let us suppose that the input signal differs slightly in frequency with respect to the VCO central frequency $\omega_0/2\pi$:

$$y_i(t) = A \sin\left[\omega_i t + \theta_i + \phi_i(t)\right]$$

When the loop is in lock, the VCO signal, in the absence of modulation ($\phi_i(t) = 0$), is expressed as

$$y_0(t) = B \cos(\omega_i t + \theta_0)$$

with

$$\theta_i - \theta_0 = \text{Arc}\sin\frac{\omega_i - \omega_0}{K}$$

In the presence of modulation, still assuming that the variations of $\phi_i(t)$ are fast enough not to modulate the VCO and, further, assuming that $\overline{\phi_i(t)} = 0$ [if this is not the case, $\overline{\phi_i(t)}$ can always be included in θ_i], the VCO signal expression becomes

$$y_0(t) = B \cos(\omega_i t + \theta_0')$$

θ_0' being the possible new phase of the VCO signal.

The signal obtained at the phase detector output is written

$$u_1(t) = K_1 \sin\left[\theta_i - \theta_0' + \phi_i(t)\right]$$

that is,

$$u_1(t) = K_1\left[\sin(\theta_i - \theta_0')\cos\phi_i(t) + \cos(\theta_i - \theta_0')\sin\phi_i(t)\right] \qquad (12.31)$$

A first consequence of Eq. 12.31 is connected with the distortion of the demodulated signal. If $\phi_i(t) = m\sin(\Omega t + \Theta)$, in addition to the distortion terms on the modulation signal odd harmonics, due to the sinusoidal characteristic of the phase detector, distortion terms appear on the even harmonics [originating in the $\cos\phi_i(t)$ term of Eq. 12.31]. This is, of course, because, as a consequence of the frequency shift $\omega_i - \omega_0$, the operating point has shifted on the phase detector characteristic, which is no longer symmetrical with respect to the modulation.

If we concentrate on the dc component $\overline{u_1(t)}$, knowing that $\overline{\phi_i(t)} = 0$ implies $\overline{\sin\phi_i(t)} = 0$, then

$$\overline{u_1(t)} = K_1 \overline{\cos\phi_i(t)}\, \sin\left(\theta_i - \theta_0'\right) \tag{12.32}$$

If we state that it is this dc component that maintains synchronism, then

$$\omega_i - \omega_0 = K_2 K_3 \overline{u_1(t)} = K \overline{\cos\phi_i(t)}\, \sin\left(\theta_i - \theta_0'\right)$$

implying

$$\theta_i - \theta_0' = \operatorname{Arc\,sin} \frac{\omega_i - \omega_0}{K\, \overline{\cos\phi_i(t)}} \tag{12.33}$$

As compared with the result obtained in the absence of modulation, it is as if the phase detector sensitivity and consequently the loop gain were multiplied by $\overline{\cos\phi_i(t)}$.

Remark. It should be noted that this apparent decrease in K_1 only occurs with respect to the loop (for the demodulated signal obtained, Eq. 12.31 should be used) and that it takes place even if $(\theta_i - \theta_0')$ remains small enough for replacing $\sin(\theta_i - \theta_0')$ by $(\theta_i - \theta_0')$. In this case, Eq. 12.32 becomes

$$\overline{u_1(t)} = K_1 \overline{\cos\phi_i(t)}\, \left(\theta_i - \theta_0'\right)$$

while the demodulated signal becomes $u_1(t) = K_1 \sin\phi_i(t)$.

We can further discuss the apparent decrease of K_1 in a few specific cases, starting with the case of a sinusoidal phase modulation: $\phi_i(t) = m\sin(\Omega t + \Theta)$. Then

$$\cos\phi_i(t) = \cos\left[\, m\sin(\Omega t + \Theta)\,\right]$$

$$= J_0(m) + 2J_2(m)\sin(2\Omega t + 2\Theta) + \cdots$$

implying that

$$\overline{\cos\phi_i(t)} = J_0(m)$$

and consequently,

$$\overline{u_1(t)} = K_1 J_0(m)\sin\left(\theta_i - \theta_0'\right)$$

In particular, when $m = 2.4$, $J_0(m) = 0$ and consequently $\overline{u_1(t)} \equiv 0$.

There is no longer an error signal, so the servodevice is no longer operative.

For m in the vicinity of 2.4, operation is possible, at least in the absence of noise. But in the presence of noise, $J_0^2(m)$ being very small, $\sigma_{\varphi_n}^2$ increases rapidly and the operating threshold of the device is very soon reached (see following section).

Let us now consider the case of a binary modulation: $\phi_i(t) = \theta f(t)$, with $f(t) = \pm 1$ depending on whether the bit transmitted is 0 or 1. Then, with the two cases assumed to be equiprobable,

$$\overline{\sin\phi_i(t)} = 0 \quad \text{and} \quad \overline{\cos\phi_i(t)} = \overline{\cos\left[\theta f(t)\right]} = \cos\theta$$

Consequently,

$$\overline{u_1(t)} = K_1 \cos\theta \sin(\theta_i - \theta_0')$$

When $\theta = \pi/2$, which corresponds to the case of antipodal modulation, then $\overline{u_1(t)} \equiv 0$, and the loop is no longer operative. In the same way as in the previous example, $\sigma_{\varphi_n}^2$ becomes very large when $\theta \to \pi/2$.

Finally, let us suppose that the modulation $\phi_i(t)$ is a Gaussian variable of null mean value and variance $\sigma_{\phi_i}^2$:

$$\overline{\sin\phi_i(t)} = 0 \quad \text{and} \quad \overline{\cos\phi_i(t)} = e^{-\frac{1}{2}\sigma_{\phi_i}^2}$$

Consequently,

$$\overline{u_1(t)} = K_1 e^{-\frac{1}{2}\sigma_{\phi_i}^2} \sin(\theta_i - \theta_0')$$

In this case, operation always remains possible, but if $\sigma_{\phi_i}^2$ is too large, the loop performances are considerably diminished, unless, as mentioned in Chapter 8, the bandwidth is widened to enable the VCO to follow the greater part of the variations of $\phi_i(t)$. In this case, the loop can no longer be used to perform a correct demodulation of the input signal (and the above calculations do not apply).

Remark 1. In three examples given above, we observe that the value of K_1 to be considered for the loop is that which would be obtained if the input signal amplitude to be used is assumed to be that of the carrier (discrete line) contained in the input signal power spectrum expression. This reflects the fact that since the loop is narrow enough for the influence of the $\phi_i(t)$ variations on the VCO phase to be eliminated, only the carrier is considered. But we have also seen that, as regards the phase detector sensitivity as a demodulation device, it is the signal amplitude A that should be used. Furthermore, the apparent K_1 decrease is only valid for sinusoidal phase detectors. For a triangular phase detector, linear over an interval $(-\pi/2, +\pi/2)$, or for a sawtooth detector, linear over an interval $(-\pi, +\pi)$, there is no K_1 decrease in the loop as long as $|\phi_i(t) + \theta_i - \theta_0| < \pi/2$ or $|\phi_i(t) + \theta_i - \theta_0| < \pi$, depending on the case [if this condition is not fulfilled, $u_1(t)$ must be calculated using a method similar to that described]. In particular, the loop remains operative with a sawtooth detector for a sinusoidal modulation with $m = 2.4$ or for an antipodal binary modulation $(\theta = \pi/2)$, although in both these cases the carrier is not a discrete line in the input signal spectrum expression.

Remark 2. In this section, we have discussed the influence of the input signal modulation, assuming the loop to be in lock when the modulation is applied. The acquisition of a modulated signal is, on the other hand, a difficult problem, so far unsolved. Experience shows that if the input signal spectrum comprises discrete components, acquisition, whether natural or aided, can occur on a spectrum component other than the carrier.

If the input signal $y_i(t)$ is accompanied by an additive noise $n(t)$ and even if the noise level is such that the VCO modulation $\varphi_n(t)$ is negligible (if this hypothesis is not verified, see Section 12.3.2 below), Eq. 12.31 must be modified to become

$$u_1(t) = K_1 \big[\sin(\theta_i - \theta_0') \cos \phi_i(t) + \cos(\theta_i - \theta_0') \sin \phi_i(t) + n'(t) \big]$$

The phase detector output noise term $K_1 n'(t)$ is unaffected by the apparent decrease of K_1. Consequently, for the loop, the equivalent phase modulation $\varphi_i(t)$ from the additive noise (see Section 7.2) becomes

$$\varphi_i(t) = \frac{n'(t)}{\cos \phi_i(t)}$$

The equivalent modulation power spectral density becomes

$$S_{\varphi_i}(f) = \frac{N_0}{A^2} \frac{1}{\big[\overline{\cos \phi_i(t)} \big]^2}$$

The apparent decrease of the loop gain K also implies a modification of the loop parameters with the result that the VCO phase variance from additive noise becomes

$$\sigma_{\varphi_n}^2 = \frac{N_0}{A^2 \big[\overline{\cos \phi_i(t)} \big]^2} \frac{\omega_n \big[1 + 4\zeta^2 \overline{\cos \phi_i(t)} \big]}{4\zeta} \tag{12.34}$$

in the case of a second-order loop with $F(s) = (1 + \tau_2 s)/\tau_1 s$.

NUMERICAL EXAMPLE. A second-order loop $[F(s)=(1+\tau_2 s)/\tau_1 s]$ such that $\zeta = 1$ and $\omega_n = 100$ rad/s is locked on an unmodulated signal with a phase error of 0.1 rad. The signal-to-noise ratio conditions are such that the VCO phase variance is 10^{-2} rad^2.

A phase modulation $(\theta = \pi/3)$, by a 10-kHz square signal, is applied to the input signal.

In the absence of modulation, we had $\theta_i - \theta_0 = 0.1$ rad, implying $\sin(\theta_i - \theta_0) = 0.0998$. The static phase error becomes (Eq. 12.33)

$$\theta_i - \theta_0' = \text{Arc} \sin \frac{0.0998}{\frac{1}{2}} = 0.201 \text{ rad}$$

The VCO phase variance was

$$\sigma_{\varphi_0}^2 = \frac{N_0}{A^2}(2B_n) = 10^{-2} \text{ rad}^2$$

implying $N_0/A^2 = 8 \times 10^{-5}$. The phase variance becomes (Eq. 12.34):

$$\sigma_{\varphi_n}^2 = \frac{8 \times 10^{-5}}{\frac{1}{4}} \times \frac{100 \times \left(1 + 4\frac{1}{2}\right)}{4} = 2.4 \times 10^{-2} \text{ rad}^2$$

In the presence of modulation, although the equivalent modulation spectral density from additive noise has been multiplied by 4, the phase variance is only multiplied by 2.4, for the loop parameters have become $\zeta = 0.707$ and $\omega_n = 70.7$ rad/s, implying $(2B_n) = 75$ Hz [as compared with $(2B_n) = 125$ Hz in the absence of modulation].

12.3.2 OPERATING THRESHOLD WITH PHASE MODULATION IN THE PRESENCE OF ADDITIVE NOISE

It is possible to use the procedure described in Section 12.1.3 to obtain an approximate analytical expression for the VCO phase variance from thermal noise when the input signal is phase modulated. This method is obviously imperfect in that it does not take into account the phenomenon of VCO cycle slipping with respect to the input signal. But it does demonstrate the diminished possibilities of the loop with respect to the causes of static phase error and the considerable increase in VCO phase variance, which also implies reduced loop performances.

The modulated input signal accompanied by an additive noise is written

$$y_i(t) = A \sin\left[\omega_i t + \theta_i + \phi_i(t)\right] + n(t)$$

The VCO signal, assuming the power spectrum of $\phi_i(t)$ to be very wide as compared with the loop bandwidth, is expressed

$$y_0(t) = B \cos \left[\omega_i t + \theta'_0 + \varphi_n(t) \right]$$

where $\varphi_n(t)$ represents the random VCO phase fluctuations due to the presence of noise $n(t)$. The spectrum-width hypothesis, mentioned above, implies that the bandwidth W preceding the phase detector is far larger than the loop equivalent noise bandwidth. Consequently (47), the phase detector output signal can be formulated

$$u_1(t) = K_1 \left\{ \sin \left[\theta_i - \theta'_0 + \phi_i(t) - \varphi_n(t) \right] + n'(t) \right\}$$

the process $n'(t)$ having the properties described in Section 7.1.2. The detector output dc component is expressed as

$$\overline{u_1(t)} = K_1 \, \overline{\sin \left[\theta_i - \theta'_0 + \phi_i(t) - \varphi_n(t) \right]}$$

Since the mean value of processes $\phi_i(t)$ and $\varphi_n(t)$ is assumed to be zero, we get

$$\overline{u_1(t)} = K_1 \, \overline{\cos \phi_i(t)} \, \overline{\cos \varphi_n(t)} \, \sin(\theta_i - \theta'_0) \qquad (12.35)$$

The detector output noise term $K_1 n'(t)$ can be interpreted, as regards the loop, as an equivalent phase modulation of the input signal $\varphi_i(t)$, so that

$$K_1 n'(t) = \left[K_1 \, \overline{\cos \phi_i(t)} \, \overline{\cos \varphi_n(t)} \, \right] \varphi_i(t)$$

Then

$$\varphi_i(t) = \frac{n'(t)}{\overline{\cos \phi_i(t)} \, \overline{\cos \varphi_n(t)}}$$

Consequently, as far as the loop is concerned, $\varphi_i(t)$ can be considered as a Gaussian random process, having a zero mean value and a two-sided spectral density

$$S_{\varphi_i}(f) = \frac{N_0}{A^2} \frac{1}{\left[\, \overline{\cos \phi_i(t)} \, \overline{\cos \varphi_n(t)} \, \right]^2} \qquad (12.36)$$

If we accept this new value for the power spectral density of the equivalent modulation due to an additive noise and if we take into account the variation of the value of K_1, and thereby the loop parameter variations, we obtain, for example for a second-order loop with integrator and correction network,

$$\sigma_{\varphi_n}^2 = \frac{N_0}{A^2} \frac{1}{\left[\overline{\cos\phi_i(t)} \ \overline{\cos\varphi_n(t)} \right]^2}$$

$$\times \frac{\omega_n \left[1 + 4\zeta^2 \ \overline{\cos\phi_i(t)} \ \overline{\cos\varphi_n(t)} \right]}{4\zeta}$$

that is, compared with the value $\sigma_{\varphi_0}^2$ derived from the linear analysis,

$$\frac{\sigma_{\varphi_n}^2}{\sigma_{\varphi_0}^2} = \frac{1}{\left[\overline{\cos\phi_i(t)} \right]^2} \frac{1}{\left[\overline{\cos\varphi_n(t)} \right]^2}$$

$$\times \frac{1 + 4\zeta^2 \ \overline{\cos\phi_i(t)} \ \overline{\cos\varphi_n(t)}}{1 + 4\zeta^2} \qquad (12.37)$$

If we state that $\overline{\cos\phi_i(t)} = \alpha_i$

$\alpha_i = J_0(m)$ for a sinusoidal modulation

$\alpha_i = \cos\theta$ for an equiprobable binary modulation

$\alpha_i = e^{-\frac{1}{2}\sigma_{\phi_i}^2}$ for a Gaussian modulation

and, bearing in mind that $\overline{\cos\varphi_n(t)} = e^{-\frac{1}{2}\sigma_{\varphi_n}^2}$, Eq. 12.37 becomes

$$\frac{\sigma_{\varphi_n}^2}{\sigma_{\varphi_0}^2} = \frac{1}{\alpha_i^2} e^{\sigma_{\varphi_n}^2} \frac{1 + 4\zeta^2 \alpha_i e^{-\frac{1}{2}\sigma_{\varphi_n}^2}}{1 + 4\zeta^2} \qquad (12.38)$$

This formula can be used to calculate easily $\sigma_{\varphi_0}^2$ versus $\sigma_{\varphi_n}^2$, for different values of α_i and of the damping factor ζ. Inversely, the variations of $\sigma_{\varphi_n}^2$ versus $\sigma_{\varphi_0}^2$ can be derived from this equation. These variations are shown in Fig. 12.5 in the case where $\zeta = 1$ and for three values of α_i corresponding to modulations of the input signal as follows:

- sinusoidal with an index $m = 1$ rad ($\alpha_i = 0.765$)
- binary with a phase deviation $\theta = 1$ rad ($\alpha_i = 0.54$)
- Gaussian with a standard deviation $\sigma_{\phi_i} = 1$ rad ($\alpha_i = 0.605$).

In the same way as for the curves in Fig. 12.2, we observe a σ_{φ_0} maximum, but it is only reached for values of σ_{φ_n} such that this method is definitely no longer utilizable [in particular, $\varphi_n(t)$ is certainly no longer Gaussian]. However, these curves can be used, if only for their qualitative value, when σ_{φ_n} is below 1 rad.

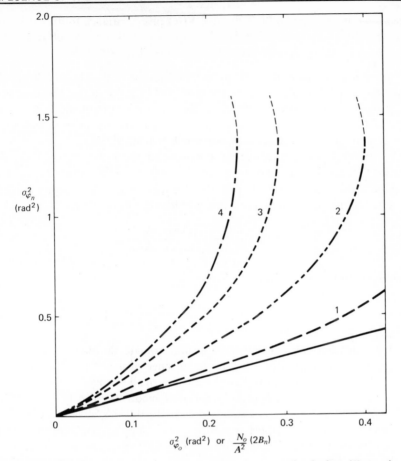

FIGURE 12.5. Departure of $\sigma^2_{\varphi_n}$ from the linear theory for a second-order loop ($K \gg \omega_n$, $\zeta = 1$) when the input signal is phase-modulated: (1) without modulation; (2) sinusoidal modulation with $m = 1$ rad; (3) Gaussian random modulation with $\sigma_{\phi_i} = 1$ rad; (4) binary modulation with $\theta = 1$ rad.

NUMERICAL EXAMPLE. If the signal-to-noise ratio conditions are such that $\sigma^2_{\varphi_o} = 0.2$ rad², a more realistic value for the VCO phase variance, in the absence of modulation, would be $\sigma^2_{\varphi_n} = 0.23$ rad², that is, an increase of $+0.6$ dB (Fig. 12.5).

For the modulations defined above, we could then have

- $\sigma^2_{\varphi_n} = 0.34$ rad² for a sinusoidal modulation (increase of $+2.3$ dB).
- $\sigma^2_{\varphi_n} = 0.52$ rad² for a Gaussian modulation (increase of $+4.2$ dB).
- $\sigma^2_{\varphi_n} = 0.7$ rad² for a binary modulation (increase of $+5.4$ dB).

Inversely, if we want to limit the real VCO phase variance to $\sigma_{\varphi_n}^2 = 0.2$ rad^2, the signal-to-noise ratio conditions have to be such that $\sigma_{\varphi_0}^2 = 0.08$ rad^2 if the input signal is modulated by a binary signal so that $\theta = 1$ rad.

Another consequence of the apparent decrease of K_1 (Eq. 12.35) is the increase in static phase errors in the loop. Let us suppose that the loop is in lock with a small difference between the input signal frequency $\omega_i / 2\pi$ and the VCO central frequency $\omega_0 / 2\pi$. In the absence of modulation and noise, we should have a phase error $(\theta_i - \theta_0)$ such that $\sin(\theta_i - \theta_0) = (\omega_i - \omega_0)/K$.

In the presence of modulation and noise, if the loop stays in lock, this means that the device continues to produce the same dc component $\overline{u_1(t)}$ to maintain the synchronism despite the decrease in K_1. Taking into account Eq. 12.35, the static phase error has become $(\theta_i - \theta_0')$ so that

$$\sin(\theta_i - \theta_0') = \frac{\omega_i - \omega_0}{K \; \overline{\cos\phi_i(t)} \; \overline{\cos\varphi_n(t)}}$$

that is, with the notation used above,

$$(\theta_i - \theta_0') = \text{Arc}\sin\left(\frac{\omega_i - \omega_0}{K} e^{\frac{1}{2}\sigma_{\varphi_n}^2} \frac{1}{\alpha_i} \right) \tag{12.39}$$

This equation can only be used to approximate the increase of a small static phase error from $(\theta_i - \theta_0)$ (in the absence of noise and modulation) to $(\theta_i - \theta_0')$. The latter quantity has also to stay small enough for the curves of Fig. 12.5 (or of Fig. 12.2 in the absence of modulation) to be used. When $(\theta_i - \theta_0')$ [and a fortiori $(\theta_i - \theta_0)$] is not small, the degradation of $\sigma_{\varphi_n}^2$ with respect to $\sigma_{\varphi_0}^2$ is very important and the cycle-slipping phenomenon can no longer be disregarded.

To close this section a little more optimistically, it should be noted that a coherent phase demodulation using a narrow phase-locked loop (that is, for a low index modulation of the input signal) has no other theoretical operating threshold than that of the loop providing for synchronization of the VCO with the carrier. This means that if the equivalent noise bandwidth of this loop $(2B_n)$ is sufficiently narrow, the demodulation continues to work correctly even if the signal-to-noise ratio at the demodulator input is very low because the bandwidth W preceding the loop is much wider than $(2B_n)$. The distinction as compared with a conventional phase demodulation (frequency discriminator followed by an integrator) regarding the theoretical operating threshold is fundamental.

Remark. We have, however, mentioned the existence of a phase detector technological threshold, related to the appearance of an undesirable output dc component when the input signal-to-noise ratio is -20 dB (bad detector) or -30 dB (good detector). But, bearing in mind the use to which the demodulated signal is to be put, such a value is rarely reached and the above conclusion remains valid.

12.3.3 OPERATING THRESHOLD WITH FREQUENCY MODULATION

When the modulation index is too high, either because it is a high-index phase modulation or because it is a frequency modulation, a loop has to be used having a sufficiently wide bandwidth for the VCO to follow the greater part of the input signal modulation and for the phase error to be fairly low. But we have seen (Section 6.4) that this produces at the phase detector output or at the loop filter output (depending on the type of loop considered) a signal that is a good duplicate of the input signal frequency modulation.

Taking into account the equivalent noise bandwidth of the loop in this case, noise protection will be distinctly less effective than in the coherent phase demodulator application, for the same signal-to-noise ratio in the preceding passband W.

We have also seen that there are two main ways of using a phase-locked loop as a discriminator, depending on the a priori information available on the modulation spectrum. If we have no information on the spectrum and if we anticipate a high spectral density and, consequently, a high frequency deviation for the maximum frequencies involved, the natural angular frequency ω_n (or the loop gain K in the case of a first-order loop) must be such that

$$\frac{\omega_n}{2\pi} > \frac{W}{2} \quad \text{or} \quad \frac{K}{2\pi} > \frac{W}{2} \qquad \begin{array}{l}(\omega_n \text{ or } K \text{ in rad/s};\\ W \text{ in Hz})\end{array}$$

The bandwidth of the bandpass filter preceding the loop W depends on the input signal frequency deviation Δf_i and the maximum frequency to be transmitted F_m. We can, for instance, use the Carson formula:

$$W = 2\Delta f_i + 2F_m$$

Since $\Delta f_i < W/2$, the phase error because of the modulation (Section 6.3) stays low, and if the phase error induced by additive noise also stays low, the linear analysis results apply (Section 7.6). The output signal-to-noise ratio formula is that obtained for a conventional discriminator; in similar fashion, the noise power at the output is limited by the passband $(-W/2, +W/2)$ or, better still, by the passband $(-F_m, +F_m)$ of the low-pass filter at the discriminator output.

If the modulation signal spectrum is such that the power spectral density and the corresponding frequency deviation are relatively low for the highest frequencies involved, a possible discriminator would be a second-order loop, followed by a low-pass filter of time constant τ_2, and such that

$$\frac{\omega_n}{2\pi} < \frac{W}{2}$$

For, if the frequency deviation at frequency F_m, $\Delta f_i(F_m)$, is small as compared with $W/2$, the corresponding phase error $\Delta\phi(F_m)$, given by (Section 6.3.2)

$$\Delta\phi(F_m) = \frac{2\pi\Delta f_i(F_m)}{2\zeta\omega_n}$$

can stay low enough for linear operating conditions to apply. If the phase error due to additive noise is also low, the output signal-to-noise ratio is identical with that obtained at the output of a standard discriminator followed by a low-pass filter performing the same filtering function as the loop plus filter unit, that is, having the transfer function $G(j\Omega)$ given in Section 7.6.4.

But we also know that when the signal-to-noise ratio in the passband W preceding a limiter followed by a standard discriminator becomes too low (that is, around $\rho = A^2/2N_0W = 10\text{–}12$ dB), a "pulse" noise appears at the discriminator output. This noise reinforces the "normal" noise as calculated by standard procedure, thus impairing the output signal-to-noise ratio. This is the well-known discriminator threshold phenomenon. Before discussing the reactions of a phase-locked loop under the same operating conditions, we shall briefly recall the mechanism underlying the pulse noise appearance in a standard discriminator.

Operating Threshold of a Standard Limiter-Discriminator (Reminder). Rice has provided a physical explanation of the phenomenon together with corresponding formulas (60). Let us consider the Fresnel diagram in Fig. 12.6, where A represents the amplitude of the useful signal and $n_1(t)$ and $n_2(t)$ are the two Gaussian random processes discussed in Section 7.1.1. Point S, which is the extremity of the vector resulting from the addition of the signal and the noise, has a random trajectory. When $n_1(t)$ and $n_2(t)$ are very small, the probability that this trajectory will encircle the origin O (case a) is very slight. This is not the case when $n_1(t)$ and $n_2(t)$ are fairly large (case b). But when the trajectory of point S encircles point O, the resulting signal equivalent phase $\varphi_i(t)$ from noise,

$$\varphi_i(t) = \operatorname{Arc} \tan \frac{n_2(t)}{n_1(t) + A}$$

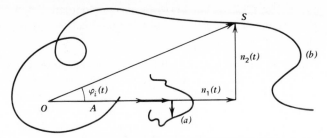

FIGURE 12.6. Signal vector and noise vector composition at the output of a bandpass filter.

steps plus or minus 2π (following the direction of the trajectory around O). The discriminator, which is sensitive to the instantaneous angular frequency

$$\frac{d}{dt}\big[\,\varphi_i(t)\,\big]=\dot{\varphi}_i(t)$$

of the signal applied, responds with a positive or negative pulse, of random height and duration but of constant area of value 2π. These pulses are generally known as clicks or spikes in technical literature. The average number of these positive N_+ or negative N_- pulses appearing per second is given by (60)

$$N_+ = N_- = \frac{W}{4\sqrt{3}}\,\text{erfc}\big[\sqrt{\rho}\,\big]\qquad\text{with }\rho=\frac{A^2}{2N_0 W}$$

This expression, in which $\text{erfc}[x]$ is the complementary error function given by

$$\text{erfc}\big[x\big]=\frac{2}{\sqrt{\pi}}\int_x^{\infty}e^{-u^2}\,du\qquad\left(\cong\frac{e^{-x^2}}{\sqrt{\pi}\,x}\text{ when }x\text{ is large}\right)$$

is valid for an ideal "rectangular" filter preceding the discriminator, when the carrier is exactly centered in the filter and in the absence of modulation. The time distribution of the pulses corresponds to a Poisson distribution and the two-sided power spectral density of this process at null frequency is given by

$$S_{\dot{\varphi}_i}(0)=8\pi^2(N_+ + N_-)$$

$$=4\pi^2\,\frac{W}{\sqrt{3}}\,\text{erfc}\big[\sqrt{\rho}\,\big]\qquad\text{in }(\text{rad/s})^2/\text{Hz}$$

The signal-to-noise ratio at the discriminator output can then be calculated, assuming that $S_{\dot{\varphi}_i}(F)\cong S_{\dot{\varphi}_i}(0)$ within the passband F_m of the output filter, taking into account both the pulse noise and the normal noise. This ratio is, in fact, the relationship between the demodulated signal and the "noise in the absence of modulation". But the modulation has the following consequences:

- An increase of the number of pulses per second (see corresponding formulas for N_+ and N_- in Ref. 60), which is normal, since the phenomenon is nonlinear.
- Loss of the $N_+ = N_-$ symmetry; when the input signal instantaneous frequency departs from the central frequency of the filter W, the clicks occur preferentially in a direction causing the output signal instantaneous absolute value to decrease. The greater the difference with respect to the central frequency, the more marked is this phenomenon. The result is that the power of the useful signal, which is multiplied by $(1 - e^{-\rho})^2$ is diminished. This reduction, which is very slight for the values of $\rho = A^2 / 2N_0 W$ generally encountered, accelerates considerably from $\rho = 1$ (0 dB). It is sometimes considered as a second discriminator operating threshold (61) and is to be considered in the same light as the phenomenon of signal suppression by the limiter (Section 9.2).

In any case, the output signal-to-noise ratio, calculated without reference to the effect of the modulation on the pulse noise, is interesting in that it is an easily measurable quantity and conserves a comparative value for different systems. If we assume the modulating signal to be a sinusoidal signal producing a peak frequency deviation equal to Δf_i and if the output filter is a low-pass filter, with a sharp cutoff at $F_m(F_m < W/2)$, we get

$$\left(\frac{S}{N}\right)_{\text{out}} = \frac{\frac{1}{2}\Delta f_i^2}{\frac{2}{3}\left(N_0/A^2\right)F_m^3 + \left(W/\sqrt{3}\,\right)\text{erfc}\left[\sqrt{\rho}\,\right]F_m}$$

that is,

$$\left(\frac{S}{N}\right)_{\text{out}} = \frac{\rho\dfrac{3}{2}\dfrac{\Delta f_i^2}{F_m^2}\dfrac{W}{F_m}}{1 + \sqrt{3}\,\dfrac{W^2}{F_m^2}\rho\,\text{erfc}\left[\sqrt{\rho}\,\right]} \tag{12.40}$$

The average number of pulses per second and the corresponding signal-to-noise ratio deterioration, which can be evaluated using the above formula (as long as $\rho > 2$ dB) depend to a great extent on the value of the bandwidth W preceding the discriminator. It is thus important for this bandwidth to be reduced to the absolute minimum, making allowances for the tolerated distortion, especially in the case of a low-pass filter with a sharp cut off at frequency F_m.

When the filter W is not an ideal rectangular filter but has a transfer function $W(j\omega)$ that is symmetrical around frequency f_0, the average number of clicks per second in the absence of modulation and when the signal frequency is equal to f_0 is given by (60)

$$N_+ = N_- = \frac{\gamma}{4\pi} \, \text{erfc}\left[\sqrt{\rho} \, \right]$$

with

$$\gamma^2 = \frac{\int_{-\infty}^{+\infty} (\omega - \omega_0)^2 \, |W(j\omega)|^2 \, d\omega}{\int_{-\infty}^{+\infty} |W(j\omega)|^2 \, d\omega}$$

Phase-Locked Loop Used as a Discriminator.   We have seen that the use of a phase-locked loop results in the same signal-to-normal-noise ratio value as that obtained with a standard discriminator, when the input signal-to-noise ratio is high. It is of interest to know whether the use of a first- or second-order loop is a means of obtaining a higher or lower pulse noise than with a standard discriminator and, consequently, whether the operating threshold of such a device would be higher or lower than that of a discriminator.

Experience and various connected theories show that the phase-locked loop output is also affected by the appearance of a pulse noise which, in this case, is the effect of the VCO cycle slipping mentioned in Section 12.2. The difficulty in comparing performances lies in the fact that the groups "band-pass filter + phase-locked loop + output filter" and "bandpass filter + standard discriminator + output filter" have to have exactly the same transfer function with respect to the modulation when the two devices are under normal operating conditions. The distortion of the demodulated signal and the output signal-to-noise ratio (with a high input signal-to-noise ratio) are then identical; these conditions have to be fulfilled for it to be possible to determine—at least experimentally—which of the two devices reaches first its operating threshold. This threshold can be defined as the signal-to-noise density ratio, preceding the input bandpass filter W, for which the output signal-to-noise ratio differs by 1 dB from the value corresponding to normal noise (conventional definition); it can also be defined as the signal-to-noise density ratio for which an average number of clicks per second is reached (definition of a more probabilist character).

Results concerning this sort of comparison are as yet fairly scarce, all the more so as it is interesting to compare simultaneously with other discriminators known as improved-threshold devices, such as the FM feedback discriminator (FMFB) or the dynamic tracking filter (DTF) (see, for example, Ref. 62 and 63 for a description of these devices).

The theoretical method most generally used to try and locate the threshold of the phase-locked loop used as a discriminator consists in inserting in the loop circuit diagram a "click generator" at the point where the output filter is connected (see Fig. 12.7). The instants corresponding to the appearance of clicks are assumed to be governed by a Poisson law, as in the Rice theory for the conventional discriminator. The one-sided power spectral density of the pulse process is then given by

$$S_{\dot{\varphi}/2\pi}(F) = 2\bar{F} \qquad (12.41)$$

where $\bar{F}$ is the mean number of slipping cycles in the loop per second. As mentioned in Section 12.2, by means of the Fokker–Planck method, we can calculate $\bar{F}$ for a first-order loop (Eq. 12.29) and, approximately, for a second-order loop with loop filter $F(s) = 1/(1 + \tau_1 s)$.

For the second-order loop with filter $F(s) = (1 + \tau_2 s)/(1 + \tau_1 s)$, an analysis (64) has been made, where it is observed that this filter can be considered as a $(1 + \varepsilon\tau_1 s)/(1 + \tau_1 s)$ filter, with $0 < \varepsilon < 1$. The results for $\bar{F}$ should thus be intermediate between those of the first-order loop ($\varepsilon = 1$) and those of the second-order loop with a simple low-pass filter ($\varepsilon = 0$). For the general second-order loop, allowances should be made for the fact that when the VCO falls out of synchronism with the input signal, slipping of several cycles generally occurs. This "burst of slipping cycles" phenomenon is particularly noticeable when $K/2$ is very large as compared with $(2B_n)$. It does not occur for a first-order loop nor, apparently, for a second-order loop with $\varepsilon = 0$ [we have seen that, for these loops, $(2B_n) = K/2$]. It can be shown (64) that for a given value of $\varepsilon = \tau_2/\tau_1$ there are two possible values for the quantity $(\tau_1/2)$ $(2B_n)$ and two possible values for the quantity $\frac{1}{2}K/(2B_n)$. If we select the lowest value for $(\tau_1/2)$ $(2B_n)$, then the $\frac{1}{2}K/(2B_n)$ solution is not much larger

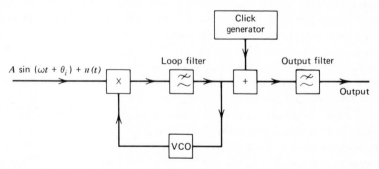

FIGURE 12.7. Model of a phase-locked loop frequency demodulator with impulse noise ("clicks").

than 1 and the "burst of slipping cycles" effect is slight. In this case, we can use for $\bar{F}$ the formula corresponding to the second-order loop with $\varepsilon = 0$, when the difference between the input signal frequency and the VCO central frequency is zero:

$$\bar{F} = \frac{1}{\pi} \left[\sqrt{\frac{1}{4\tau_1^2} + \frac{K}{\tau_1}} - \frac{1}{2\tau_1} \right] e^{-2\rho_L} \qquad (12.42)$$

with

$$\rho_L = \frac{A^2}{N_0(2B_n)}$$

These formulas are, of course, based on the assumption that the preceding filter passband W is wide as compared with $(2B_n)$. If the output filter in Fig. 12.7 is a low-pass filter with a sharp cutoff frequency F_m, chosen lower than the loop -3 dB cutoff frequency, then the "signal-to-noise in the absence of modulation" ratio formula is given by

$$\left(\frac{S}{N} \right)_{\text{out}} = \frac{\frac{1}{2}\Delta f_i^2}{\frac{2}{3}(N_0/A^2)F_m^3 + 2\bar{F}F_m} \qquad (12.43)$$

For a second-order loop where $\varepsilon = 0$, Eq. 12.43 gives an operating threshold in the vicinity of $\rho_L = +4$ dB when $\zeta = \sqrt{2}/2$, $\tau_1 = 1/(2K)$ and $F_m = K/2\pi$ (see Fig. 6.19). This threshold is obviously better than for a standard discriminator (Eq. 12.40), since we have assumed $W \gg (2B_n)$. However, to obtain a valid comparison, in the case of the standard discriminator, W would have to be reduced to the point where the modulation signal filtering is equivalent to that performed by the loop. Furthermore, Eqs. 12.40 and 12.43 use the "average number of clicks per second" formula, in the absence of modulation. Since the mechanism of click appearance is not the same, it is by no means certain that, in the presence of modulation, the result would be as favorable for the phase-locked loop with loop filter $1/(1 + \tau_1 s)$.

However, the fact that the number of clicks is less for a phase-locked loop discriminator than for a standard discriminator would appear to be confirmed by a computation (65) based on another method also proposed by Rice. This calculation proves that when the loop gain K approaches infinity, the operating principle is identical with that of a standard discriminator. The threshold gain would be about 2 dB for a modulation index of $\Delta f_i/F_m = 3$ and nearly 4 dB for $\Delta F_i/F_m = 12$.

Another interesting analysis (66) comprises a physical explanation of the threshold improvement phenomenon observed at the output of a discriminator using a first-order loop. For the unmodulated input signal $A \sin(\omega t + \theta_i)$

accompanied by an additive noise $n(t)$, we observe a click when the VCO $B\cos[\omega t + \theta_i + \varphi_n(t)]$ slips one cycle with respect to signal $A\sin(\omega t + \theta_i)$, the function of the loop being to follow the input signal despite the additive noise. If we consider the signal applied to the input as being the composite signal expressed (see Fig. 12.6) as

$$\sqrt{[A + n_1(t)]^2 + n_2^2(t)}\ \sin\left[\omega t + \theta_i + \operatorname{Arc\,tan} \frac{n_2(t)}{n_1(t) + A}\right]$$

we can say that we gain one click with respect to the standard discriminator each time this composite signal undergoes a 2π phase jump without the VCO following. When the loop gain K is very large as compared with W, the first-order loop tends to follow all the phase jumps of the composite signal and performances are identical with those of a standard discriminator. On the other hand, when K is not very large as compared with W, we can well imagine that the VCO will not follow all the phase jumps of the composite signal (especially the very fast ones). This results in a gain in performance: there are less clicks at the first-order loop output than at the standard discriminator output. It is worth noting that when the input signal-to-noise ratio $\rho = A^2/2N_0W$ is not too low, the trajectories of point S of the vector representing the composite signal that produce the 2π phase jumps pass very close to point O. The composite signal amplitude $A' = \sqrt{[A + n_1(t)]^2 + n_2^2(t)}$ at this instant is very low, with the result that instantaneous loop gain $K_{inst} = K(A'/A)$ is also low, which means that the loop cannot follow the phase jump. Under these conditions, we also note that threshold improvement is more apparent when the loop is not preceded by a limiter than otherwise since in this case the loop scarcely benefits from the instantaneous loop gain reduction. This latter point would appear to be confirmed by experience (66).

Also of interest in this connection are several experimental studies (see, for example, Ref. 67), which indicate a threshold improvement with respect to a standard discriminator even more evident if the loop uses a triangular phase detector or, better still, a sawtooth detector. Even though these characteristics tend to become sinusoidal when the signal-to-noise ratio is low (Sections 7.1.4 and 7.1.5), for a second-order loop, the natural angular frequency value can be lower, for the linearity condition is easier to maintain with regard to the phase error due to modulation. Thus if the signal-to-noise ratio for a sawtooth detector exceeds 0 or a few positive decibels (Fig. 7.6), the loop noise bandwidth $(2B_n)$ can be set at a lower value and the number of clicks will be reduced.

To summarize our discussion of this important problem, which is still the subject of numerous investigations, we can state that if the loop bandwidth is large in comparison with the preceding bandwidth, the gain with respect to the operating threshold of the standard discriminator is slight or nonexistant. On the other hand, in the case of a loop where the bandwidth is not very large as compared with the preceding bandwidth, it is possible to obtain a better operating threshold than that of a standard discriminator, but not necessarily better than those of the improved-threshold devices. In all cases, to obtain a valid comparison, the output signal distortion must be equal and output signal-to-noise ratio identical for above threshold operating conditions.

APPENDIX A
DERIVED LOOPS

A.1 WORKING PRINCIPLE OF A MIXER

The mixer is a nonlinear device the working principle of which is similar to that of a phase detector: it is generally a diode, sometimes a transistor, device effecting the multiplication of the signals applied. In fact, it rarely operates as a true multiplier, especially in the high-frequency zone where it is most often used. But, if the amplitude of the local oscillator signal (reference signal, without noise) is sufficiently high as compared with the input signal amplitude, two or four diode devices exist that operate as multipliers by the sign of the reference. The output "low-frequency" component (that is, at the difference frequency) is identical with that which would be obtained with a true multiplier.

Let us call $y_i(t)$ the input signal and $y_R(t)$ the reference signal:

$$y_i(t) = A \sin(\omega_i t + \theta_i)$$

$$y_R(t) = B \cos(\omega_R t + \theta_R)$$

The multiplier output is

$$y(t) = AB \sin(\omega_i t + \theta_i) \cos(\omega_R t + \theta_R)$$

or

$$y(t) = (AB/2)\{\sin[(\omega_i - \omega_R)t + \theta_i - \theta_R] + \sin[(\omega_i + \omega_R)t + \theta_i + \theta_R]\}$$

The $(\omega_i + \omega_R)$ angular frequency term is suppressed by filtering (a mixer is used in most cases to translate down the frequency; consequently, $\omega_i - \omega_R \ll \omega_i$ so $\omega_R \cong \omega_i$ and $\omega_i - \omega_R \ll \omega_i + \omega_R$), which leaves

$$y(t) = \frac{AB}{2} \sin[(\omega_i - \omega_R)t + \theta_i - \theta_R]$$

If the mixer is not a true multiplier, the output signal amplitude is not $AB/2$. But if the mixer works as a multiplier by the sign of the reference signal, we have seen in Section 2.1.3 that the output signal amplitude is proportional to the input signal amplitude. Generally speaking, we shall designate $y(t)$ as follows:

$$y(t) = kA \sin \left[(\omega_i - \omega_R)t + \theta_i - \theta_R \right]$$

where the coefficient k represents the mixer conversion loss, usually expressed in decibels.

Let us now suppose the input signal $y_i(t)$ to be accompanied by a stationary Gaussian noise $n(t)$, having a null mean value and a one-sided power spectral density N_0 (in W/Hz or in dBm/Hz) in a bandwidth W centered around frequency f_i:

$$y_i(t) = A \sin (\omega_i t + \theta_i) + n(t)$$

It can be shown that if $W/2$ is not only small as compared with f_i but also smaller than $(f_i - f_R)$, the mixer output signal can be expressed as

$$y(t) = kA \sin \left[(\omega_i - \omega_R)t + \theta_i - \theta_R \right] + kn'(t)$$

where $n'(t)$ is a stationary Gaussian noise, having a null mean value and a one-sided power spectral density N_0 in a frequency bandwidth W centered around frequency $(f_i - f_R)$. This means that the noise spectrum is simply frequency translated from f_i to $(f_i - f_R)$ and that the signal-to-noise ratio remains unchanged.

A.2   INDIRECT OR HETERODYNE LOOP

Let us consider the circuit shown in Fig. A.1, where a frequency translation is induced in the input signal by means of the signal from the VCO. The signal resulting from the frequency translation, after filtering and amplifica-

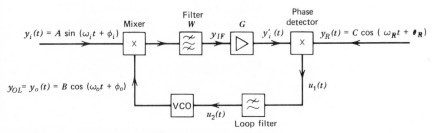

FIGURE A.1.   Block diagram of a heterodyne phase-locked loop.

tion if necessary, is applied to the phase detector, which also receives a reference signal. The phase detector output error signal, after filtering in the loop filter, is applied to the VCO modulation input.

The signal obtained at the output of the bandpass filter centered on the frequency $(f_i - f_0)$ following the mixer is expressed as

$$y_{IF}(t) = \frac{AB}{2} \sin\left[(\omega_i - \omega_0)t + \phi_i - \phi_0\right]$$

After amplification of gain G at the frequency $(f_i - f_0)$ the signal obtained becomes the phase detector input signal:

$$y_i'(t) = G\frac{AB}{2} \sin\left[(\omega_i - \omega_0)t + \phi_i - \phi_0\right]$$

$$= A'\sin\left[(\omega_i - \omega_0)t + \phi_i - \phi_0\right]$$

The phase detector output signal, assumed to operate as a multiplier, is expressed, after elimination of the $(f_i - f_0 + f_R)$ frequency term, as

$$u_1(t) = \frac{A'C}{2} \sin\left[(\omega_i - \omega_0 - \omega_R)t + \phi_i - \phi_0 - \theta_R\right]$$

Let us write $A'C/2 = G(ABC/4) = K_1$. The quantity K_1 is the phase detector sensitivity for an input signal of amplitude $A' = G(AB/2)$. Thus

$$u_1(t) = K_1 \sin\left[(\omega_i - \omega_0 - \omega_R)t + \phi_i - \phi_0 - \theta_R\right]$$

Furthermore, let us call $f(t)$ the loop filter impulse response:

$$u_2(t) = u_1(t) * f(t)$$

and

$$\frac{d\phi_0}{dt} = K_3 u_2(t)$$

This three-equation system is similar to that obtained considering the circuit in Fig. A.2, which is consequently an equivalent circuit.

For this circuit, the signal $y_i''(t)$ applied to the phase detector is expressed as

$$y_i''(t) = G\frac{AC}{2} \sin\left[(\omega_i - \omega_R) + \phi_i - \theta_R\right]$$

$$= A''\sin\left[(\omega_i - \omega_R)t + \phi_i - \theta_R\right]$$

After multiplication by $y_0(t)$ and elimination of the $(f_i - f_R + f_0)$ frequency

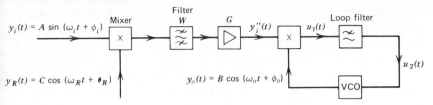

$y_i(t) = A \sin (\omega_i t + \phi_i)$

Mixer

Filter
W

G

$y_i''(t)$

$u_1(t)$

Loop filter

$u_2(t)$

$y_R(t) = C \cos (\omega_R t + \theta_R)$

$y_o(t) = B \cos (\omega_o t + \phi_o)$

VCO

FIGURE A.2. Equivalent model of a heterodyne phase-locked loop.

term, the signal obtained is

$$u_1(t) = \frac{A''B}{2} \sin \left[(\omega_i - \omega_R - \omega_0)t + \phi_i - \theta_R - \phi_0 \right]$$

This time

$$K_1 = \frac{A''B}{2} = G\frac{ABC}{4}$$

is the sensitivity of a phase detector operating with signals at the frequency $(f_i - f_R)$, instead of f_R. The $u_1(t)$ expression becomes

$$u_1(t) = K_1 \sin \left[(\omega_i - \omega_R - \omega_0)t + \phi_i - \theta_R - \phi_0 \right]$$

Furthermore,

$$u_2(t) = u_1(t) * f(t)$$

and

$$\frac{d\phi_0}{dt} = K_3 u_2(t)$$

The circuits in Figs. A.1 and A.2 are thus equivalent since the same equations govern their operation. But the second of these circuits is that of the standard phase-locked loop, and, consequently, the operation of the heterodyne loop is identical with that of the standard loop.

One of the advantages of the indirect loop is that the phase detector works with signals at frequency f_R, which can be lower than the VCO frequency f_0. This is of interest when f_0 has to be high enough for the VCO modulation possibilities to be compatible with those of the input signal.

Furthermore, after acquisition, $\omega_i - \omega_0 = \omega_R$. The frequency of the signal applied to the phase detector input is exactly equal to that of the reference signal; it is consequently extremely stable if the reference signal is derived from a crystal oscillator. This means that a narrow bandpass filter can be used preceding the phase detector, thus improving signal-to-noise ratio conditions for this critical device. However, care must be taken to ensure that this bandpass filter is large enough with respect to the loop filter; otherwise, the transfer function of its equivalent low-pass filter will disturb the loop response. In addition, even if this filter is relatively large, its presence can considerably modify the loop operation during acquisition, causing what is known as a false acquisition. This phenomenon is particularly prevalent when the number of poles is fairly high.

A.3 LONG LOOP

Let us consider the circuit in Fig. A.3, where the local oscillator signal applied to the mixer is obtained by frequency multiplication from the VCO signal. We shall assume that the filtering accompanying the frequency multiplying circuits is large enough to let through the VCO signal modulation in its entirety without amplitude or phase distortion.

The local oscillator signal is expressed as

$$y_0'(t) = B' \cos\left(N\omega_0 t + N\phi_0\right)$$

Consequently, signal $y_i'(t)$ applied at the phase detector input is written

$$y_i'(t) = G\frac{AB'}{2} \sin\left[\left(\omega_i - N\omega_0\right)t + \phi_i - N\phi_0\right]$$

$$= A' \sin\left[\left(\omega_i - N\omega_0\right) + \phi_i - N\phi_0\right]$$

The phase detector output signal, after suppression of the $(\omega_i - N\omega_0 + \omega_0)$ angular frequency term, is expressed as

$$u_1(t) = \frac{A'B}{2} \sin\left\{\left[\omega_i - (N+1)\omega_0\right]t + \phi_i - (N+1)\phi_0\right\}$$

that is,

$$u_1(t) = K_1 \sin\left\{\left[\omega_i - (N+1)\omega_0\right]t + \phi_i - (N+1)\phi_0\right\}$$

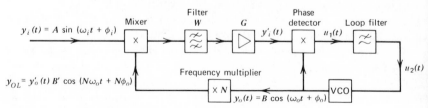

FIGURE A.3. Block diagram of a long-loop type of a phase-locked loop.

with

$$K_1 = \frac{A'B}{2} = G\frac{ABB'}{4}$$

The VCO command signal is given by $u_2(t) = u_1(t) * f(t)$, $f(t)$ being the loop filter impulse response.

Finally, the VCO phase is related to the command signal $u_2(t)$ by the equation $d\phi_0/dt = K_3 u_2(t)$, K_3 being the modulation sensitivity of the VCO the central angular frequency of which is ω_0.

The equations giving $u_1(t)$, $u_2(t)$, and $d\phi_0/dt$ are similar to those obtained using the equivalent circuit shown in Fig. A.4.

In this case, the signal obtained at the phase detector output, after elimination of the $[\omega_i + (N+1)\omega_0]$ angular frequency term, is expressed as

$$u_1(t) = \frac{1}{2} AG\frac{BB'}{2}\sin\left\{\left[\omega_i - (N+1)\omega_0\right]t + \phi_i - (N+1)\phi_0\right\}$$

that is,

$$u_1(t) = K_1 \sin\left\{\left[\omega_i - (N+1)\omega_0\right]t + \phi_i - (N+1)\phi_0\right\}$$

while

$$u_2(t) = u_1(t) * f(t) \quad \text{and} \quad \frac{d\phi_0}{dt} = K_3 u_2(t)$$

The long loop shown in Fig. A.3 is thus the equivalent of a direct loop at frequency f_i; the equivalent VCO is obtained by following the VCO at frequency f_0 by a frequency multiplier by $(N+1)$.

It is then as if the modulation sensitivity (and the frequency deviation) of the VCO were multiplied by $(N+1)$. But the essential advantage, as compared with a direct loop at frequency f_i, is that the long loop phase detector operates at frequency f_0, which is well below the minimum required for the

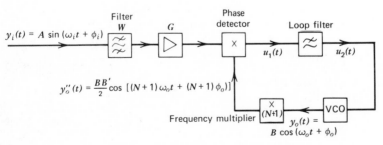

FIGURE A.4. Equivalent model of a long-loop type of a phase-locked loop.

direct loop, implying a relative facility of construction.

Moreover, when the loop is in lock, $(N+1)\omega_0 = \omega_i$. The angular frequency of the signal obtained at the mixer output and passing through the bandpass filter preceding the phase detector is

$$\omega_{IF} = \omega_i - N\omega_0 = \omega_i - \frac{N}{N+1}\omega_i$$

that is,

$$\omega_{IF} = \frac{\omega_i}{N+1}$$

In particular, for a drift $\Delta\omega_i$ of the input signal angular frequency, the frequency drift in the bandpass filter is $\Delta\omega_i/(N+1)$, which means that a filter almost as narrow as for an indirect loop can be used, thus improving considerably the signal-to-noise ratio conditions at the phase detector input. The remarks noted previously relating to the bandwidth of this filter are also valid in this case.

APPENDIX B
EXAMPLE OF
A ROUGH DRAFT OF
A COHERENT RECEIVER

B.1 TERMS OF THE PROBLEM

We are required to estimate the value of the main elements and the performances of a receiver aboard a satellite that is part of a radiolocalization system.

This receiver is designed for the reception of signals sequentially transmitted by a large number of beacons that the satellite has to locate by performing certain measurements:

- the Doppler effect influence on the carrier frequency,
- the time required for signal propagation.

The beacons are equipped with coherent transponders which, when interrogated by the satellite, transmit the signals necessary to these measurements.

The Doppler effect measurement consists in measuring the frequency of the carrier received by the satellite. This frequency is

$$f = f_0 \frac{c+v}{c-v}$$

where f_0 is the nominal frequency of the carrier, c is the propagation speed of the electromagnetic waves and v the relative speed (or range-rate) of the satellite with respect to the beacon interrogated.

From the measurement of f and knowledge of f_0 and c we can derive the value of the relative speed v:

$$v = c \frac{f-f_0}{f+f_0}$$

Measurement of the signal propagation time consists in measuring the phase-shift of a subcarrier (tone) of frequency F_0 modulating the carrier. If we call D the distance (or range) between the beacon interrogated and the satellite at the instant of measurement, the received subcarrier phase-lags, with respect to the signal of the same frequency kept aboard the satellite, by a quantity

$$\Delta\phi = 2\pi F_0 \frac{2D}{c}$$

From the measurement of $\Delta\phi$ and knowledge of F_0 and c, we can derive the value of distance D:

$$D = \frac{c}{2} \frac{\Delta\phi}{2\pi F_0}$$

B.1.1 PURPOSE OF THE RECEIVER

The function of the receiver aboard the satellite is

- To provide for the measurement of the frequency f of the carrier received with a degree of precision such that the speed v can be evaluated with an accuracy of 2 mm/s.
- To perform the demodulation of the signal received in order to obtain the signal of the subcarrier at frequency F_0.
- To provide for the measurement of the phase-shift $\Delta\phi$ of the signal at frequency F_0 with a degree of precision such that the distance D can be evaluated with an accuracy of 2 m.

B.1.2 INITIAL NUMERICAL DATA

The main available data are as follows:

- $f_0 = 2.1$ GHz
- $F_0 = 1$ MHz
- If we call f_D the quantity $f - f_0$, the geometrical relations related to the satellite movement are such that

$$|f_D| \leqslant 58 \text{ kHz}$$

- In the same way, if we call F_D the quantity $F - F_0$, F being the instantaneous frequency of the subcarrier received, then

$$|F_D| \leqslant 28 \text{ Hz}$$

- Still on the basis of the movement of the satellite, the frequency of the signals received varies versus time. The maximum value for the slope of this variation is given by

$$\left|\frac{\Delta f}{\Delta t}\right| \leqslant 104 \text{ Hz/s}$$

$$\left|\frac{\Delta F}{\Delta t}\right| < 5.1 \times 10^{-2} \text{ Hz/s}$$

- The receiver is connected at a point called antenna access. At this point the total power received P_r is such that

$$-133 \text{ dBm} \leqslant P_r \leqslant -123 \text{ dBm}$$

- The noise temperature of the antenna evaluated at the antenna access is $T_{ant} = 250°K$. The noise figure of the receiver preamplifier is $NF = 4$ dB. The losses in the circuits located between the antenna access and the preamplifier input are estimated at 1.1 dB.
- The subcarrier is transmitted as a square-wave signal at the frequency F_0, modulating the carrier in phase with a modulation index $\theta = \pi/4$.
- Taking into account the beacon transponder circuits, we can consider that the carrier received is also phase modulated by a random process $\alpha(t)$, the two-sided power spectral density of which is known:

$$S_\alpha(f) = \nu |H_1(j2\pi f)|^2$$

where $H_1(j2\pi f)$ is the transfer function of a second-order loop (with integrator and phase-lead correction) the parameters of which are $\omega_{n1} = 200$ rad/s and $\zeta_1 = 1$ and ν a constant, of value 5.6×10^{-5} rad^2/Hz.
- We have available aboard the satellite a clock signal at frequency $f_C = 5$ MHz. Its frequency stability, from both long- and short-term point-of-view, has no influence on the measurement precision. It should be remembered that the signal at frequency F_0 is also available; the phase shift $\Delta\phi$ is measured with respect to this signal.

B.2   FIRST APPROACH

With this type of problem it is often useful to make a preliminary rough estimate, which is a means of evaluating rapidly the main characteristics and, in some cases, of detecting at a relatively early stage the major difficulties.

B.2.1 PROCESSING OF THE CARRIER

B.2.1.1 Intermediate Frequency Signal-to-Noise Ratio. The first quantity required is the one-sided density of the additive noise at the receiver input. If we call L the losses in the circuits between the antenna access and the preamplifier input, the total equivalent noise temperature at the antenna access is

$$T_{eq} = T_{ant} + T_0(L \times NF - 1)$$

that is,

$$T_{eq} = 250 + 293(1.29 \times 2.5 - 1) \cong 900°K$$

which implies, k being Boltzmann's constant, namely, $k = 1.38 \times 10^{-23}$ J/°C, that

$$N_0 = kT_{eq} = 1.24 \times 10^{-20} \text{ W/Hz}$$

that is,

$$N_0 = -169 \text{ dBm/Hz}$$

Bearing in mind the value of F_0 and $|f_D|$, the minimum bandwidth for the amplifiers at intermediate frequency (IF) would be

$$B_{min} = 2(10^6 + 58 \times 10^3) = 2.116 \text{ MHz}$$

The signal-to-noise ratio in this bandwidth (disregarding the power lost outside the bandwidth) would be, for the minimum value of the signal received,

$$\left(\frac{S}{N}\right)_{IF} = \frac{P_r}{N_0 B_{min}} \Rightarrow -133 + 169 - 63.3 = -27.3 \text{ dB}$$

It is obviously impossible under these conditions to undertake a conventional demodulation, and the receiver will have to use coherent techniques: a phase-locked loop will provide a duplicate of the carrier (carrier loop) in order to perform the demodulation of the signal received and the measurement of f_D. A second-order loop will be used since the value of f_D is such as to necessitate a high loop gain K value.

B.2.1.2 Equivalent Noise Bandwidth of the Carrier Loop. The equivalent noise bandwidth $(2B_n)$ should be as narrow as possible with respect to the additive noise, but this involves two difficulties:

1. if the natural angular frequency ω_n is too small, the phase error due to

the frequency ramp $|\Delta f/\Delta t|$ may be too high.

2. in the same way, if ω_n is too small, the phase error due to the random modulation $\alpha(t)$ may be excessive.

If we are to tolerate a phase error due to the frequency ramp $|\Delta f/\Delta t|_{max}$ of 0.1 rad, ω_n must be greater than $\omega_{n\,min}$ so that (see Section 5.3.2)

$$\omega_{n\,min}^2 = \frac{2\pi}{0.1}\left|\frac{\Delta f}{\Delta t}\right|_{max} = 6535 \ (\text{rad/s})^2$$

that is,

$$\omega_{n\,min} \cong 80 \ \text{rad/s}$$

Assuming $\zeta = 1$, using our knowledge of phenomena similar to that mentioned in (2) above (see Chapter 8), the equivalent noise bandwidth would be $(2B_n)_{min} \cong 100$ Hz.

The power of the carrier P_c is equal to half the total power P_r ($\pi/4$ modulation by a square-wave signal). Under these conditions the VCO phase variance $\sigma_{\varphi_0}^2$ from the additive noise is

$$\sigma_{\varphi_0}^2 = \frac{N_0}{2P_c}(2B_n)_{min} = \frac{N_0}{P_r}(2B_n)_{min}$$

that is,

$$\sigma_{\varphi_0}^2 = 2.5\times 10^{-4}\times 10^2 = 2.5\times 10^{-2} \ \text{rad}^2$$

The phase error variance σ_φ^2 due to the additive noise is equal to $\sigma_{\varphi_0}^2$. If we assume an identical quantity for the phase error due to the modulation $\alpha(t)$, the total phase error variance σ_ϕ^2 will be

$$\sigma_\phi^2 = 2\sigma_\varphi^2 = 5\times 10^{-2} \ \text{rad}^2$$

With this value, we are sure that the carrier loop will remain in the linear zone.

B.2.1.3 Precision of the Range Rate Measurement. The measurement of the VCO frequency (which carries the Doppler effect f_D to be measured) is performed by measuring the phase increase during a certain period of time T (counting time):

$$f_{\text{VCO}} = \frac{\phi_0(t+T)-\phi_0(t)}{2\pi T}$$

For a time interval T sufficiently large as compared with $1/\omega_n$, the two quantities $\phi_0(t + T)$ and $\phi_0(t)$ are independent (see Eq. 7.22). Consequently,

$$\sigma_{f_{\mathrm{vco}}} = \frac{\sqrt{2}\,\sigma_{\phi_0}}{2\pi T}$$

If we then suppose, very roughly, that the VCO phase variance due to the random modulation $\alpha(t)$ is comparable to that due to the additive noise, $\sigma_{\phi_0}^2 \cong 2\sigma_{\varphi_0}^2 = 5 \times 10^{-2}$ rad^2, then

$$\sigma_{f_{\mathrm{vco}}} = \frac{\sqrt{2}\,\times 0.223}{2\pi T} = \frac{0.05}{T}\ \mathrm{Hz}$$

For counting time T of 1 s, $\sigma_{f_{\mathrm{vco}}} = 5 \times 10^{-2}$ Hz, and for $T = 5$ s, $\sigma_{f_{\mathrm{vco}}} = 10^{-2}$ Hz. The corresponding error on the range-rate measurement v is

$$\sigma_v \cong \frac{c}{2f_0}\,\sigma_{f_{\mathrm{vco}}}$$

Therefore,

$$\sigma_v \cong \frac{3 \times 10^8 \times 5 \times 10^{-2}}{2 \times 2 \times 10^9}\,\frac{1}{T} = \frac{3.6}{T}\ \mathrm{mm/s}$$

We can thus anticipate a range-rate measurement precision of between 3.6 mm/s (for $T = 1$ s) and 0.71 mm/s (for $T = 5$ s).

B.2.1.4 Carrier Loop Acquisition. Before we can attempt to make the Doppler measurement the carrier loop has to be in lock.

The possible frequency variation range of the signal applied at the receiver input is $2|f_D|_{\max} = 116$ kHz. The quantity $|f_D|_{\max}$ is certainly greater than the natural acquisition range given by Eq. 10.17, and an acquisition auxiliary device will be necessary (if natural acquisition could occur, the time required is given by Eq. 10.14:

$$T_{\mathrm{acq}} = \frac{4\pi^2 f_{D\max}^2}{2\zeta\omega_n^3} = 1.3 \times 10^5\ \mathrm{s})$$

If acquisition is performed by sweeping the VCO central frequency, bearing in mind the value of σ_ϕ^2, we can anticipate a sweeping speed of around $\mathcal{R} = \frac{1}{2}\omega_n^2$ rad/s^2 (see Chapters 10 and 11); that is,

$$R = \frac{\mathcal{R}}{2\pi} = \frac{\omega_n^2}{4\pi} = 520\ \mathrm{Hz/s}$$

Since the search range is 116 kHz, acquisition could not be completed in less than $116 \times 10^3/520 = 223$ s.

In order to obtain a reasonable acquisition time (a few seconds, 2–3, to be precise), we have to gain a factor 10^2 on R, that is, a factor 10 on ω_n. With $\omega_{n\,\mathrm{acq}} = 800$ rad/s, $\sigma_{\varphi_0}^2 = \sigma_\varphi^2 = 0.25$ rad^2 [but the error due to the additive noise is practically the only error to be considered: for such a value of ω_n, the phase error due to the random modulation $\alpha(t)$ is negligible]. Figures 11.9 and 11.10 show that the probability of successful acquisition for $\mathcal{R} = \frac{1}{2}\omega_{n\,\mathrm{acq}}^2$ is no longer equal to 1. Thus, even at this early stage, we can be prepared to encounter difficulties with the carrier signal acquisition problem. Furthermore, once acquisition is complete, we shall have to return to a natural angular frequency value closer to 80 rad/s.

B.2.2   PROCESSING OF THE SUBCARRIER

B.2.2.1   Video Frequency Signal-to-Noise Ratio   Given the value of σ_ϕ^2 for $\omega_n = 80$ rad/s, the demodulation will be effected by multiplying the modulated signal by the carrier loop VCO signal, in an analog multiplier or a multiplier by the sign of the VCO signal.

If the bandwidth W of the IF amplifier preceding the demodulator is wide enough, the demodulated signal is a square signal of frequency F_0 and peak amplitude $K_1\sin\theta$. This signal is accompanied by an additive noise, the two-sided density of which is $K_1^2(N_0/2P_r)$ from $-W/2$ to $+W/2$ (see Sections 7.1.2 and 7.1.3). On the basis of the above hypothesis and the value of F_0, $W/2$ is equal to several megahertz, at least 5 MHz, for instance. The signal-to-noise ratio at the demodulator output $(S/N)_\mathrm{dem}$ is given by

$$\left(\frac{S}{N}\right)_\mathrm{dem} = \frac{K_1^2\sin^2\theta}{K_1^2(N_0/2P_r)W} = \frac{P_r\sin^2\theta}{N_0(W/2)} \Rightarrow 36 - 3 - 67 = -34 \text{ dB}$$

B.2.2.2   Precision of the Range Measurement   It is obviously impossible to connect the phase meter used for the distance measurement directly to the demodulator output under the conditions evaluated above. The F_0 signal will have to be filtered to remove most of the accompanying noise.

The frequency variation range resulting from the Doppler effect influence on the signal at F_0 is $2|F_D|_\mathrm{max} = 56$ Hz. Even supposing we could use a filter having an equivalent noise bandwidth $B_\mathrm{eq} = 56$ Hz (the parasitic phase-shift introduced by this filter versus the Doppler effect would have to be carefully compensated for), the signal-to-noise ratio would be

$$\left(\frac{S}{N}\right)_{B_{eq}} = \frac{\frac{1}{2}(16/\pi^2)K_1^2\sin^2\theta}{K_1^2(N_0/P_r)B_{eq}}$$

$$= \frac{8}{\pi^2}\frac{P_r\sin^2\theta}{N_0 B_{eq}} \Rightarrow -1 - 133 - 3 + 169 - 17.5 \text{ dB}$$

$$= 14.5 \text{ dB}$$

The error variance due to the noise involved during the measurment of the phase difference $\Delta\phi$ would be $\sigma^2_{\Delta\phi}$ with

$$\sigma_{\Delta\phi} = \frac{1}{\sqrt{2(S/N)_{B_{eq}}}} = \frac{1}{\sqrt{56.4}} = 0.133 \text{ rad}$$

to which quantity corresponds a standard deviation of the range measurement,

$$\sigma_D = \frac{c}{2}\frac{\sigma_{\Delta\phi}}{2\pi F_0} = 3.18 \text{ m}$$

The theoretical value, estimated without a safety margin, is below the performance objective and we shall now see what performances would be expected of a phase-locked loop ensuring a better filtering of the subcarrier F_0 (subcarrier loop).

If we tolerate a phase error due to the maximum subcarrier frequency ramp

$$\left|\frac{\Delta F}{\Delta t}\right|_{\max} = 5.1 \times 10^{-2} \text{ Hz/s}$$

of $2\pi \times 10^{-3}$ rad (that is, an error of 0.15 m on the distance measurement), ω'_n must be greater than $\omega'_{n\min}$ so that

$$\omega'^2_{n\min} = \frac{2\pi}{2\pi \times 10^{-3}}\left|\frac{\Delta F}{\Delta t}\right|_{\max} = 51 \text{ (rad/s)}^2$$

that is,

$$\omega'_{n\min} \geqslant 7.14 \text{ rad/s}$$

If we take $\omega'_n = \omega'_{n\min}$ and $\zeta' = \sqrt{2}/2$, $(2B'_n) = 7.6$ Hz. The phase error variance due to noise is

$$\sigma^2_{\Delta\phi} = \sigma^2_{\varphi'_0} = \frac{K_1^2(N_0/P_r)}{(16/\pi^2)K_1^2\sin^2\theta}(2B'_n) = \frac{\pi^2}{16}\frac{N_0}{P_r\sin^2\theta}(2B'_n)$$

that is,

$$\sigma_{\Delta\phi}^2 = 0.625 \times (5 \times 10^{-4}) \times 7.6 = 2.38 \times 10^{-3} \text{ rad}^2$$

This implies for the distance measurement standard deviation that

$$\sigma_D = \frac{c}{2} \frac{\sigma_{\Delta\phi}}{2\pi F_0} = 1.16 \text{ m}$$

which is an improvement of $+8.8$ dB over the conventional filtering solution.

B.2.2.3 Subcarrier Loop Acquisition. For $\omega_n' = \omega_{n\min}'$ defined above, the natural acquisition time is far too long. Acquisition will thus have to be performed with a higher value for ω_n', followed by a commutation.

Making allowances for the Doppler effect $|F_D|_{\max}$ and for a stability of 10^{-5} of the VCO central frequency, the maximum possible frequency variation is $\Delta F = |F_D|_{\max} + 10^{-5}F_0 = 38$ Hz. The acquisition time is then given by Eq. 10.14:

$$T_{\text{acq}}' = \frac{(2\pi\Delta F)^2}{2\zeta_{\text{acq}}' \omega_{n\,\text{acq}}'^3}$$

If this time is required to stay below 1 s, with $\zeta_{\text{acq}}' = \sqrt{2}/2$, $\omega_{n\,\text{acq}}'$ will have to be such that

$$\omega_{n\,\text{acq}}'^3 \geqslant \frac{(2\pi\Delta F)^2}{\sqrt{2}} = 40.3 \times 10^3 \text{ (rad/s)}^3$$

that is,

$$\omega_{n\,\text{acq}}' \geqslant 34.3 \text{ rad/s}$$

B.3 SECOND APPROACH

The conclusions reached after the first approach, given the specified requirements, would appear to indicate that it is possible to construct a coherent receiver using a carrier loop and a subcarrier loop to filter the carrier and subcarrier signals in such a way as to obtain a range-rate measurement precision better than 2 mm/s and a range measurement precision better than 2 m.

The major difficulties would appear to be connected with the acquisition of the signals by the loops. If we wish to avoid a response duration for each beacon of more than 5–6 s, the following points are essential:

- For the carrier loop: acquisition must take place with the loop having a "wide bandwidth," assisted by sweeping, followed by switching to a "narrow bandwidth."
- For the subcarrier loop: the acquisition can be natural, the loop having a "wide bandwidth" followed by a switch to a "narrow bandwidth."

In the second approach, we shall examine these various problems in greater detail and calculate the main parameters of the loops and the Receiver circuits.

B.3.1 CARRIER LOOP

B.3.1.1 Value of the Parameters During the Measurement Phase (Narrow Band)

a. Phase-locked loop phase error. The three sources of error in the carrier loop are:

- input signal frequency variation (Doppler effect)
- receiver additive noise
- input signal parasitic modulation

When the input signal frequency is affected by a linear drift of slope $\Delta f/\Delta t$, the corresponding phase error is given by Eq. 5.15, where the third term of the second member very quickly becomes negligible (the closer ζ is to 1, the quicker this term will disappear). This leaves, assuming $K \gg \omega_n$,

$$\phi(t) = \frac{2\pi(\Delta f/\Delta t)}{\omega_n^2} + \frac{2\pi(\Delta f/\Delta t)t}{K}$$

We have seen that the choice of $\omega_n \geqslant 80$ rad/s makes it possible to keep the first error term at about 0.1 rad. If we specify the same value for the second term after $t = 10$ s (the exact beacon response duration is not yet known), we need to have

$$K \geqslant 2\pi \times 10.4 \times 10^3 \text{ rad/s}$$

that is,

$$K \geqslant 6.53 \times 10^4 \text{ rad/s}$$

But $K = 6.53 \times 10^4$ rad/s would be barely sufficient as regards the condition $K \gg \omega_{n\,\mathrm{acq}}$ during the acquisition phase. We shall therefore take $K \geqslant 100\, \omega_{n\,\mathrm{acq}}$, that is, using the value of $\omega_{n\,\mathrm{acq}}$ proposed in Section B.2.1.4,

$$K \cong 10^5 \text{ rad/s}$$

Under these conditions the second error term above will remain below 0.065 rad as long as $t \leqslant 10$ s.

We shall now try and find an optimum value of ω_n that would minimize the phase error variance due to the other two sources. This variance is given by (see Section 8.3.2)

$$\sigma_\phi^2 = \frac{N_0}{2P_c} \int_{-\infty}^{+\infty} |H(j2\pi f)|^2 \, df$$

$$+ \nu \int_{-\infty}^{+\infty} |H_1(j2\pi f)|^2 |1 - H(j2\pi f)|^2 \, df$$

The result of the first integral is straightforward and leads to (Eq. 7.18)

$$\frac{N_0}{P_r} \frac{\omega_n(1 + 4\zeta^2)}{4\zeta}$$

The calculation of the second integral is tedious and, in a subsequent stage of the design, should be undertaken on a computer, to explore on either side of the values of ω_n and ζ determined below in a more simple way. Taking $\zeta = 1$, we can calculate

$$\int_{-\infty}^{+\infty} |H_1(j2\pi f)|^2 |1 - H(j2\pi f)|^2 \, df = \frac{\omega_n}{4a} \frac{4a^2 + 12a + 5}{(a+1)^3} \quad \text{with } a = \frac{\omega_n}{\omega_{n1}}$$

Consequently,

$$\sigma_\phi^2 = \frac{N_0}{P_r} \frac{5}{4} \omega_n + \nu \frac{\omega_{n1}}{4} \frac{4a^2 + 12a + 5}{(a+1)^3}$$

If there is an optimum value for ω_n, it corresponds to the value a solution of

$$\frac{\partial}{\partial a} \left(\frac{\sigma_\phi^2}{\omega_{n1}} \right) = 0$$

that is,

$$\frac{5}{4} \frac{N_0}{P_r} - \nu \frac{a^2 + 4a + 0.75}{(a+1)^4} = 0$$

Case of the minimum signal received. Based on the values $N_0/P_r = 2.5 \times 10^{-4}$ and $\nu = 5.6 \times 10^{-5}$ the plot of the functions $\frac{5}{4}(N_0/P_r)(a+1)^4$ and $\nu(a^2 + 4a + 0.75)$ shows that they have no point of intersection. The variance σ_ϕ^2 is an increasing function of a and ω_n should be as small as possible.

Case of the maximum signal received. With $N_0/P_r = 2.5 \times 10^{-5}$, a root is found to exist in the vicinity of $a = 0.6$; bearing in mind the value $\omega_{n1} = 200$ rad/s, for optimum results, when the signal received is maximum, we should keep as close as possible to $\omega_n = 120$ rad/s and $\zeta = 1$.

b. VCO Phase Variance. The range-rate measurement is related to the VCO signal phase value $\phi_0(t)$, which is affected by the three sources of error mentioned in Section B.3.1.1a, above.

The linear drift of slope $\Delta f/\Delta t$ of the input signal frequency produces, once the transients have disappeared, a VCO constant frequency error of

$$\frac{2\pi(\Delta f/\Delta t)}{K} \text{ rad/s} \quad \text{or} \quad \frac{\Delta f/\Delta t}{K} \text{ Hz}$$

With the values $\Delta f/\Delta t = 104$ Hz/s and $K = 10^5$ rad/s, this error remains below 1.04×10^{-3} Hz; the corresponding error on the range-rate measurement $(7.42 \times 10^{-2}$ mm/s$)$ is well below the specified value.

The other two causes produce a random error directly related to the VCO phase variance value $\sigma_{\phi_0}^2$. This is given by (see Section 8.3.2)

$$\sigma_{\phi_0}^2 = \frac{N_0}{P_r} \int_{-\infty}^{+\infty} |H(j2\pi f)|^2 df$$

$$+ \nu \int_{-\infty}^{+\infty} |H_1(j2\pi f)|^2 |H(j2\pi f)|^2 df$$

The result of the first integral is known. The calculation of the second is tedious (see remarks above). Assuming $\zeta = 1$, we can calculate

$$\int_{-\infty}^{+\infty} |H_1(j2\pi f)|^2 |H(j2\pi f)|^2 df = \frac{\omega_n}{4} \frac{5a^2 + 19a + 5}{(a+1)^3} \quad \text{with } a = \frac{\omega_n}{\omega_{n1}}$$

Consequently,

$$\sigma_{\phi_0}^2 = \frac{N_0}{P_r} \frac{5}{4} \omega_n + \nu \frac{\omega_n}{4} \frac{5a^2 + 19a + 5}{(a+1)^3}$$

The calculation of $(\partial/\partial a)(\sigma_{\phi_0}^2/\omega_{n1})$ shows that this quantity is always positive (for $a > 0$). Thus, as regards the range-rate measurement, the lowest possible value should be taken for ω_n.

c. Choice of the Parameters. During the Measurement Phase, the parameters will be chosen so that when the signal received is minimum,

$$K = 10^5 \text{ rad/s}, \qquad \omega_n = 80 \text{ rad/s}, \qquad \zeta = 1$$

When the signal received is maximum, the value of the parameters should vary as little as possible (for the range-rate precision) or the value of ω_n should not much exceed $\omega_n = 120$ rad/s (for the servo-device phase error).

It would therefore be advisable to precede the carrier loop by a limiter (easier to construct than an AGC).

B.3.1.2 Value of the Parameters During the Acquisition Phase (Wide Bandwidth). If we specify a VCO frequency stability of $\pm 10^{-5}$, assuming the VCO frequency to be below 100 MHz, provision must be made for a search range of

$$\Delta f_{\text{search}} \leqslant 2|f_D|_{\max} + 2 \times 10^{-5} \times 10^8 = 118 \text{ kHz}$$

We shall raise this figure to obtain

$$\Delta f_{\text{search}} = 125 \text{ kHz}$$

The two important questions now posed are

- how much time can be allowed for the acquisition?
- what will be the probability of success P?

To try and answer these two questions we shall use the results of Section 11.2. As these results correspond to laboratory conditions of experiment, it is advisable to fix an a priori high value for P in order to make allowances for the very different operating conditions of the receiver under investigation.

a. $p = 10^{-2}$. If we fix the parameter p at 10^{-2} (see Section 11.2.2), we get a probability of success $P = 0.995$ for the conditions below:

$\dfrac{\mathfrak{R}}{\omega_n^2}$	$\sigma_{\varphi_0}^2$	ω_n	$\dfrac{\mathfrak{R}}{2\pi}$	$\sigma_{\varphi_n}^2$	$\dfrac{\mathfrak{R}}{\omega_n^2}e^{\frac{1}{2}\sigma_{\varphi_n}^2}$	p'
(rad)	(rad^2)	(rad/s)	(kHz/s)	(rad^2)	(rad)	
0.25	0.4	1280	65.2	0.57	0.332	4×10^{-3}
0.365	0.32	1024	60.9	0.42	0.450	1.6×10^{-3}
0.45	0.25	800	45.8	0.31	0.525	$7\text{–}8 \times 10^{-4}$
0.535	0.20	640	34.9	0.24	0.602	2×10^{-4}

obtained as follows:

- The first two columns are derived from Fig. 11.10.
- The third column is derived from the second, using the formula

$$\sigma_{\varphi_0}^2 = \frac{N_0}{P_r}\omega_n \frac{1+4\zeta^2}{4\zeta} \qquad \text{with } \zeta = 1$$

- The fourth column corresponds to the use of columns 1 and 3.
- Column 5 is obtained from column 2 using the curve $\zeta = 1$ of Fig. 12.2.
- Column 6 is obtained from columns 1 and 5.
- Column 7 corresponds to the use of the column 6 results together with the curves of Fig. 11.12.

b. $p = 10^{-3}$. If we fix the parameter p at 10^{-3}, we obtain a probability of success $P = 0.995$ for the following conditions:

$\dfrac{\mathcal{R}}{\omega_n^2}$	$\sigma_{\varphi_0}^2$	ω_n	$\dfrac{\mathcal{R}}{2\pi}$	$\sigma_{\varphi_n}^2$	$\dfrac{\mathcal{R}}{\omega_n^2}e^{\frac{1}{2}\sigma_{\varphi_n}^2}$	p'
(rad)	(rad^2)	(rad/s)	(kHz/s)	(rad^2)	(rad)	
0.225	0.4	1280	58.7	0.57	0.299	3.2×10^{-4}
0.305	0.32	1024	50.9	0.42	0.376	1.5×10^{-4}
0.375	0.25	800	38.2	0.31	0.438	$7-8 \times 10^{-5}$
0.45	0.2	640	29.3	0.24	0.507	2.2×10^{-5}

c. Choice of the parameters. The choice of $p = 10^{-3}$ is more judicious, although resulting in lower search speeds $\mathcal{R}/2\pi$ than for $p = 10^{-2}$, since it involves lower values for the parameter p' (characterizing the probability of an accidental sweeping restart, which, in our case, would be disastrous).

If we require acquisition to take place as quickly as possible, ω_n must be maximum. However, a value of $\sigma_{\varphi_0}^2 = 0.4$ rad^2 (implying $\sigma_{\varphi_n}^2 = 0.57$ rad^2) would appear to be a little excessive. We shall adopt

$$\omega_{n\,\text{acq}} = 10^3 \text{ rad/s}, \qquad \zeta_{\text{acq}} = 1$$

providing for a sweeping speed $\mathcal{R}/2\pi = 50$ kHz/s, which corresponds to a maximum searching time

$$T_{\text{search}} = \frac{125 \times 10^3}{50 \times 10^3} = 2.5 \text{ s}$$

The decision level δ (see Section 11.2.2) will be chosen at $0.375\,K_{1\,\text{min}}$ and the time constant τ of the decision filter will be 8.5 ms.

Once the acquisition is complete, the loop filter elements will have to be commuted to obtain the measurement-phase parameter values ($\omega_n = 80$ rad/s; $\zeta = 1$). An experimental study of the device—or a method similar to that described in Section B.3.2.2 below—will show whether it is possible to commute directly from $\omega_{n\,\text{acq}} = 10^3$ rad/s to $\omega_n = 80$ rad/s or whether an intermediate value of ω_n is necessary.

N.B.   The acquisition conditions were determined for the minimum signal received. When the signal applied to the loop input is higher than the minimum, the loop natural angular frequency is higher while $\sigma_{\varphi_n}^2$ is lower; these two factors improve simultaneously the probability P at a predetermined sweeping speed.

B.3.1.3   Determination of the Filter and the Limiter Preceding the Loop.

The use of a bandpass filter preceding the loop improves the signal-to-noise ratio at the phase detector input (advantage of indirect loops—see Appendix A). Providing the bandwidth of this filter is narrow enough to obtain an output signal-to-noise ratio greater than 1, a limiter can be used, by means of which the loop parameters can be kept reasonably constant.

If an indirect loop is chosen (see Section B.3.3.1) the filter must not be too narrow or the servo-device stability will be too precarious.

During the acquisition phase, for the minimum signal received, we have chosen $\omega_{n\,acq} = 10^3$ rad/s and $\zeta_{acq} = 1$. Bearing in mind the value $K = 10^5$ rad/s, the time constants τ_1 and τ_2 of the loop filter will be chosen so that (Eqs. 3.41 and 3.52)

$$\tau_{1\,acq} = \frac{K}{\omega_{n\,acq}^2} = \frac{10^5}{10^6} = 0.1 \text{ s}$$

$$2\zeta_{acq}\omega_{n\,acq} = 2 \times 10^3 = \frac{1 + K\tau_{2\,acq}}{\tau_{1\,acq}} \cong \frac{10^5\tau_{2\,acq}}{0.1}$$

Therefore,

$$\tau_{2\,acq} = 2 \text{ ms}$$

The corresponding Bode plot (Fig. 4.5) gives, for the angular frequency value corresponding to a gain of unity $(\omega = K\,(\tau_2/\tau_1) = 2 \times 10^3$ rad/s), a phase difference

$$90° + \text{Arc} \tan (2 \times 103 \times 0.1) - \text{Arc} \tan (2 \times 10^3 \times 2 \times 10^{-3})$$

that is,

$$90° + \text{Arc} \tan 200 - \text{Arc} \tan 4 = 103.75°$$

The phase safety margin, for the loop supposed perfect during the acquisition phase, is 76.25°. The passband of the phase detector is generally very wide and this component has no influence in this respect. If we take 6.25° to cover the phase shift induced by the VCO at the angular frequency $\omega = K\,(\tau_2/\tau_1)$, this corresponds to a VCO cutoff frequency at -3 dB with respect to the modulation, f_{cVCO}, such that

$$\frac{\omega}{2\pi f_{cVCO}} \leqslant \tan 6.25° = 0.11$$

or

$$f_{cVCO} \geqslant 2.9 \times 10^3 \text{ Hz}$$

Since this performance is not difficult to achieve, we can allow $10°$ for the preceding filter phase shift, still keeping a comfortable $60°$ margin for servo-device stability.

If we use a two-pole filter of the Butterworth type, having a cutoff angular frequency (at -3 dB) of the equivalent low-pass filter ω_c, the phase-shift involved at angular frequency ω is β so that

$$\tan \beta = \frac{\sqrt{2}\,(\omega/\omega_c)}{1-(\omega/\omega_c)^2}$$

which leads to

$$\frac{\sqrt{2}\,(\omega/\omega_c)}{1-(\omega/\omega_c)^2} = \tan 10° = 0.1763$$

We also obtain $\omega/\omega_c = 0.1228$; for $\omega = K(\tau_2/\tau_1) = 2 \times 10^3$ rad/s this leads to $\omega_c = 16.3 \times 10^3$ rad/s, that is, $f_c = \omega_c/2\pi = 2.6$ kHz.

The equivalent noise bandwidth of the bandpass filter W_c is then

$$W_c = 2f_c \frac{\pi}{2\sqrt{2}} = 5.8 \text{ kHz}$$

The signal-to-noise ratio at the output of filter W_c is

$$\left(\frac{S}{N}\right)_{W_c} = \frac{P_r \cos^2 \theta}{N_0 W_c}$$

thus, for the minimum, signal received,

$$\left(\frac{S}{N}\right)_{W_c} \Rightarrow -133 - 3 + 169 - 37.6 = -4.6 \text{ dB}$$

For the maximum signal received, this ratio will be $+5.4$ dB taking into account the expected $+10$ dB dynamic.

The use of a limiter (followed by a zonal filter) between the filter W_c and the phase detector, leads to the following values for the suppression factor (see Fig. 9.2):

$$\alpha = 0.475 \qquad \text{for} \left(\frac{S}{N}\right)_{W_c} = -4.6 \text{ dB}$$

$$\alpha = 0.91 \qquad \text{for} \left(\frac{S}{N}\right)_{W_c} = +5.4 \text{ dB}$$

The zonal filter output dynamic is thus reduced to 5.65 dB instead of 10 dB at the limiter input. This will cause a certain variation of the loop parameters (see the following section). For this variation to be reduced, better signal-to-noise ratio conditions are necessary at the limiter input, that is, a narrower filter W_c. At this stage in the design, it is impossible to choose W_c narrower if we want to leave an adequate safety margin for the servo device during the acquisition phase when the input signal is high.

B.3.1.4   Calculation of the Carrier Loop Elements

a. Sensitivity K_1 of the Phase Detector.   The detector we intend to use can accept a $+3$ dBm signal (in a 50 Ω resistive impedance) at the input "signal" when the signal applied to the input "reference" is a sinusoidal $+13$-dBm signal (in a 50 Ω resistive impedance). The detector operates in this case as a multiplier by the sign of the reference signal. If the input signal is itself sinusoidal, the slope K_1 of the detector (indicated by the constructor) is then 0.25 V/rad.

In the calculation below, we shall assume that the limiter behaves like an AGC device having a gain of 0 dB for the minimum signal received and -4.35 dB for the maximum. Although this procedure is based on false premises, the numerical results provide a good approximation (see Section 9.2).

When the input signal is at its minimum value, the signal-to-noise ratio at the detector input is $(S/N)_{W_c} = -4.6$ dB. Since the noise is greater than the signal, it is the noise power which must remain below $+3$ dBm. For this condition to be fulfilled 95% of the time, the noise power N_c applied at the "signal" input of the detector should not exceed -3 dBm.

If we call G_c the total gain of the receiving chain preceding the phase detector, then

$$N_c = G_c N_0 W_c = -3 \text{ dBm}$$

Consequently,

$$G_c = 128.4 \text{ dB}$$

For the minimum signal received, the power of the useful signal applied to the detector is $P_c = G_c P_r \cos^2 \theta$ that is,

$$P_c \Rightarrow 128.4 - 133 - 3 = -7.6 \text{ dBm}$$

The phase detector sensitivity is then

$$K_{1\min} = \frac{0.25}{3.4} = 73.8 \text{ mV/rad}$$

For the maximum signal received, making allowances for the limiter, we shall have

$$P_c \Rightarrow -7.6 + 5.65 = -1.95 \text{ dBm}$$

implying

$$K_{1\max} = \frac{0.25}{1.77} = 141.4 \text{ mV/rad}$$

b. VCO modulation sensitivity K_3. We want to obtain a minimum loop gain $K_{\min} = 10^5$ rad/s but $K_{\min} = K_{1\min} K_2 K_3$. We thus require

$$K_2 K_3 = \frac{10^5}{73.8 \times 10^{-3}} = 1.36 \times 10^6 \text{ rad/s/V}$$

or

$$K_2 K_3 = 216 \text{ kHz/V}$$

Restricting the VCO modulation sensitivity to a value of around 5.4 kHz/V (bearing in mind the $\pm 10^{-5}$ stability required of the VCO), we should have to add an operational gain amplifier $K_2 = 40$. This will give

$$K_{1\min} = 73.8 \text{ mV/rad}, \qquad K_2 = 40, \qquad K_3 = 5.4 \text{ kHz/V}$$

c. Loop filter elements. In the measurement phase, for the minimum signal received, we require the parameters ω_n and ζ to have the following values:

$$\omega_{n\min} = 80 \text{ rad/s}, \qquad \zeta_{\min} = 1$$

The time constant τ_1 will consequently be chosen (Eq. 3.41) so that

$$\tau_1 = \frac{K_{\min}}{\omega_{n\min}^2} = \frac{10^5}{(80)^2} = 15.625 \text{ s}$$

The time constant τ_2 is given by (Eq. 3.52)

$$\frac{1 + K_{\min} \tau_2}{\tau_1} = 2\zeta_{\min} \omega_{n\min} = 160 = \frac{1 + 10^5 \tau_2}{15.625}$$

that is,

$$\tau_2 \cong 0.025 \text{ s}$$

These time constants can be obtained with a passive filter (R_1, R_2, C) such that

$$R_1 = 1.56 \text{ M}\Omega, \qquad R_2 = 2.5 \text{ k}\Omega, \qquad C = 10 \text{ }\mu\text{F}$$

For the maximum signal received, we shall have

$$\omega_{n\,\text{max}}^2 = \frac{K_{\text{max}}}{\tau_1}$$

that is,

$$\omega_{n\,\text{max}} = 110.7 \text{ rad/s}$$

Furthermore,

$$\zeta_{\text{max}} = 1.38$$

During the acquisition phase, for the minimum signal received, we require parameters ω_n and ζ to have the following values:

$$\omega_{n\,\text{acq}} = 10^3 \text{ rad/s}, \qquad \zeta_{\text{acq}} = 1$$

The time constants τ_1 and τ_2 will consequently be chosen so that: $\tau_{1\,\text{acq}} = 0.1$ s and $\tau_{2\,\text{acq}} = 0.002$ s (see Section B.3.1.3).

These time constants are obtainable with a passive filter (R_1, R_2, C) having the same capacitance value as during the measurement phase (to avoid transients which could disturb the commutation), giving

$$R_{1\,acq} = 9.8 \text{ k}\Omega, \qquad R_{2\,acq} = 200 \text{ }\Omega, \qquad C = 10 \text{ }\mu\text{F}$$

For the maximum signal received, $\omega_{n\,\text{acq}} = 1384$ rad/s and $\zeta_{\text{acq}} = 1.38$, values that are quite safe as regards the servo-device stability, given the choice made for W_c.

B.3.2 SUBCARRIER LOOP

B.3.2.1 Operating Principle of the Coherent Demodulator

a. Bandwidth of the Filter Preceding the Demodulator. The signal received is modulated by a square-wave signal of frequency F_0. Two extreme options are possible for the bandwidth of the filter W_m preceding the demodulator phase detector:

- Option 1:   filter as little as possible the modulated signal before applying it to the demodulator.
- Option 2:   filter the modulated signal so as to let through only the first component of the signal spectrum on either side of the carrier.

Let us first analyze the first option. If we assume W_m to be sufficiently wide, the signal obtained at the demodulator output is a square signal of frequency F_0 and amplitude $K_1 \sin \theta$, if we call K_1 the sensitivity of the demodulation phase detector. This signal would be accompanied by a noise having a one-sided density $K_1^2(N_0/P_r)$ from $f=0$ to $f=W_m/2$.

Two options are then possible for the subcarrier loop:

- No filtering is performed around F_0 before applying the signal to the subcarrier loop detector, assumed to operate as a multiplier by the sign of the VCO signal at 1 MHz.

   The detector will then have a triangular characteristic of slope $K_1' = (2/\pi)K_1 \sin \theta$ and the two-sided density of the equivalent modulation by the noise will be $(N_0/2P_r \sin^2 \theta)\cdot(\pi^2/4)$ (see Section 7.1.4a).
- A relatively narrow filtering is performed around F_0 before applying the signal to the detector. This signal applied is then a sinusoidal signal of amplitude $(4/\pi)K_1 \sin \theta$. The detector will then have a sinusoidal characteristic of slope $K_1' = (8/\pi^2)K_1 \sin \theta$ and the two-sided density of the equivalent modulation by the noise will be $(N_0/2P_r \sin^2 \theta)\cdot(\pi^2/8)$.

As regards the second option for the filter W_m, the signal applied to the demodulator is expressed, if $F_0 < W_m/2 < 3F_0$, as

$$y(t) = A \cos \theta \sin \omega_0 t + \frac{2}{\pi} A \sin \theta \sin (\omega_0 + \Omega_0)t$$

$$- \frac{2}{\pi} A \sin \theta \sin (\omega_0 - \Omega_0)t$$

At the output of the demodulator of sensitivity K_1, a sinusoidal signal is obtained at frequency F_0, of amplitude $(4/\pi)K_1 \sin \theta$, accompanied by a noise having a one-sided density $K_1^2(N_0/P_r)$ from $f=0$ to $f=W_m/2$.

Whether filtering is performed around F_0 or not, the subcarrier loop phase detector will have a sinusoidal characteristic of slope $K_1' = (8/\pi^2)K_1 \sin \theta$ and the two-sided spectral density of the equivalent modulation will be $(N_0/2P_r \sin^2 \theta)\cdot(\pi^2/8)$ (provided $W_m/2 \leqslant 2F_0$).

Consequently, the choice of a filter having a bandwidth $W_m \cong 4$ MHz preceding the demodulator provides the best operating conditions for the subcarrier loop, without having to use a filter centered on F_0 (if it is decided to use such a device, it would be with the intention of improving the subcarrier loop detector operating conditions). Furthermore, a relatively narrow filter W_m improve signal-to-noise ratio conditions at the demodulator input. From this point of view, it would be advisable for W_m to be only slightly higher than 2 MHz. However, the parasitic phase shift induced in the subcarrier F_0 could be quite high. It would thus be preferable to widen this filter enough so that it only attenuates perceptibly the $3F_0$ modulation components and beyond.

b. Demodulated Signal Amplitude. In a bandwidth $W_m \cong 4$ MHz, the signal-to-noise ratio is

$$\left(\frac{S}{N}\right)_{W_m} \cong \frac{P_r}{N_0 W_m} \Rightarrow -133 + 169 - 66 = -30 \text{ dB}.$$

The signal applied to the demodulator thus consists essentially of noise. If we decide to use for the demodulation a phase detector similar to that used for the carrier loop, the total gain G_m of the receiver chain preceding the demodulator must be such that the noise power N_m does not exceed -3 dBm. Consequently

$$N_m = G_m N_0 W_m = -3 \text{ dBm}$$

implying

$$G_m = 100 \text{ dB}$$

The total power of the useful signal applied to the demodulator (disregarding the part of the power suppressed by filtering W_m) will be $P_m = G_m P_r = -33$ dBm. To this value corresponds a demodulator sensitivity K_{1m}:

$$K_{1m} = \frac{0.25}{63.1} = 4 \text{ mV/rad}$$

The demodulated signal amplitude will be

$$A' = \frac{4}{\pi} K_{1m} \sin\theta = 3.6 \text{ mV}$$

(which corresponds to a power of -38.9 dBm in a 50 Ω resistive load). The one-sided density N_0' of the noise accompanying this signal is

$$N_0' = K_{1m}^2 \frac{N_0}{P_r} = 40 \times 10^{-10} \text{ V}^2/\text{Hz}$$

in a bandwidth from 0 to $W_m/2 = 2$ MHz, that is a noise power of 8×10^{-3} V^2 (or -11 dBm in a 50 Ω resistor).

B.3.2.2 Analysis of the Wide-to-Narrow Bandwidth Commutation. We saw in the course of the first approach discussion that by choosing $\omega'_{n\,acq} = 34.3$ rad/s and $\zeta'_{acq} = \sqrt{2}\,/2$, we can guarantee a natural acquisition time for the subcarrier below 1 s. We also saw that choosing $\omega'_n = 7.14$ rad/s and $\zeta' = \sqrt{2}\,/2$ for the measurement phase produced an interesting distance measurement performance. The natural acquisition time is generally considered as the time required for the phase plane trajectories to reach the separatrix (Section 10.2.1). It is useful to know, at least approximately, after how long we can proceed to the loop parameter commutation and after how long we can then start the distance measurement, if we keep the above values for the parameters.

The ultimate aim is to reduce the phase error as far as possible. Since the signal-to-noise ratio conditions are very good, it is obvious that the value $\omega'_{n\,acq}$ should be kept as long as possible since this is a means of reaching quickly a given value of the phase error.

The least favorable case is that where, upon completion of natural acquisition, the trajectory has just crossed the separatrix. We then use the curves of Fig. 10.12 to calculate the time required for the trajectory to stabilize in the linear zone (defined, for example, by $|\phi'| < \pi/6$), after which the linear equations can be used to determine (more accurately than with the curves of Fig. 10.12) the time necessary before the measurement can be performed (assuming this can be undertaken when $|\phi'| < 2\pi \times 10^{-3}$ rad). The least favorable phase plane departure point is defined by ($\dot{\phi}'_0 = 2.7\,\omega'_{n\,acq}$; $\phi'_0 = -\pi$). After a time lapse of around $t_1 = 25/4\omega'_{n\,acq}$, the trajectory reaches the point ($\dot{\phi}'_1 = -1.95\,\omega'_{n\,acq}$; $\phi'_1 = +\pi/6$) and is such as to remain in the linear operating zone of the wide loop.

We can then calculate the loop response by adding the response to a phase step $\theta = \phi'_1 = \pi/6$ to the response to an angular frequency step

$$\Delta\omega = \dot{\phi}'_1 + 2\zeta'_{acq}\omega'_{n\,acq}\sin\phi'_1$$

that is,

$$\Delta\omega \cong \omega'_{n\,acq}(-1.95 + 0.74) = -1.21\,\omega'_{n\,acq}$$

The expressions in Chapter 5 can be used to determine that after a time $t_2 = 5/\omega'_{n\,acq}$ the trajectory reaches the point ($\dot{\phi}'_2 = 0.04\,\omega'_{n\,acq}$; $\phi'_2 = 0.01$).

If the loop bandwidth commutation is performed at this instant, the initial phase plane conditions corresponding to the new parameters are

$$\frac{\dot{\phi}'_2}{\omega'_n} = \frac{\dot{\phi}'_2}{\omega'_{n\,acq}}\frac{\omega'_{n\,acq}}{\omega'_n}$$

and ϕ'_2, that is,

$$\frac{\dot{\phi}'_2}{\omega'_n} = 0.04 \frac{34.3}{7.14} = 0.192$$

and $\phi'_2 = 0.01$.

After checking that with this initial point in the narrow-loop phase plane, the trajectory will remain entirely in the linear zone, we can calculate the response of the new loop as the response to a phase step $\theta = 0.01$ rad together with the response to an angular frequency deviation:

$$\Delta\omega = \dot{\phi}'_2 + 2\zeta'\omega'_n \sin\phi'_2 = (0.192 + 0.014)\omega'_n$$

The expressions in Chapter 5 can be used to ensure that after a time $t_3 = 4/\omega'_n$, the phase error in the loop will be below $2\pi \times 10^{-3}$ rad (that is a maximum error of 0.15 m for the distance measurement).

Consequently, if we call t_0 the instant where the carrier loop acquisition is completed, we can state that

- At instant $t_0 + 1$ s, the natural acquisition process with $\omega'_{n\,acq} = 34.3$ rad/s and $\zeta'_{acq} = \sqrt{2}/2$ is such that all the trajectories have reached or gone beyond the separatrix.
- At instant $t_0 + 1\,s + 25/4\omega'_{n\,acq} = t_0 + 1.183$ s, all the trajectories have reached the linear operating zone.
- At instant $t_0 + 1.183\,s + 5/\omega'_{n\,acq} = t_0 + 1.329$ s, we can perform the loop parameter commutation and at instant $t_0 + 1.329\,s + 4/\omega'_n = t_0 + 1.889$ s proceed to the distance measurement with an error (caused by the end of the acquisition phenomenon) below 15 cm.

B.3.2.3   Calculation of the loop elements

a. Calculation of the loop gain. The phase error due to the linear slope of the input signal frequency F (Doppler effect) is given by (see Section B.3.1.1)

$$\phi'(t) = \frac{2\pi}{\omega'^2_n}\left|\frac{\Delta F}{\Delta t}\right| + \frac{2\pi}{K'}\left|\frac{\Delta F}{\Delta t}\right| t$$

The choice of $\omega'_n = 7.14$ rad/s (see Section B.2.2.2) is a means of ensuring that the first error term is below $2\pi \times 10^{-3}$ rad. If we want the second term to be also below $2\pi \times 10^{-3}$ rad after a time $t = 5$ s (the exact beacon response duration is not known), K' must be chosen so that

$$K' \geqslant 5.1 \times 10^{-2} \times 5 \times 10^3 = 255 \text{ rad/s}$$

Another condition for K' is related to the frequency difference ΔF that can exist between the VCO signal and the input signal. Since acquisition is not performed by sweeping the VCO frequency, any frequency difference ΔF appears as a phase error $2\pi\Delta F/K'$. We have seen that the difference $|\Delta F|$ is less than 38 Hz. If we take $K' = 10^4$ rad/s, the phase error will remain below 2.4×10^{-2} rad, that is an error on the distance measurement of 0.57 m, compatible with the required precision.

b.   Phase Detector Sensitivity.   The demodulator output noise power is -11 dBm. If we use a detector identical with those used for carrier loop and the demodulator, the maximum possible amplification will be $+8$ dB. The power of the useful signal applied to the detector will then be $-38.9 + 8 = -30.9$ dBm for the minimum signal received, leading to $K'_1 = 0.25/49.5 \cong 5$ mV/rad.

To obtain $K' = 10^4$ rad/s, the sensitivity K'_3 and the amplifier gain K'_2 would have to be

$$K'_2 K'_3 = \frac{10^4}{5 \times 10^{-3}} = 2 \times 10^6 \text{ rad/s/V}$$

that is,

$$K'_2 K'_3 = 318 \text{ kHz/V}$$

Given the frequency stability required of the VCO at 1 MHz, it is difficult to obtain a modulation sensitivity greater than $K'_3 = 250$ Hz/V. This would imply $K'_2 = 1273$. In other words, taking into account the signal-to-noise ratio at the detector input $(-27.9$ dB), this means that the VCO drift, due to the phase detector "zero" drift multiplied by this value for K'_2, would exceed the frequency stability required.

K'_1 must therefore be increased and the detector operating conditions must be improved, using a bandpass filter centered on F_0 between the demodulator and the phase detector of the subcarrier loop. If we choose a filter having a bandwidth $W_{F_0} = 20$ kHz, we gain simultaneously 20 dB on the signal-to-noise ratio conditions and a factor 10 on K'_1. Under these conditions, the gain to be provided for between demodulator and detector is 28 dB and we shall have

$$K'_1 = 50 \text{ mV/rad}, \qquad K'_2 = 128, \qquad K'_3 = 250 \text{ Hz/V}$$

N.B.   The stability required of the phase detector "zero" is $2\pi \times 10^{-3}$ rad, to avoid a distance error greater than 0.15 m. This will certainly prove to be one of the difficult development points.

If the parasitic phase shift is such as to prevent us from reducing W_{F_0} below 20 kHz, the signal-to-noise ratio conditions are such that a limiter (or a standard AGC) would not be very effective to maintain the loop parameters reasonably constant for the measurement phase. A coherent AGC will thus have to be used: by means of a second phase detector, inserted as a coherent amplitude detector, we can detect the amplitude of the signal at F_0 and adjust accordingly the video amplifier gain from 28 dB to 18 dB when the signal received passes from minimum to maximum value. Under these conditions the loop parameters will remain constant.

c. *Calculation of the Loop Filter Elements.* During the measurement phase, we require $\omega_n' = 7.14$ rad/s and $\zeta' = \sqrt{2}/2$. The time constant τ_1' is given by Eq. 3.41:

$$\tau_1' = \frac{K'}{\omega_n'^2} \cong \frac{10^4}{51} = 196 \text{ s}$$

The time constant τ_2' is given by Eq. 3.52, leading to

$$\frac{1 + 10^4 \tau_2'}{196} = 10.1$$

that is,

$$\tau_2' = 0.2 \text{ s}$$

These values can be obtained by choosing the following values for the loop filter elements R_1', R_2', and C':

$$R_1' = 1.96 \text{ M}\Omega, \qquad R_2' = 2 \text{ k}\Omega, \qquad C' = 100 \text{ }\mu\text{F}$$

During the acquisition phase, for the minimum signal received, we require $\omega_{n\,acq}' = 34.3$ rad/s, and $\zeta_{acq}' = \sqrt{2}/2$. The time constant $\tau_{1\,acq}'$ is given by

$$\tau_{1\,acq}' = \frac{K'}{\omega_{n\,acq}'} = \frac{10^4}{1.18 \times 10^3} = 8.5 \text{ s}$$

The time constant $\tau_{2\,acq}'$ is derived from

$$\frac{1 + 10^4 \tau_{2\,acq}'}{8.5} = 48.5$$

that is,

$$\tau_{2\,acq}' = 41.2 \text{ ms}$$

ROUGH DRAFT OF A COHERENT RECEIVER

Keeping the same capacitance C' for the loop filter, as for the measurement phase, we shall take

$$R'_{1\,\text{acq}} = 85 \text{ k}\Omega, \qquad R'_{2\,\text{acq}} = 412 \ \Omega$$

When the signal received is higher than its minimum value, natural acquisition will take place more quickly than expected; then, with the help of the coherent AGC, the parameters will take the values calculated above for the end of the acquisition process.

B.3.3 GENERAL ORGANIZATION OF THE RECEIVER

B.3.3.1 Frequency Determination. We have seen in Section B.3.1.4 that the gain of the receiving chain preceding the phase detector of the carrier loop should be around 128 dB.

The gain of the preamplifier at 2100 MHz can be about 30 dB. Making allowances for the conversion losses of the first mixer, the intermediary frequency (IF) gain should be about 105 dB.

As it is rather difficult to obtain this amplification with a single IF amplifier, two devices should be used (amplifier IF_1 and amplifier IF_2) with two local oscillators (LO_1 and LO_2).

The choice of frequencies IF_1 and IF_2 is based on a compromise between several considerations (protection with respect to the image frequency, technological problems involved in the construction of amplifiers, filters and local oscillators, protection against interference, and so on). A decision in this respect has to be made at this stage of the design, but, as specifically coherent techniques are not involved, we shall not develop further this aspect.

On the other hand, we now have to analyze the choice of the type of carrier loop, which affects the receiver structure.

The use of a short loop seems preferable (see Appendix A). With a long loop, the LO_1 signal would be derived from the VCO by frequency multiplication. When the long loop was in lock, it would be the LO_1 signal (the frequency of which in the vicinity of 2100 MHz) that would "carry" the greater part of the Doppler frequency deviation to be measured. A second frequency multiplication chain would then have to be used, starting from the clock signal at f_C, to translate down the LO_1 frequency to a frequency of a few Megahertz, permitting the Doppler measurement to be made. It is preferable to use a short loop and a single frequency multiplication chain: that which derives the LO_1 signal from the clock signal.

Once the decision has been made regarding the short loop principle, we have to choose between the direct or indirect type (or heterodyne—see Appendix A).

The stability required of the VCO central frequency is $\pm 10^{-5}$ or better (Section B.3.1.2). This supposes the use of a crystal-stabilized oscillator (VCXO). The maximum frequency of a device of this type (quartz oscillating at fundamental mode) is restricted to 25 or 30 MHz and the maximum frequency deviation is limited to 10 or 20 kHz, depending on the stability required. Consequently, we shall not be able to use the VCXO at its nominal frequency, in view of the value of f_D: the VCXO will have to be followed by a frequency multiplier, which will give an equivalent VCO of frequency comprised between about 60 and 150 MHz (if the multiplication factor is chosen within 3 to 6).

If we use a direct loop, the central frequency of the IF_2 amplifier should be equal to the VCO central frequency. An IF_2 between 60 and 150 MHz has the disadvantage of implying a relatively high IF_1. Furthermore, the carrier loop and demodulator phase detectors would be difficult to construct.

On the other hand, given the equivalent VCO frequency, it is fairly easy to use this as LO_2, that is, construct an indirect loop. The IF_1 can then be chosen between 20 or 30 MHz and 150 MHz and the IF_2 can be chosen between a few megahertz and 20 or 30 MHz. In any case, this frequency should be a multiple of f_C, since the simplest solution consists in deriving the indirect loop reference signal from the clock signal at f_C. This choice also means that filter W_c can be used, which produces a noticeable improvement in signal-to-noise ratio conditions at the input of the "carrier loop" phase detector, as was noted in Section B.3.1.3.

The first possible value for IF_2 is $f_C = 5$ MHz. However, bearing in mind the modulation of the signal received, this value would seem rather low to perform the demodulation under favourable conditions. As the value immediately above in the LO_1 multiplication chain is, for example, $3f_C$, it would seem preferable to take 15 MHz as value of the IF_2, no major difficulties being involved in the construction of phase detectors operating at this frequency.

B.3.3.2 Gain Determination. If the preamplifier gain is 30 dB, making allowances for the losses $(-1.1$ dB) between the antenna access and the preamplifier and the conversion losses of the first mixer $(-6$ dB), we need an IF gain (towards the carrier loop) of $128.4 - 30 + 1.1 + 6 = 105.5$ dB.

The IF_2 gain should be limited, because of the problems of radiation of the $3f_C$ reference signal in the equipment. A value of 50 dB should be obtainable without involving expensive precautionary measures.

If the second mixer conversion losses are assumed to be -6 dB, the IF_1 gain should be $105.5 - 50 + 6 = 61.5$ dB.

The filter W_c should be located as close as possible to the second mixer. The IF_2 preceding the demodulator will then be completely independent of that of the carrier loop. Its gain should be $50 - (128.4 - 100) = 21.6$ dB.

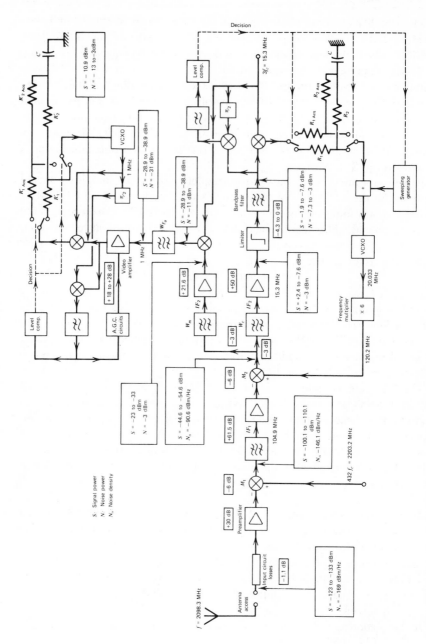

FIGURE B.I. General block diagram of the coherent receiver.

The gain of the video amplifier placed between the demodulator and the subcarrier loop varies between 18 and 28 dB under the influence of the coherent AGC (Section B.3.2.3.).

Given the acquisition devices necessary for the carrier and subcarrier loops, we arrive at the synoptic diagram of Fig. B.1, on which are indicated the respective signal and noise levels at the most characteristic points.

B.3.4. RECEIVER PERFORMANCES

B.3.4.1 Range-Rate Measurement. We have seen (Section B.2.1.3) that the standard deviation of the range-rate measurement is given by

$$\sigma_v = \frac{c}{2f_0}\sigma_{f_{vco}} = \frac{c}{2f_0}\frac{\sqrt{2}\ \sigma_{\phi_0}}{2\pi T} = \frac{0.016}{T}\sigma_{\phi_0}$$

We shall calculate, for the two extreme values of the received signal power, the value of σ_{ϕ_0} and the value of σ_ϕ, using the expressions given in Section B.3.1.1.

P_r	ω_n	ζ	$a = \dfrac{\omega_n}{\omega_{n1}}$	$\dfrac{N_0}{P_r}(2B_n)$	$\nu\dfrac{\omega_n}{4}\dfrac{5a^2+19a+5}{(a+1)^3}$	$\sigma_{\phi_0}^2$	σ_{ϕ_0}
(dBm)	(rad/s)			(rad^2)	(rad^2)	(rad^2)	(rad)
-133	80	1	0.4	2.5×10^{-2}	5.5×10^{-3}	3.05×10^{-2}	0.175
-123	110	1.38	0.55	4.3×10^{-3}	7×10^{-3}	1.13×10^{-2}	0.106

P_r	$\dfrac{N_0}{P_r}(2B_n)$	$\nu\dfrac{\omega_{n1}}{4}\dfrac{4a^2+12a+5}{(a+1)^3}$	σ_ϕ^2	σ_ϕ
(dBm)	(rad^2)	(rad^2)	(rad^2)	(rad)
-133	2.5×10^{-2}	1.07×10^{-2}	3.57×10^{-2}	0.189
-123	4.3×10^{-3}	$9.6\ \times10^{-3}$	1.39×10^{-2}	0.118

At the end of this second approach, we can state that the receiver contribution to the range-rate measurement precision, characterized by the standard deviation σ_v, is comprised between 2.8 mm/s (for $T = 1$ s) and 0.56 mm/s (for $T = 5$ s), when the signal received is at its minimum value. Thus, to comply with specifications, the counting time T must exceed 1.4 s.

B.3.4.2   Range Measurement.   We have seen (Section B.2.2.2.) that the standard deviation of the distance measurement is given by

$$\sigma_D = \frac{c}{2}\frac{\sigma_{\Delta\phi}}{2\pi F_0} = \frac{c}{2}\frac{\sigma_{\varphi_0'}}{2\pi F_0}$$

$$= \frac{c}{2}\frac{1}{2\pi F_0}\left[\frac{\pi^2}{16}\frac{N_0}{P_r \sin^2\theta}(2B_n')\right]^{1/2}$$

with $(2B_n') = 7.6$ Hz.

For the two extreme values of the received signal power, we can calculate the value of $\sigma_{\varphi_0'}$ and σ_D:

P_r	$(2B_n')$	$\dfrac{\pi^2}{16}\dfrac{N_0}{P_r\sin^2\theta}$	$\sigma_{\varphi_0}^2$	$\sigma_{\varphi_0'}$	σ_D
(dBm)	(Hz)	(rad^2/Hz)	(rad^2)	(rad)	(m)
-133	7.6	3.08×10^{-4}	2.35×10^{-3}	4.85×10^{-2}	1.16
-123	7.6	3.08×10^{-5}	2.35×10^{-4}	1.53×10^{-2}	0.37

B.3.4.3   Beacon Response Duration.   The carrier loop signal search time is 2.5 s. If we reserve 1 s after acquisition for the wide-to-narrow bandwidth commutation and disappearance of the corresponding transients, the range-rate measurement can start 3.5 s after the beacon starts transmitting. Similarly, subcarrier signal acquisition can start 3.5 s after the beginning of transmission, finishing 2 s later, with the wide-to-narrow bandwidth commutation completed. The distance measurement can start 5.5 s after the beginning of transmission. If the range-rate measurement is completed 6 s after the beginning of the transmission, performances will be as follows:

- $\sigma_v \leqslant 1.12$ mm/s, which allows a *margin of $+5$ dB* with respect to the specifications.
- $\sigma_D \leqslant 1.16$ m, which allows a *margin of $+4.8$ dB* with respect to the specifications.

A possible improvement would consist in only applying the modulation signal after the range-rate measurement has started and suppressing it before the end of the measurement. This would allow an *additional margin of $+3$ dB*, both for the carrier acquisition and the range-rate measurement performance. We can also modulate the carrier by the F_0 signal with a modulation index of $\pi/3$ instead of $\pi/4$, which the carrier loop can easily withstand, given the values of σ_ϕ^2 given in B.3.4.1. This will allow an *additional margin of $+1.8$ dB*, both for the subcarrier acquisition and for the distance measurement performance.

REFERENCES

1. "The Phase Locked Loop I.C. as a Communication System Building Block," National Semiconductor Corp., AN-46, June 1971.

2. "Monolithic Phase Locked Loop," Exar Integrated Systems Inc., X-R 215, April 1972.

3. "Double Balanced Mixers," Hewlett Packard, Technical Data, October 1967.

4. "Analog Multipliers and Multiplier/Dividers," a short-form Catalog of Analog Circuit Modules, Analog Devices, 1971.

5. "Quarter-Square Multiplier/Dividers," General Catalogue, Burr-Brown, 1970.

6. L. M. Robinson, "Tanlock: a Phase-Lock Loop of Extended Tracking Capability," Proc. 1962 IRE Convention on Military Electronics, Los Angeles, California.

7. M. Balodis, "Laboratory Comparison of Tanlock and Phaselock Receivers," Proceedings of the National Telemetring Conference, Los Angeles, California, 1964.

8. G. A. McKay, "An Extended Phase Detector for Phase-Locked Receivers," 1967 Conference Record IEEE Region III Convention.

9. A Acampora and A. Newton, "Use of Phase Subtraction to Extend the Range of a Phase-Locked Demodulator," *RCA Review*, **27**, No. 4, December 1966, p. 577-599.

10. P. Vovelle, "Les Oscillateurs à Fréquence Contrôlée," *Onde Electrique*, No. 466, Janvier 1966, p. 105-110.

11. S. C. Gupta and R. J. Solem, "Optimum Filters for Second- and Third-Order Phase-Locked Loops by an Error Function Criterion," *IEEE Transactions on Space Electronics and Telemetry*, **SET-11**, June 1965, p. 54-62.

12. R. C. Tausworthe and R. B. Crow, "Improvements in Deep-Space Tracking by use of Third-Order Loops," Proceedings of the International Telemetring Conference, 1972, Los Angeles, California.

13. J. G. Truxal, *Automatic Feedback Control System Synthesis*, McGraw-Hill Book Company, New York, 1955.

14. C. S. Weaver, "A New Approach to the Linear Design and Analysis of Phase-Locked Loops," *IRE Transactions on Space Electronics and Telemetry*, **SET-5**, No. 4, December 1959, p. 166-178.

15. B. D. Martin, "The Pioneer IV Lunar Probe: a Minimum Power FM/PM System Design," J. P. L. Technical Report No. 32–215, March 15, 1962.

16. F. M. Gardner, S. S. Kent, and R. D. Dasenbrock, "Theory of Phaselock Techniques," REC-TR-22A, Resdel Engineering Corporation, Pasadena, California, 1964, Chapter 10-4.

17. W. B. Davenport Jr. and W. L. Root, *An Introduction to the Theory of Random Signals and Noise*, McGraw-Hill Book Company, New York, 1958, Chapter 8.5.

18. A. J. Viterbi, *Principles of Coherent Communication*, McGraw-Hill Book Company, New York, 1966, Appendix A, p. 295.

19. J. M. Wozencraft and I. M. Jacobs, *Principles of Communication Engineering*, John Wiley and Sons Inc., New York, 1965, p. 186.

20. W. B. Davenport Jr. and W. L. Root, *An Introduction to the Theory of Random Signals and Noise*, McGraw-Hill Book Company, New York, 1965, Chapter 13.

21. M. Schwartz, *Information Transmission, Modulation and Noise*, McGraw-Hill Book Company, New York, 1970, p. 455.

22. A. Pouzet, private communication; see also "Characteristics of Phase Detectors in presence of noise," Proceedings of the 1972 International Telemetering Conference, Los Angeles, California.

23. A. J. Viterbi, *Principles of Coherent Communication*, McGraw-Hill Book Company, New York, 1966, p. 36, Table 2.2.

24. R. Jaffe and E. Rechtin, "Design and Performance of Phase-Lock Circuits Capable of Near-optimum Performance over a Wide Range of Input Signal and Noise Levels," *IRE Transactions on Information Theory*, **IT-1**, March 1955, p. 66-76.

25. S. C. Gupta and R. J. Solem, "Optimum Filters for Second- and Third-Order Phase-Locked Loops by an Error Function Criterion," *IEEE Transactions on Space Electronics and Telemetry*, **SET-11**, June 1965, p. 54-62.

26. H. E. Rowe, *Signals and Noise in Communication Systems*, Bell Telephone Laboratories Series, D. Van Nostrand Company, Princeton, 1965.

27. P. Debray, "Etude d'une boucle de phase en présence d'un bruit blanc et gaussien," unpublished CNES Internal Document, September 1966.

28. J. A. Develet, "Fundamental Accuracy Limitations in a Two-Way Coherent Doppler Measurement System," *IEEE Transactions on Space Electronics and Telemetry*, **SET-7**, September 1961, p. 80-85.

29. A. J. Viterbi, "Optimum Coherent Demodulation for Continuous Modulation Systems," *Proceedings of the National Electronics Conference, 1962*, Vol. 18, p. 498.

30. W. C. Lindsey, "Optimum and Sub-Optimum Frequency Demodulation," Technical Report No. 32-637, Jet Propulsion Laboratory, June 15, 1964.

31. R. M. Gray and R. C. Tausworthe, "Frequency-Counted Measurements and Phase-Locking to Noisy Oscillators," *IEEE Transactions on Communications Technology*, **COM-19**, No. 1, February 1971, p. 21-30.

32. A. Blanchard and J. C. Durand, "Influence de la stabilité à court terme des oscillateurs sur les mesures précises de vitesse par effet Doppler," *Colloque International l'Espace et la Communication*, Editions CHIRON, Paris, April 1971.

33. F. M. Gardner, S. S. Kent and R. D. Dasenbrock, "Theory of Phase-Lock Techniques," REC-TR-22A, Resdel Engineering Corporation, Pasadena, California, 1964, Chap. 9-4.

34. J. C. Springett and M. K. Simon, "An Analysis of the Phase Coherent-Incoherent Output of the Bandpass Limiter," *IEEE Transactions on Communication Technology*, **COM-19**, No. 1, February 1971, p. 42-49.

35. R. C. Tausworthe, "Theory and Practical Design of Phase-Locked Receivers," J. P. L. Technical Report No. 32-819, February 1966.

36. W. B. Davenport Jr., "Signal-to-Noise Ratios in Band-Pass Limiters," *Journal of Applied Physics*, **24**, No. 6, June 1953, p. 720-727.

37. J. C. Springett, "A note on Signal-to-Noise and Signal-to-Noise Spectral Density Ratios at the Output of a Filter-Limiter Combination," J. P. L. Space Programs Summary No. 37-36, Vol. IV.

38. A. J. Viterbi, "Acquisition and Tracking Behavior of Phase-Locked Loops," Jet Propulsion Laboratory, External Publication No. 673, July 1959.

39. R. W. Sanneman and J. R. Rowbotham, "Unlock Characteristics of the Optimum Type II Phase-Locked Loop," *IEEE Transactions on Aerospace and Navigational Electronics*, **ANE-11**, March 1964, p. 15-24.

40. C. R. Cahn, "Piecewise Linear Analysis of Phase-Lock Loops," *IRE Transactions on Space Electronics and Telemetry*, **SET-8**, No. 1, March 1962, p. 8-13.

41. E. N. Protonotarios, "Pull-in Performance of a Piecewise Linear Phase-Locked Loop," *IEEE Transactions on Aerospace and Electronic Systems*, **AES-5**, No. 3, May 1969, p. 376-386.

42. A. Jay Goldstein, "Analysis of the Phase-Controlled Loop with a Sawtooth Comparator," *The Bell System Technical Journal*, **41**, March 1962, p. 603-633.

43. F. M. Gardner, *Phaselock Techniques*, John Wiley and Sons, New York, 1966, Chap. 4-3.

44. S. Vialle, "Résultats obtenus en balayant en fréquence une boucle d'asservissement de phase ouverte," unpublished CNES Internal Doucument, May 1969.

45. J. P. Frazier and J. Page, "Phase-Lock Loop Frequency Acquisition Study," *IRE Transactions on Space Electronics and Telemetry*, **SET-8**, September 1962, p. 210-227.

46. S. Vialle, "Acquisition d'un signal de fréquence inconnue à l'aide d'une boucle d' asservissement de phase," unpublished CNES Internal Document, November 1969.

47. A. J. Viterbi, *Principles of Coherent Communication*, McGraw-Hill Book Company, New York, 1966, Chapter 4.

48. A. J. Viterbi, *Principles of Coherent Communication*, McGraw-Hill Book Company, New York, 1966, Section 2.7.

49. J. K. Holmes, "On a Solution to the Second-Order Phase-Locked Loop," *IEEE Transactions on Communication Technology*, **COM-18**, No. 2, April 1970, p. 119-126.

50. W. C. Lindsey and M. K. Simon, "The Effect of Loop Stress on the Performance of Phase-Coherent Communication Systems," *IEEE Transactions on Communication Technology*, **COM-18**, No., 5, October 1970, p. 569-588.

51. J. A. Develet Jr., "An Analytic Approximation of Phase-Lock Receiver Threshold," *IEEE Transactions on Space Electronics and Telemetry*, **SET-9**, March 1963, p. 9-12.

52. R. C. Tausworthe, "Theory and Practical Design of Phase-Locked Receivers," J. P. L. Technical Report No. 32-819, February 1966.

53. H. L. Van Trees, "Functional Techniques for the Analysis of the Non-Linear Behavior of Phase-Locked Loops," *Proc. IEEE*, **52**, No. 8, August 1964, p. 894-911.

54. W. C. Lindsey and R. C. Tausworthe, "A Survey of Phase-Locked Loop Theory," J. P. L. Technical Report.

55. E. A. Bozzoni et al., "An Extension of Viterbi's Analysis of the Cycle Slipping in a First-Order Phase-Locked Loop," *IEEE Transactions on Aerospace and Electronic Systems*, **AES-6**, No. 4, July 1970, p. 484-489.

56. A. Blanchard, "Influence d'un fading bref sur la Mesure de l'Effet Doppler à l'aide d'un Récepteur à Commande Automatique de Phase," *Colloque International sur l'Electronique et l'Espace*, Paris, April 1967.

57. J. R. Rowbotham and F. W. Sanneman, "Random Characteristics of the Type II Phase-Locked Loop," *IEEE Transactions on Aerospace and Electronic Systems*, **AES-3**, No. 4, July 1967, p. 604-612.

58. B. M. Smith, "The Phase-Locked Loop with Filter: Frequency of Skipping Cycles," *Proc. IEEE*, **54**, Febuary 1966, p. 296-297.

59. D. Sanger and R. Tausworthe, "Experimental Study of the First-Slip Statistics of the Second-Order Phase-Locked Loop," J. P. L. Space Programs Summary 37-43, Vol. III, 1967.

60. S. O. Rice, "Noise in FM Receivers," in *Time Series Analysis*, M. Rosenblatt, Ed., John Wiley & Sons, New York, 1963, Chap. 25.

61. M. Schwartz, W. R. Bennett, and S. Stein, *Communication Systems and Techniques*, McGraw-Hill Book Company, New York, 1966, p. 134.

62. D. T. Hess, "Equivalence of FM Threshold Extension Receivers," *IEEE Transactions on Communication Technology*, **COM-16**, No. 5, October 1968, p. 746-748.

63. J. H. Roberts, "The Frequency Feedback Receiver as a Low Threshold demodulator in FMFDM Satellite Systems" and "The Dynamic Tracking Filter as a Low Threshold demodulator in FMFDM Satellite Systems," Colloquium on Threshold Extension Techniques, IEEE Electronics Division, *Colloquium Digest No. 1968/9*, March 1968.

64. S. C. Gupta, J. W. Bayless, and D. R. Hummels, "Threshold Investigation of Phase-Locked Discriminators," *IEEE Transactions on Aerospace and Electronic Systems*, **AES-4**, No. 6, November 1968, p. 855-863.

65. P. W. Osborne and D. L. Schilling, "Threshold Analysis of Phase-Locked Loop Demodulators Using Most Likely Noise," *IEEE Transactions on Communication Technology*, **COM-19**, No. 1, February 1971, p. 31-41.

66. D. T. Hess, "Cycle Slipping in a First Order Phase-Locked Loop," *IEEE Transactions on Communication Technology*, **COM-16**, No. 2, April 1968, p. 255-260.

67. J. C. Lindenlaub et al., "A Study of the Extended Linear Range Phase Lock Loop," Purdue University, Lafayette, Indiana, August 1968.

INDEX

Accidental sweeping restart, 296, 300
Accidental sweeping stop, *see* False alarm
    probability
Acquisition, 241-278
  duplication of, 280
  input signal frequency, 245-278
  input signal phase, 243-245
  necessary condition for, 298
  range, 253, 260, 265, 266, 271, 277, 279,
    320, 356
  time, 260, 262, 271, 273, 276, 277, 356,
    359, 372
Aided acquisition, 279-300, 329, 356
  decision filter, *see* Decision filter
  decision level, *see* Decision level
  frequency dispersion, 286, 294
  probability of success, *see* Probability of
    successful acquisition
  requirements, 363
Amplitude detector, 224. *See also* Coherent
    amplitude detector
Antenna access, 353-354
  noise equivalent temperature, 353-354
Antipodal binary modulation, 328-329
Autocorrelation function, VCO phase, 169-
    170, 304-305
  VCO signal, 168, 170
Automatic frequency control (A.F.C.), 30,
    245, 280
  discriminator, 280
Automatic gain control (A.G.C.), 167, 186,
    218-226, 245, 367, 375
  reduction factor, *see* Suppression factor
  see also Coherent automatic gain control
Auxiliary phase detector, 290-292

Balanced modulator, 20
Beat signal frequency, 246, 253, 267, 283,
    299
Binary modulation, 328-329, 332-333
Bode diagram, 60, 61, 64, 65, 67, 69, 71,
    365

Booton's method, *see* Quasi-linearization
    method

Cascaded loops, 212
Click noise spectral density, 337, 340
Clicks, 336-342
  average number per second, 337-339
  time distribution, 337
Coherent amplitude detector, 225, 283, 289-
    292
  sensitivity, 290
Coherent automatic gain control, 225, 280,
    375
Coherent phase demodulation, *see* Phase
    demodulation
Comb filter, 281
Command signal, 4-5, 135
Commutation, loop bandwidth, 372-373
  loop filter components, 279
  loop parameters, 357, 364
Convolution product, 45
Cycle slipping, 277, 319-324, 334, 339-343
  mean frequency of, 322, 340-341
  mean time between, 321-324
  probability of, 322-324

Damping factor, 55, 66
  choice, 37, 67, 68, 70, 83, 89, 92, 93, 96,
    99, 108, 118, 128, 136, 160, 161, 165
  measurement, 125
  optimization, 208, 209
  variations, 58-59, 133, 222-224, 234-235,
    316, 329-330, 332
Decision delay, 286, 287, 289
Decision filter, 283, 284, 289
  bandwidth determination, 296, 300
  time constant, 295, 364
Decision level, 283, 286, 287, 298
  determination, 294, 300, 364
Dirac delta function, 81, 169
Direct loop, 376-377
Discriminator, 30

385